Advances in Industrial Control

For further volumes:
www.springer.com/series/1412

Pål Liljebäck · Kristin Y. Pettersen ·
Øyvind Stavdahl · Jan Tommy Gravdahl

Snake Robots

Modelling, Mechatronics, and Control

Springer

Pål Liljebäck
Applied Cybernetics
SINTEF ICT
Trondheim, Norway
and
Department of Engineering Cybernetics
Norwegian University of Science & Technology
Trondheim, Norway

Kristin Y. Pettersen
Department of Engineering Cybernetics
Norwegian University of Science & Technology
Trondheim, Norway

Øyvind Stavdahl
Department of Engineering Cybernetics
Norwegian University of Science & Technology
Trondheim, Norway

Jan Tommy Gravdahl
Department of Engineering Cybernetics
Norwegian University of Science & Technology
Trondheim, Norway

ISSN 1430-9491 ISSN 2193-1577 (electronic)
Advances in Industrial Control
ISBN 978-1-4471-2995-0 ISBN 978-1-4471-2996-7 (eBook)
DOI 10.1007/978-1-4471-2996-7
Springer London Heidelberg New York Dordrecht

Library of Congress Control Number: 2012938883

Printed on acid-free paper

Springer is part of Springer Science+Business Media (www.springer.com)

To our families

Series Editors' Foreword

The series *Advances in Industrial Control* aims to report and encourage technology transfer in control engineering. The rapid development of control technology has an impact on all areas of the control discipline. New theory, new controllers, actuators, sensors, new industrial processes, computer methods, new applications, new philosophies, . . . , new challenges. Much of this development work resides in industrial reports, feasibility study papers and the reports of advanced collaborative projects. The series offers an opportunity for researchers to present an extended exposition of such new work in all aspects of industrial control for wider and rapid dissemination.

Over the years the *Advances in Industrial Control* series has been very fortunate in publishing monographs that were often seminal for the development of new areas in control systems theory and industrial technology. These monographs were often written by young researchers making their way in the industrial control field or were a report of a substantial research project that was now ready for holistic presentation and dissemination. For a monograph series that spans two decades, it is actually quite easy to find examples of this type of motivational text. From the early years of the series, *Iterative Learning Control for Deterministic Systems* by Kevin L. Moore (ISBN 978-3-540-19707-2, 1992) and *Autotuning of PID Controllers* by Cheng-Ching Yu (ISBN 978-3-540-76250-8, 1999) are good examples. In more recent years we can cite *Control of Fuel Cell Power Systems* by Jay T. Pukrushpan, Anna G. Stefanopoulou, and Huei Peng (ISBN 978-1-85233-816-9, 2004), *Predictive Functional Control* by Jacques Richalet and Donal O'Donovan (ISBN 978-1-84882-492-8, 2009) and finally *Internet-Based Control Systems* by Shuang-Hua Yang (ISBN 978-1-84996-358-9, 2011) as typical *Advances in Industrial Control* monographs that are studied as key texts for their respective topics.

Clearly *Snake Robots* by Pål Liljebäck, Kristin Y. Pettersen, Øyvind Stavdahl, and Jan Tommy Gravdahl is going to be a much read, studied, and cited monograph in this particular field of robot development. After a truly fascinating introductory chapter that examines among other topics, biological snake motion, the monograph is structured into two parts. Part I investigates and reports on modelling, technology, and control for snake robot locomotion in a planar (flat) environment (Chaps. 2–8).

Part II moves on to snake locomotion in a cluttered environment with stationary objects (Chaps. 9–13). This group of chapters introduces and explores the concept of "obstacle-aided" locomotion. So much of mobile robot technology is concerned with obstacle avoidance, so it is interesting to see a robot application that exploits the contact with objects (obstacles) in the environment to *aid* locomotion. The monograph closes with a concluding chapter, three short technical appendices, a useful Glossary of technical terms, and an exhaustive Index.

The authors have succeeded in writing a well-structured text that is both a scientific and an engineering monograph. The structure and contents of the monograph can be accessed in several different ways. For example, the monograph presents mathematical models that elucidate snake motion per se; this describes important *fundamental* scientific principles in the field. Alternatively, the monograph can be used as a source for up-to-date survey and review material on snake-robotic engineering and technology; apart from the thorough historical review in Chap. 1, each subsequent chapter opens with a section that creates the context and reviews the past literature relating to the work to be presented by the authors. One of the attractive features of the monograph is the way in which the authors use the chapter sequence to work through increasingly complex issues in the understanding, control and technology of snake robots. Such careful attention to structure allows the expert researcher and the researcher new to the field rapidly to assess the importance of the material presented and its relation to past developments. Finally, the monograph presents the authors' own research and development in the field. This research covers the full spectrum of mathematical modelling, control design, simulation studies and fascinating experimental demonstrator prototypes. Closing the monograph is a chapter on the future research and technological challenges for snake robot locomotion (Chap. 14).

The series Editors have no doubt that the control and robotics community will find much of interest in this monograph. The monograph's progress through the historical record for the field, the proofs and descriptions of fundamental snake robot principles and the practical demonstrations using robot prototypes will ensure this new entry to the *Advances in Industrial Control* series becomes a key reference and source text for snake-robot locomotion research and development.

Industrial Control Centre
Glasgow
Scotland, UK

M.J. Grimble
M.A. Johnson

Preface

The purpose of this book is to present theoretical and practical topics related to snake robots. Snake robots are robotic mechanisms designed to move like biological snakes. The advantage of such mechanisms is their ability to move and operate in challenging environments where human presence is unwanted or impossible. Future applications of these mechanisms include search and rescue operations, inspection and maintenance in industrial process plants, and subsea operations. Research on snake robots has been conducted for several decades. For instance, the world's first snake robot was developed in Japan already in 1972. There are, however, still many theoretical and practical aspects of snake robot locomotion which have not yet been addressed in the snake robot literature. Current literature is characterised by numerous different approaches to modelling, development, and control of these mechanisms, but a unified theoretical foundation of snake robots has not yet been established.

In this book, we attempt to target these limitations of current literature on snake robots. The main goal of the book is to contribute to the mathematical foundation of the control theory of snake robots, and also stimulate and support future research on these fascinating mechanisms. To this end, the book is a complete treatment of snake robotics, with topics ranging from mathematical modelling techniques, mechatronic design and implementation, and control design strategies. In particular, several new approaches to modelling snake robot locomotion are presented. Moreover, numerous properties of snake robot dynamics are derived using nonlinear system analysis tools, and several new control strategies for snake robots are proposed. The book also describes the development of two snake robots that are employed to experimentally validate many of the theoretical results. Whereas previous literature has mainly focused on flat surface locomotion, a distinct feature of the book is the strong focus on locomotion in uneven and cluttered environments. The organisation of the book is detailed in Sect. 1.5.

Although the results presented in this book are new and based on recent conference and journal publications, they are presented at an initial level which is accessible to audiences with a standard undergraduate background in control theory or mechatronics. The book is written in a clear and easily understandable manner with numerous figures and pictures which help illustrate and visualise the material. The target audience of this book includes academic researchers and graduate students with an interest in snake robots or underactuated systems in general. The book may also be used for self-study or as a reference by engineers and applied mathematicians, and by anyone who would like to find out more about the exciting field of snake robotics. We believe the book will be particularly useful to new researchers taking on a topic related to snake robotics since the book provides an extensive overview of the snake robot literature and also represents a suitable starting point for research in this area.

We are indebted to a number of people who have been integral to the completion of this book. We express our sincere gratitude to Professor Scott David Kelly (University of North Carolina at Charlotte), Professor Shugen Ma (Ritsumeikan University), and Professor Ole Morten Aamo (Department of Engineering Cybernetics at NTNU) for their feedback to the material in this book in conjunction with their participation in the doctoral dissertation of Pål Liljebäck.

Furthermore, we gratefully acknowledge all the support we have received from our friends and colleagues at the Department of Engineering Cybernetics at NTNU. In particular, we thank Idar Haugstuen for our cooperation in conjunction with his M.Sc. project on snake robots in 2009/2010. Moreover, we thank Christian Holden for providing useful feedback to the material in this book, and we thank Alexey Pavlov for our talks and his many useful suggestions regarding our research. For their untiring help and efforts with the experimental systems considered in this book, we thank Terje Haugen, Per Inge Snildal, and Glenn Angel at the mechanical workshop of the department. We also thank Stefano Bertelli for his positive spirit and for his help with documenting the experimental results.

We owe many thanks to our friends and colleagues at SINTEF Applied Cybernetics for their support and for contributing to a positive and stimulating work environment. In particular, we thank the Research Director, Sture Holmstrøm, for his enthusiasm and his willingness to financially support research on snake robots. We are thankful to Aksel A. Transeth, Sigurd Fjerdingen, and Erik Kyrkjebø for their positivity and for our many interesting discussions related to snake robots over the last years. A special recognition goes to Aksel A. Transeth for contributing to the research underlying this book with his knowledge and expertise of snake robots. We thank Geir Mathisen, Espen Helle, and Knut Vidar Skjersli for their work on circuit boards and software for the experimental systems considered in the book. We are grateful to Anders Beitnes for conceiving the idea of a self-propelled fire hose, which initiated research on snake robots at SINTEF and NTNU. We also thank Wheeko and Kulko for their outstanding and obedient performance during experiments, and for never complaining about late work hours.

Finally, we express our deepest gratitude to the Norwegian University of Science and Technology (NTNU) and SINTEF for providing the resources and environment that made it possible to write this book, and to the Research Council of Norway for supporting our research on snake robots.

Trondheim, Norway

Pål Liljebäck
Kristin Y. Pettersen
Øyvind Stavdahl
Jan Tommy Gravdahl

Contents

Chapter 1
Introduction

1.1 Background and Motivation

A snake robot is a robotic mechanism designed to move like a biological snake. Inspired by the robustness and stability of biological snake locomotion, snake robots carry the potential of meeting the growing need for robotic mobility in unknown and challenging environments. These mechanisms typically consist of many serially connected joint modules capable of bending in one or more planes. The many degrees of freedom of snake robots make them difficult to control but provide potential locomotion skills in cluttered and irregular environments which surpass the mobility of more conventional wheeled, tracked, and legged robots.

P. Liljebäck et al., *Snake Robots*, Advances in Industrial Control,
DOI 10.1007/978-1-4471-2996-7_1,

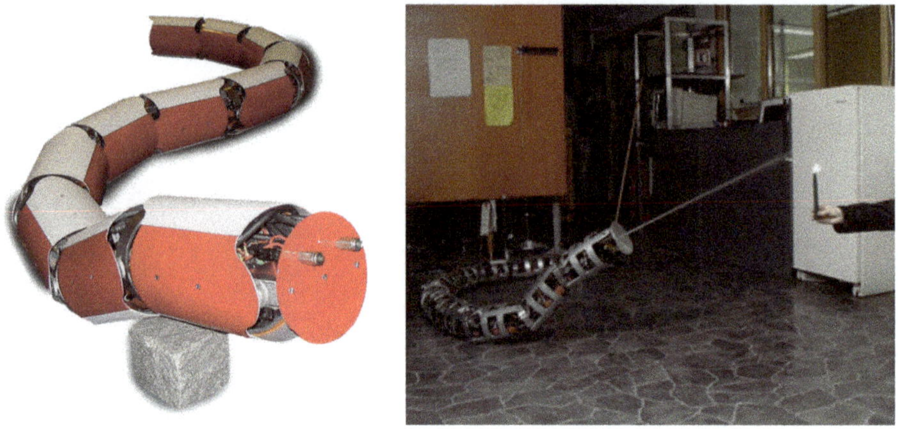

Fig. 1.1 The water hydraulic snake robot *Anna Konda*

Before we motivate research on snake robots in general, let us begin by describing how the authors became involved in and excited about this research activity. Research on snake robots at the Norwegian University of Science and Technology (NTNU) has spawned from a research project at SINTEF.[1] The project was initiated in 2003 after several major city fires in Trondheim, which launched an initiative to bring the fire department in closer relation with the research community in Trondheim to stimulate efforts that would improve fire safety. A specific idea which spurred from this initiative was the vision of a self-propelled fire hose as a robotic tool to aid human firefighters. This idea is clever in that the high-pressure water inside the hose can be employed as a *hydraulic* medium in the propulsion mechanism, a *fire extinguishing* medium, and a *cooling* medium for cooling the robot in environments with extreme temperatures. The resulting system would be a robotic fire hose that could move in extreme environments with the agility of a biological snake or, in other words, a water hydraulic snake robot. The Applied Cybernetics department at SINTEF was brought in to investigate this idea further and so began the research activity on snake robots at SINTEF and NTNU.

Researchers at SINTEF and NTNU quickly realised that developing such a mechanism represents a highly interdisciplinary task with challenges ranging from heat-resistant materials and water hydraulic joint actuation to control design and human–machine interaction. To show the feasibility of the concept, it was decided to develop a simple technology demonstrator in the form of a water hydraulic snake robot. The robot, which was named *Anna Konda*, is shown in Fig. 1.1 and is described in more detail in Liljebäck et al. (2006), Transeth et al. (2010). Anna Konda can move over relatively flat surfaces and can spray water through nozzles in its head. The robot is, however, far from ready for operating in harsh environments.

[1] SINTEF is a Norwegian research organisation which is tightly coupled with NTNU both geographically and through joint research activities.

The work on the Anna Konda robot helped us identify several major research challenges. The critical and most significant research challenge was, and still is, the serpentine propulsion mechanism of this system. After the development of Anna Konda, the research on snake robots at NTNU and SINTEF has therefore targeted snake robot locomotion in general without concern about the specific application of the robot. Although fire fighting was the initial motivation behind this research, the scope of the current research activities extends beyond merely fire intervention tasks since snake robots can potentially be used in many other applications where robust robotic mobility is required. A very relevant and important application of future snake robots involves search and rescue missions in earthquake areas. In particular, the earthquake in Haiti in January 2010 and the Tsunami which struck Japan in March 2011 are natural disasters in recent time which illustrate the need for technology that can be employed to efficiently locate survivors inside the debris of collapsed buildings. Other potential applications of snake robots include inspection and intervention operations in hazardous environments of industrial plants, manipulator tasks in tight spaces, and space and subsea operations. Figure 1.2 illustrates some of these applications.

In a global perspective, research on snake robots has been conducted for several decades. The research field was pioneered about 40 years ago by Professor Shigeo Hirose at Tokyo Institute of Technology, who developed the world's first snake robot as early as 1972 (see Hirose 1993). The robot, which is shown in Fig. 1.7, was equipped with passive wheels mounted tangentially along its body. The wheels enabled the robot to travel forward on a flat surface by controlling the joints according to a periodic body wave motion similar to the body waves displayed by biological snakes. In the decades following the pioneering research by Professor Hirose, several agile and impressive snake robots have been developed by research communities around the world in efforts to mimic the motion capabilities of their biological counterpart. However, the locomotive capabilities of current snake robots are still limited to fairly simple and controlled lab environments, and the world has not yet seen practical applications of snake robot locomotion. Nonetheless, researchers working with snake robots are motivated by the vision of a robotic propulsion mechanism with robust and agile mobility in challenging environments.

Development and control of snake robots is generally quite challenging for two primary reasons. First of all, a snake robot has many degrees of freedom, which means that the physical mechanism will contain a complex interconnection of sensors, actuators, and control logic. Moreover, the many degrees of freedom represent complex nonlinear dynamics which is challenging to analyse from a control design perspective. Second, the dependence on environment interaction is more complicated for a snake robot than for more conventional mobile robots. In particular, the propulsion mechanism of a wheeled, tracked, or legged robot is achieved with a separate and dedicated part of the robot. A snake robot, on the other hand, has no separate part which is dedicated to propulsion. Being essentially a smooth and flexible manipulator arm, the propulsion mechanism of a snake robot is rather an integrated part of the entire body, which means that propulsion requires synchronised motion of the entire robot in order to produce appropriate environment interaction forces.

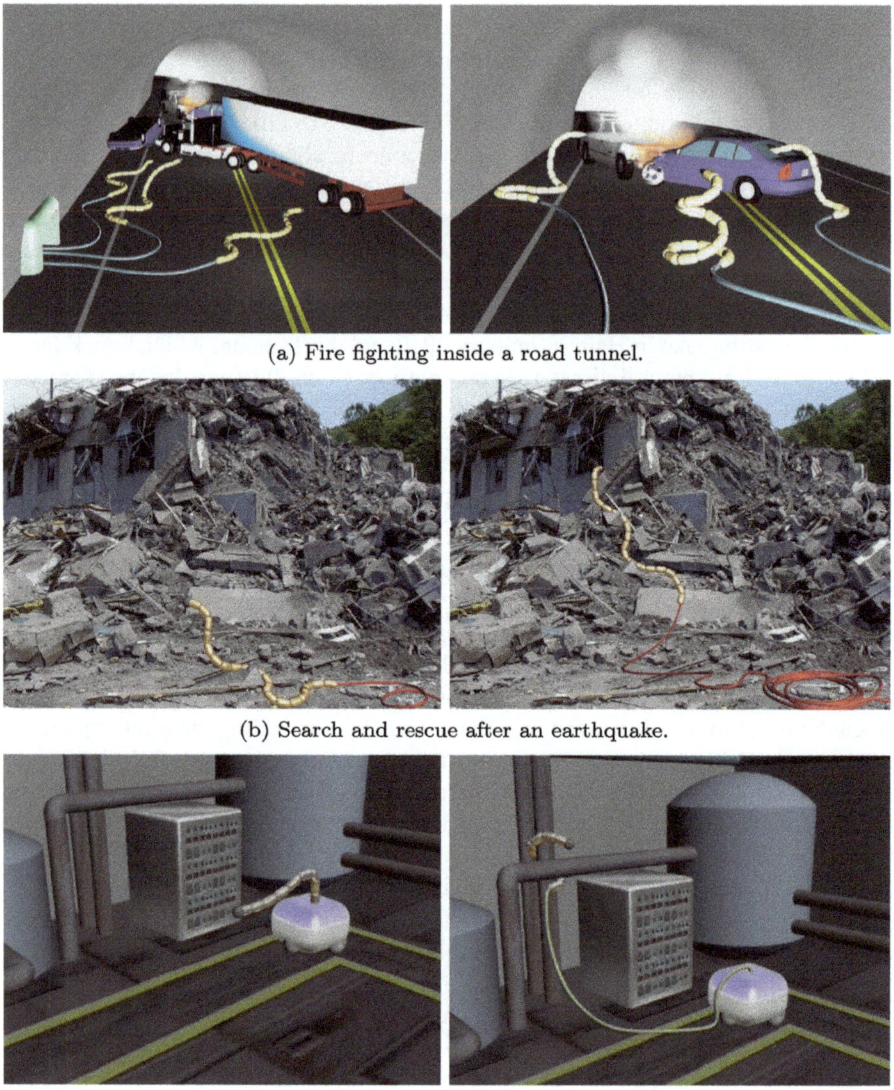

(a) Fire fighting inside a road tunnel.

(b) Search and rescue after an earthquake.

(c) Inspection and maintenance inside a process plant.

Fig. 1.2 Examples of future applications of snake robots

Motion based on such environment interaction is challenging both with respect to control design and mechanical implementation.

This book targets some of the challenges discussed above, and it is motivated by the long-term goal of developing snake robots which can move in unknown and challenging environments in order to support human intervention tasks. The main goal of the book is to create a foundation for future research by presenting the fundamentals of snake robot locomotion. To this end, the focus of the book is primarily directed towards control design. Efficient control strategies are vital to future applications of snake robots, and they are also instrumental in the development of these mechanisms. In particular, a control strategy with a solid mathematical foundation will immediately reveal what sensory capabilities, ground friction properties, actuator forces, etc., which are required in the physical robot to achieve a specific control objective. It is our hope that the results presented in this book will stimulate and support future research on these fascinating mechanisms.

1.2 Biological Snakes

This book is inspired by the robust motion capabilities of biological snakes. These amazing creatures are optimal in the sense that they have emerged through millions of years of evolution. In the following, we present aspects of biological snakes that we consider relevant to modelling, mechatronics, and control of snake robots. The material is based on Bauchot (1994), Hirose (1993), Hu et al. (2009), Mattison (2002).

1.2.1 The Anatomy of Snakes

The typical appearance of the skeletal structure of a snake is shown in Fig. 1.3 and consists of vertebrae, ribs, and a skull. Snakes can have between 130 and 500 vertebrae, with ribs attached to each one. The vertebrae constitute a column of movable joints that run through the body of the snake, and protects the spinal cord, which runs through a channel along the top of the vertebral column. The ribs attached to each side of a vertebra protect the internal organs.

The mechanical interconnection of the vertebrae is interesting. As illustrated in Fig. 1.4, two vertebrae are connected in a ball and socket arrangement. The magnitude of the relative rotational motion between two vertebrae is quite limited. In particular, the relative rotation between two vertebrae about the vertical axis ranges between 10° and 20°, while the relative rotation about the horizontal axis is limited to only a few degrees. These limitations may appear contradictory to the flexibility that snakes are known for, but this flexibility is, in fact, produced by the sum of the small movements of many vertebrae. Moreover, limiting the range of the relative movements leads to increased strength in the connection between the vertebrae. To prevent damage to the spinal cord due to twisting of the vertebrae about the axis tangential to the body, each vertebra has a number of wing-like projections that interlock loosely with their counterparts on the adjacent vertebrae. This limits the amount of twisting.

The body shape of a snake is changed with the help of muscles that are arranged diagonally along each side of the snake. The ends of these muscles are attached to

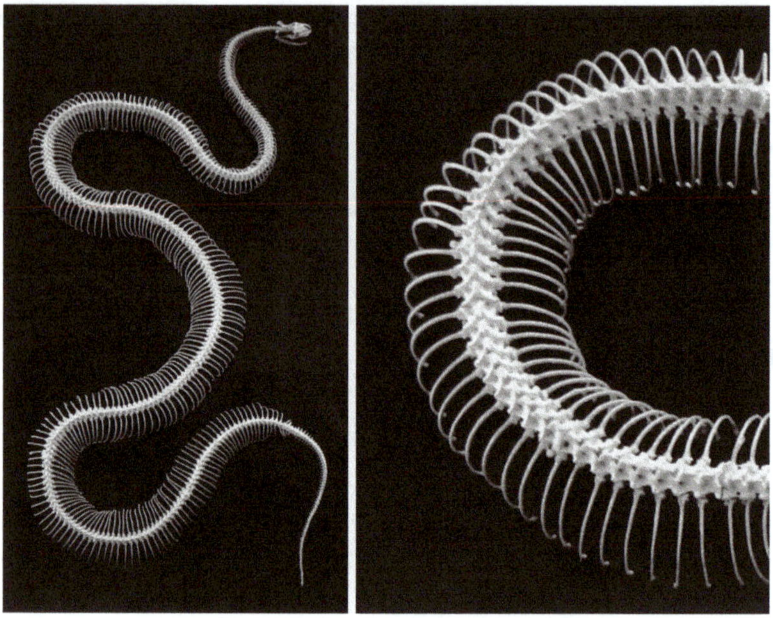

Fig. 1.3 The skeleton of a snake consisting of vertebrae, ribs, and a skull

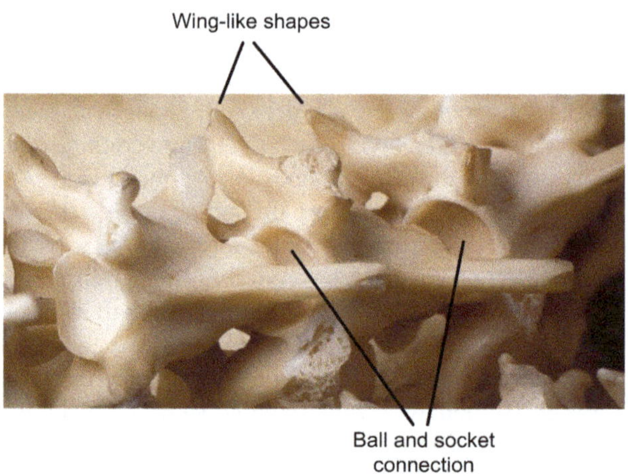

Fig. 1.4 Close-up of vertebrae from a snake

ribs, sometimes joining adjacent ribs, but mostly joining ribs that are some distance apart. The pattern of contraction and relaxation of these muscles determines the type of locomotion that is performed. For instance, if muscles on one side of the snake are contracted at the same time as the equivalent muscles on the other side are relaxed, then the body will be bent. If, on the other hand, opposite sets of muscles are contracted or relaxed simultaneously, then the snake will, to some extent, be able to shorten or extend its body at this location.

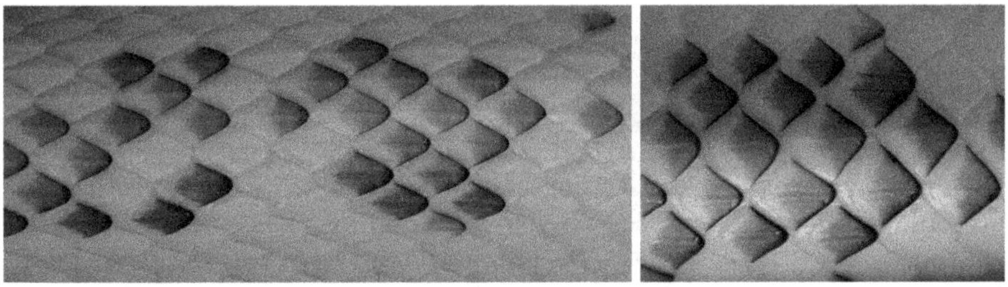

Fig. 1.5 The skin of a snake is completely covered by scales, which are formed from thickened areas of the skin

The skin of a snake is completely covered with scales. The scales are formed from thickened areas of the skin and are therefore integrated with the skin. The typical appearance of snake scales is shown to the left of Fig. 1.5, while the right shows the skin when it is stretched, thereby pulling the scales apart. The areas of skin between the scales allow the snake to flex its body while maintaining a smooth coverage of the scales. An important purpose of the scales is to form a physical protection from general wear and tear when the snake moves across rough surfaces. At the same time, the use of small units of armour allows greater flexibility than would large bony plates. Another feature of the scales is that they give the snake *anisotropic* ground friction properties, i.e. the scales give the snake a larger friction coefficient in the transversal direction of the snake body compared to in the tangential direction. Studies of biological snakes and simulation studies have indicated that this difference in the friction coefficients is important during forward gliding motion. The assumption on the importance of this friction property is proved in this book (Sect. 4.4).

1.2.2 The Locomotion of Snakes

Snakes are almost unique among the terrestrial vertebrates in their lack of legs. However, the lack of legs do not appear to have placed restrictions on the ability of snakes to move around. On the contrary, snake locomotion is stable, robust, and versatile. The speed of snake locomotion is, however, relatively slow, although certain species can move at speeds up to 11 km/h. Some snakes display specialised forms of motion. For instance, certain snakes can jump to heights of up to 1 m by curving their body into a vertical S-shape to serve as a spring, and then jump by stretching their body. Other snakes are able to glide through the air by throwing themselves from trees and forming their body in an aerodynamically favourable manner. In the following, the four most common types of biological snake locomotion are presented.

Lateral Undulation

Lateral undulation, also called serpentine crawling, is the fastest and most common form of snake locomotion and is illustrated in Fig. 1.6(a). During lateral undulation, continuous waves are propagated backwards along the snake body from head to tail. During this wave motion, the sides of the snake body push against irregularities in the surface, thereby pushing the snake forward. This form of locomotion is therefore not suitable on slippery surfaces. As the snake progresses, every point along the body passes the same point on the ground, and there is never any static contact between the ground and any point along the body. During swimming, the same wave motion is produced, but the body then pushes against the resistance of the water. The weight distribution of a snake during lateral undulation is not uniform, but rather distributed so that the peaks of the body wave curve are slightly lifted from the ground.

Concertina Locomotion

Concertina locomotion, which is illustrated in Fig. 1.6(b), is often employed in narrow spaces where the available range of motion is limited. The motion is carried out by first extending the front part of the body forward, while the back part is curved several times to provide an anchor against the narrow environment. Once the head and front part of the body are fully extended, they are subsequently used to provide an anchor in the same way so that the back part of the body can be drawn up. The sequence is then repeated.

The principle behind concertina locomotion relies on the difference between the large static friction forces at the anchor points and the low kinetic friction forces in the part of the body which is extended. The motion pattern is not very efficient in terms of energy consumption, but is often needed in order to traverse tight spaces.

Rectilinear Crawling

Rectilinear crawling is a slow form of locomotion often employed by heavy-bodied snakes. Also snakes in the final stages of stalking their pray use rectilinear crawling to avoid alerting their intended victim. During rectilinear crawling, the snake uses the edges of the scales on its underside as anchor points to pull itself forward in a more or less straight line. The operation consists of stretching forward and hooking the edges of the scales over small irregularities, then pulling the body up to this point. Alternate parts of the body will be stretching and pulling at the same time. The motion pattern is illustrated in Fig. 1.6(c).

Sidewinding

Sidewinding is a form of locomotion which is usually employed by snakes that live in areas of loose sand, e.g. desert snakes. The motion resembles concertina motion

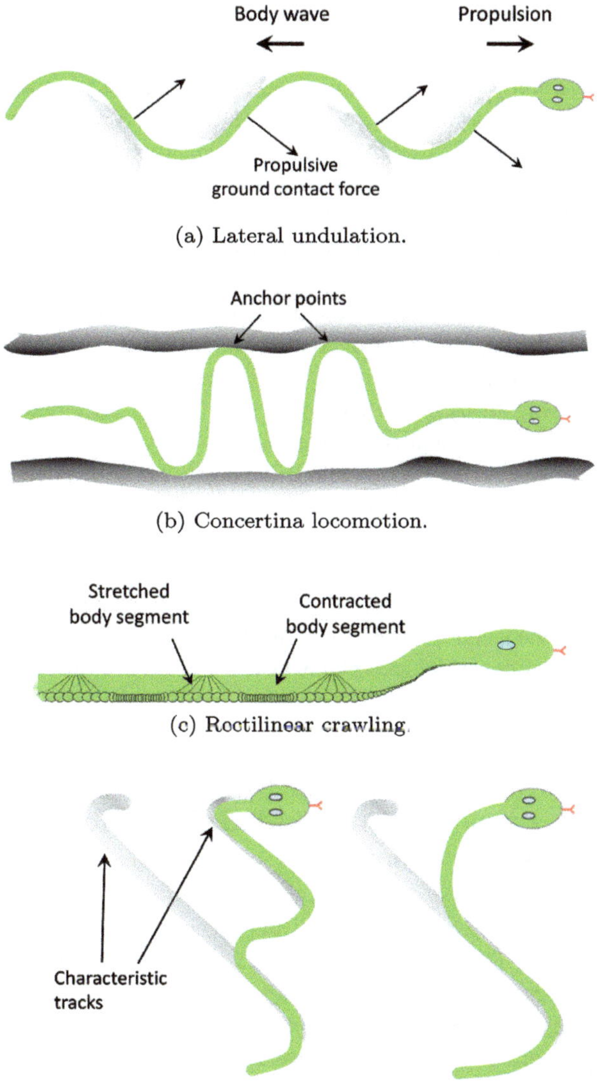

(a) Lateral undulation.

(b) Concertina locomotion.

(c) Rectilinear crawling.

(d) Sidewinding.

Fig. 1.6 Different forms of biological snake locomotion

in that one part of the body acts as an anchor, while another part is moved forward. Starting from a resting position, the head and neck are raised off the ground and thrown sideways, while the rest of the body provides an anchor against the ground. Once the head and fore part of the body are again on the ground, they in turn act as an anchor while rest of the body repeats the same motion. The snake moves at about 45° with respect to its heading and leaves a trail of characteristic markings in the sand, as illustrated in Fig. 1.6(d).

The Control System of Snakes

The employed locomotion method of snakes sometimes depends on the size of the snake and sometimes on the substrate over (or through) which it is moving. In fact, an interesting difference between snake locomotion and legged forms of locomotion is that the basic repeating motion that leads to propulsion of legged animals to a large extent depends on the progression speed of the animal. On the other hand, the basic repeating motion that leads to propulsion of snakes largely depends on the environment, and not on the speed.

Considering the large number of muscles involved in the motion of a snake and also the large number of contact points that are sensed by its nervous system, it is fair to say that the coordination of snake movements is both impressive and complex. Investigations of the electrical activity that accompanies the muscular contraction during movement show that the motor response is segmentary. Nerve impulses are propagated backwards along the snake body through the bone marrow. These impulses successively activate local muscle groups, which bend the snake body. Musculature is, in other words, successively, and not simultaneously, active, and only for a few elements at a time. The bending motion at a point along the snake body is also influenced by the sensory information transmitted by the skin. Simply speaking, the snake produces a relatively simple motor command which is modulated by local reflexes. This explains how every point in the body is able to follow the same trajectory.

1.3 Previous Work on Modelling, Mechatronics, and Control of Snake Robots

In the following, we provide an overview of previous literature on snake robot locomotion. The review is structured according to the title of this book by first considering research efforts related to modelling and analysis of snake robots, followed by research on physical implementation of these mechanisms, and finally considering previous control design efforts for snake locomotion. The scope of this book, which we present in Sect. 1.4, is motivated and justified based on this literature review.

1.3.1 Previous Work on Modelling and Analysis of Snake Robots

Previous literature on modelling and analysis of snake robot locomotion is summarised in Table 1.1. The table separates works that consider snake locomotion from a planar (2D) perspective and works that also include three-dimensional aspects of the motion. A more detailed description of this literature is presented in the following.

Table 1.1 Previous work on modelling and analysis of snake robots

Biomechanical studies of biological snakes	
2D perspective	Gray (1946), Moon and Gans (1998), Ma (1999)
3D perspective	Hirose (1993), Hu et al. (2009)
Flat surface locomotion with sideslip constraints	
2D perspective	Hirose (1993), Krishnaprasad and Tsakiris (1994), Kelly and Murray (1995), Ostrowski (1996), Ostrowski and Burdick (1998), Ishikawa (2009), Hatton and Choset (2009a), Prautsch and Mita (1999), Ute and Ono (2002), Matsuno and Mogi (2000), Matsuno and Sato (2005)
3D perspective	Ma et al. (2003), Tanaka and Matsuno (2008b), Date and Takita (2005)
Flat surface locomotion without sideslip constraints	
2D perspective	Ma (2001), Ma and Tadokoro (2006), Saito et al. (2002), Li and Shan (2008), Kane and Lecison (2000), Grabec (2002), Hicks (2003), Mehta et al. (2008), Chernousko (2005), Nilsson (2004), Hu et al. (2009)
3D perspective	Shapiro et al. (2007), Ma et al. (2004), Transeth et al. (2008a)
Robotic fish and eel-like mechanisms	
2D perspective	McIsaac and Ostrowski (2003a), Kanso et al. (2005)
3D perspective	Boyer et al. (2006), Zuo et al. (2008), Morgansen et al. (2001, 2002, 2007), Vela et al. (2002a)
Locomotion in environments with obstacles	
2D perspective	Shan and Koren (1993), Bayraktaroglu and Blazevic (2005), Date and Takita (2007)
3D perspective	Chirikjian (1992), Chirikjian and Burdick (1995), Yamada and Hirose (2006a), Shan and Koren (1995), Tanev et al. (2005), Transeth et al. (2008b)

Biomechanical Studies of Biological Snakes

A complete treatment of previous studies of biological snakes is beyond the scope of this book. The biomechanical studies that we consider to be most relevant to this book are summarised in the following.

One of the earliest analytical studies of snake locomotion was given in Gray (1946), where mathematical descriptions of the forces acting on a snake are proposed and used to derive properties of snake locomotion. One of Gray's conclusions was that forward motion of a planar snake requires the existence of external forces acting in the direction normal to the snake body.

Hirose (1993) studied biological snakes and modelled the snake body as a continuous curve that could not move sideways (sideslip constraints). A well-known result by Hirose is the formulation of the *serpenoid curve*, which is a mathematical description of lateral undulation (the most common form of snake locomotion). This mathematical description is elaborated in Sect. 4.7. Hirose also investigated adaptive functions of biological snakes (i.e. sinus-lifting, the α-adaptive principle, and the l-adaptive principle) and proposed mathematical descriptions of how external factors, such as ground friction and temperature, affect the shape of a snake during locomotion. Furthermore, Hirose investigated locomotion efficiency inside a maze, i.e. when the snake touches a wall on each side.

An alternative description of lateral undulation, named the *serpentine curve*, was proposed in Ma (1999), where a mathematical model of the muscle characteristics of snakes is employed to derive the resulting form of the body shape during lateral undulation. Ma showed that snake locomotion according to the serpentine curve has a higher locomotive efficiency than locomotion according to the serpenoid curve. The locomotive efficiency during slip-free motion was defined as the ratio between the tangential and normal direction friction forces on the snake body.

Other interesting studies of snake locomotion include the work in Moon and Gans (1998), which considers the mechanism by which muscular activity of a snake produces curvature and propulsion. In particular, the muscular activity is studied as a snake interacts with pegs in order to push itself forward. A more recent study given in Hu et al. (2009) investigates the frictional properties of snake skin both mathematically and experimentally. In particular, the study shows that the friction coefficient of a snake in the transversal direction of the body is larger than the friction coefficient in the tangential direction. This property is important during forward gliding motion. The study also shows that the weight distribution of a snake during lateral undulation is not uniform, but rather distributed so that the peaks of the body wave curve are slightly lifted from the ground. This is often referred to as sinus-lifting.

Modelling of Flat Surface Locomotion with Sideslip Constraints

As noted in e.g. Gray (1946), each part of a biological snake conducting lateral undulation follows the path traced out by the head. This phenomenon is partially explained by the frictional anisotropy of snake skin studied in e.g. Hu et al. (2009) but is also caused by irregularities on the surface that provide grip and enable the snake to glide forward without slipping sideways. To mimic this motion, many models of snake robots have been developed under the explicit assumption that the body cannot move sideways (sideslip constraints). This assumption introduces nonholonomic constraints (Bloch et al. 2003) in the equations of motion of the robot. In practice, such conditions are usually achieved by installing passive wheels along the body of the snake robot.

Several works attack the motion control problem of wheeled snake robots with tools from differential geometry. Early approaches of such a form are presented in Kelly and Murray (1995), Krishnaprasad and Tsakiris (1994), which model the kinematics of wheeled snake robots and analyse the relationship between body shape changes and the resulting displacement of the robot. These works also assess the controllability of such mechanisms. Similar approaches are considered in Ostrowski and Burdick (1998), Ostrowski (1996), where also the dynamics of wheeled snake robots is considered, and where system symmetries are utilised to arrive at reduced forms of the model. Modelling and controllability analysis of the kinematics of a three-linked wheeled snake robot is also considered in Ishikawa (2009). Furthermore, Hatton and Choset (2009a) introduce the concept of a body velocity integral in order to easily approximate the net displacement of a snake robot during a gait. The method requires that the system coordinates are properly chosen.

A model of the 2D dynamics of a wheeled snake robot is developed in Prautsch and Mita (1999) from Lagrange's equations of motion, and in Ute and Ono (2002) from first principles. The works in Matsuno and Mogi (2000), Matsuno and Sato (2005) present models of the 2D kinematics and dynamics of snake robots, respectively, where some, but not all, of the links are wheeled. The wheel-less links correspond to links that are lifted from the ground. Lifting some of the wheeled links is sometimes desirable from a control perspective to make the motion of the robot less constrained. A model of the 3D kinematics of a snake robot that describe the lifting of the links more accurately is presented in Ma et al. (2003). Furthermore, Tanaka and Matsuno (2008b) present a model of the 3D dynamics of a snake robot consisting of a grounded base part and a lifted head part (for manipulation purposes), where some, but not all, of the links in the base part are wheeled.

Continuum models of snake robot dynamics, where the snake is treated as a continuous curve that cannot move sideways, are presented in Date and Takita (2005), Hirose (1993). The model in Hirose (1993) is planar, while the model in Date and Takita (2005) considers the 3D dynamics of the continuous snake robot.

Modelling of Flat Surface Locomotion Without Sideslip Constraints

In addition to the many models of snake robots with sideslip constraints, there are also many models that do not enforce such constraints, but instead only assume that the links exhibit *anisotropic* ground friction properties similar to biological snakes. With anisotropic ground friction properties, the friction coefficients describing the friction force in the tangential and normal direction of a link, respectively, are different. Models based on such ground friction properties are generally more complex to analyse than models based on sideslip constraints since there is no longer a direct connection between the body shape changes and the resulting displacement of the robot.

Ma (2001) employs the Newton–Euler formulation to develop a 2D model of the dynamics of a snake robot with anisotropic ground friction properties. The ground friction model includes both static and dynamic Coulomb ground friction forces. The model of the robot is formulated in two ways, where the first form gives the propulsion of the robot and the joint torques based on knowledge of the body shape changes, whereas the second form gives the propulsion and body shape changes of the robot based on knowledge of the joint torques. The model is extended in Ma and Tadokoro (2006) to also describe snake locomotion on a slope.

Another model of planar wheel-less snake robot dynamics is developed in Saito et al. (2002) from first principles. The model considers both viscous and Coulomb ground friction forces. Simulations with the model are carried out to derive properties of snake robot dynamics. The model from Saito et al. (2002) is employed in Li and Shan (2008) to study the controllability of the joints of a snake robot under the assumption that one joint is passive. However, the analysis does not consider the position of the robot.

Models of planar snake robot dynamics with anisotropic viscous ground friction are presented in Grabec (2002), Hicks (2003), Kane and Lecison (2000). The work Hicks (2003) exploits symmetries in the system (cyclic coordinates) to transform the model to a reduced form where the shape dynamics is decoupled from the displacement dynamics of the snake robot and investigates general requirements for the propulsion of a three-linked snake robot. A friction model that includes both viscous and Coulomb friction forces is proposed and analysed in Mehta et al. (2008).

A model that considers *isotropic* Coulomb ground friction forces (both static and dynamic friction) is presented in Chernousko (2005). Isotropic ground friction is also assumed in Nilsson (2004), where a continuum approach along with energy arguments are employed to analyse planar snake locomotion under isotropic friction conditions. Shapiro et al. (2007) model the frictional contact forces between a snake robot and a compliant surface. The dynamics of planar snake locomotion is described in terms of a continuum model in Hu et al. (2009), where the snake is treated as a continuous curve influenced by Coulomb friction forces from the ground. The model is employed to study the effect of anisotropic ground friction properties on the propulsion of snakes.

The 3D dynamics of a snake robot during locomotion across flat surfaces is considered in Ma et al. (2004), Transeth et al. (2008a). The model in Ma et al. (2004) is developed from the Newton–Euler formulation and includes both static and dynamic Coulomb ground friction forces. The model is employed to study sinus-lifting during lateral undulation. Transeth et al. (2008a) model snake robot dynamics by use of the framework of nonsmooth dynamics. The model, which represents a hybrid system, describes the normal direction contact forces from the ground and the Coulomb ground friction forces by use of set-valued force laws.

Modelling of Robotic Fish and Eel-Like Mechanisms

Research on robotic fish and eel-like mechanisms is relevant to research on snake robots since these mechanisms are very similar. A complete treatment of robotic underwater locomotion is beyond the scope of this review. However, a representative part of previous research related to modelling of such mechanisms is presented in the following.

A model of eel-like motion is developed in McIsaac and Ostrowski (2003a) based on tools from differential geometry that were also considered in some of the works concerning wheeled snake robots described above. However, the model does not place sideslip constraints on the robot. Instead, the eel-like mechanism is propelled by hydrodynamic forces modelled by a viscous friction model. The dynamics of eel-like motion is also considered in Kanso et al. (2005), where a model reduction is proposed to allow the net motion of the robot to be described as a sum of geometric and dynamic phases over closed curves in the shape space, in Boyer et al. (2006), where a continuum model is formulated based on beam theory, and in Zuo et al.

(2008), where first principles are employed to model the dynamics of a swimming snake robot. The works in Morgansen et al. (2001, 2002, 2007), Vela et al. (2002a), model the dynamics of a robotic fish influenced by lift and drag forces in an inviscous fluid. The controllability of the fish-like mechanism is also assessed in these works.

Modelling of Locomotion in Environments with Obstacles

In Chirikjian and Burdick (1995), Chirikjian (1992), the kinematics of snake robots is modelled in terms of a continuous backbone curve that captures the macroscopic geometry of the robot. Gaits for the backbone curve, which determine the shape of the snake robot, are specified with respect to environment constraints and the desired locomotion trajectory of the robot. The approach is original in that the problem of locomotion in cluttered environments is attacked at a purely kinematic level. The work by Chirikjian and Burdick is extended in Yamada and Hirose (2006a), where a continuum kinematics model is presented that explicitly handles the case of backbone curves that can be bent, but not twisted. This condition is in line with most physical snake robots, which are generally able to bend but not twist their body. The kinematic constraints imposed on a snake robot due to external obstacles are modelled in Shan and Koren (1993, 1995). These works also analyse how obstacles around a snake robot affect its degrees of freedom.

The only known works that consider the dynamics of snake robots in environments with obstacles (i.e. where obstacle contact forces are considered) are presented in Bayraktaroglu and Blazevic (2005), Date and Takita (2007), Tanev et al. (2005), Transeth et al. (2008b). In Bayraktaroglu and Blazevic (2005), a dynamic simulation software called *WorkingModel* is used to simulate a planar snake robot interacting with circular obstacles. Contact forces are calculated from a spring-damper approximation. A similar approach is employed in Tanev et al. (2005), where the simulation software *Open Dynamics Engine* (ODE) is used to model a snake robot interacting with various forms of obstacles. Date and Takita use the multi-body dynamics simulation software *Autolev* to study the motion of a snake robot during contact with a single peg, where the contact with the peg is modelled as a spring-damper system. The works in Bayraktaroglu and Blazevic (2005), Date and Takita (2007), Tanev et al. (2005) do not provide the equations underlying the dynamics of the snake robot due to the use of general-purpose simulation software. On the other hand, the model proposed in Transeth et al. (2008b) is, to our best knowledge, the only work which explicitly presents the equations of motion underlying the obstacle interaction dynamics of a snake robot. The model, which represents a hybrid system, is formulated within the framework of nonsmooth dynamics. A timestepping method is used to simulate the dynamics of the robot, which means that the system equations are discretised with a time step determined by a fixed error criterion, and trajectories of the system are approximated without tracking events (i.e. obstacle impacts). Transeth et al. (2008b) also introduce the term *obstacle-aided locomotion*, which involves using external obstacles as push points to aid the propulsion instead of avoiding them.

Table 1.2 Previous work on implementation of physical snake robots

Snake robots without contact force sensors	
With passive wheels	Endo et al. (1999), Togawa et al. (2000), Ma et al. (2001), Wiriyacharoensunthorn and Laowattana (2002), Mori and Hirose (2002), Miller (2002), Ye et al. (2004b, 2007), Yamada et al. (2005), Crespi and Ijspeert (2008), Yu et al. (2008, 2009), Kamegawa et al. (2009)
Without passive wheels	Yim (1994), Yim et al. (2002), Worst and Linnemann (1996), Dowling (1997, 1999), Nilsson (1998), Ohno and Hirose (2001), Saito et al. (2002), Brunete et al. (2006), Chen et al. (2007), Wright et al. (2007), Kuwada et al. (2008), Yamada and Hirose (2008, 2009), Ohashi and Hirose (2010)
With active propulsion	Kimura and Hirose (2002), Yamada and Hirose (2006b), Taal et al. (2009), Fjerdingen et al. (2009), Kamegawa et al. (2004), Masayuki et al. (2004), Granosik et al. (2006), Gao et al. (2008), McKenna et al. (2008), Ijspeert et al. (2007), Hara et al. (2007)
Snake robots with contact force sensors	
With passive wheels	Hirose (1993), Chen et al. (2008)
Without passive wheels	Bayraktaroglu (2008), Andruska and Peterson (2010), Liljebäck et al. (2006), Fjerdingen et al. (2008)
With active propulsion	Taal et al. (2009)

1.3.2 Previous Work on Implementation of Physical Snake Robots

Previous literature that considers development of physical snake robots is summarised in the following. The review is structured according to the focus of this book on snake robot locomotion based on measurements of environment contact forces, which we consider important for body shape adaptation in cluttered environments. In particular, we have chosen to separate the works that consider snake robots *with* contact force sensors from the works that do *not* include such sensor capabilities in the robot design. This partitioning illustrates, as seen in Table 1.2, that previous research on environment sensing for snake robots is limited. The referred works are summarised in Table 1.2, which separates between snake robots with passive wheels, which are advantageous during motion across flat surfaces, snake robots without such passive wheels, and snake robots equipped with active propulsion.

Snake Robots Without Contact Force Sensors

Hirose developed the world's first snake robot as early as 1972 (Hirose 1993). The robot, which is shown in Fig. 1.7, was equipped with passive wheels to realise the anisotropic ground friction property that enables forward locomotion on flat surfaces.

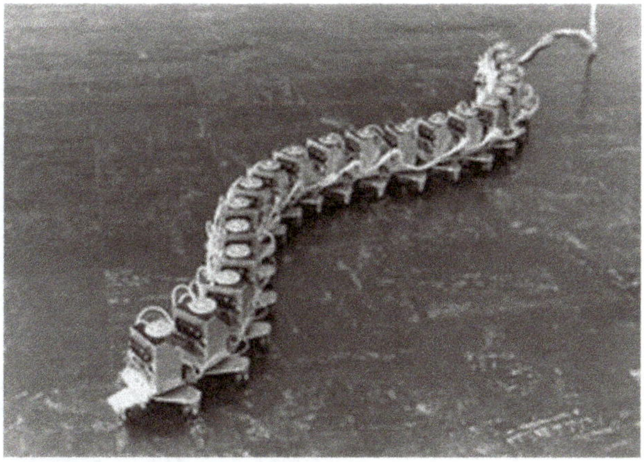

Fig. 1.7 The snake robot *ACM III*, which was the world's first snake robot developed by Prof. Shigeo Hirose in 1972. Courtesy of Tokyo Institute of Technology

Fig. 1.8 The snake robot *ACM R3* developed at Tokyo Institute of Technology. The robot is covered with passive wheels. Courtesy of Tokyo Institute of Technology

Several other snake robots with passive wheels have been proposed over the years, such as the robots presented in Endo et al. (1999), Togawa et al. (2000), Ma et al. (2001), Wiriyacharoensunthorn and Laowattana (2002), Mori and Hirose (2002) (see Fig. 1.8), Miller (2002) (see Fig. 1.9), Ye et al. (2004b, 2007), Yamada et al. (2005) (see Fig. 1.10), Crespi and Ijspeert (2008), Yu et al. (2008, 2009), and Kamegawa et al. (2009). Some of the robots can only display planar motion, while other robots can move their links both horizontally and vertically. Some robots have shielded joint modules that enable motion in environments with e.g. mud and dust, and even motion under water (see Yamada et al. 2005 and Fig. 1.10), while other robots have modules with exposed electronic components which only allow them to move in clean lab environments. A common feature of these mechanisms, however, is that they are generally only able to move across relatively flat surfaces since passive wheels do not move very well in a cluttered environment. Such mechanisms are therefore suitable for motion control experiments on relatively flat surfaces, but not for practical applications of snake robots in more challenging environments.

Fig. 1.9 The snake robot *S5* developed by Dr. Gavin Miller. The robot has passive wheels on its underside. Courtesy of Dr. Gavin Miller

Snake robots without passive wheels, i.e. robots that basically consist of straight links interconnected by motorised joints, are presented in Yim (1994), Yim et al. (2002), Worst and Linnemann (1996), Dowling (1997, 1999), Nilsson (1998), Ohno and Hirose (2001), Saito et al. (2002), Brunete et al. (2006), Chen et al. (2007), Wright et al. (2007), Kuwada et al. (2008), Yamada and Hirose (2008, 2009), Ohashi and Hirose (2010) (see Fig. 1.11). Despite its lack of wheels, the snake robot in Saito et al. (2002) maintains an anisotropic ground friction property since the underside of each link has edges, or grooves, that run parallel to the link. This robot can therefore move forward by lateral undulation through purely planar motion. Robots whose ground friction properties are isotropic and, on the other hand, can move forward during lateral undulation by resorting to sinus-lifting, i.e. by slightly lifting the peaks of the body wave curve from the ground (see e.g. Ohno and Hirose 2001; Yamada and Hirose 2008). However, robots with isotropic friction are mostly used for studying gaits other than lateral undulation, such as gaits based on sidewinding, inchworm motion, or lateral rolling. A notable feature of the works presented in Wright et al. (2007) (see Fig. 1.12) and Yamada and Hirose (2009) (see Fig. 1.13) is the focus on development of small, light-weight, and strong joint actuation mechanisms, which are important for many future applications of snake robots.

Fig. 1.10 The snake robot *ACM R5* developed at Tokyo Institute of Technology. The robot is covered by passive wheels and can swim under water. Courtesy of Tokyo Institute of Technology

Fig. 1.11 The snake robot *ACM R7* developed at Tokyo Institute of Technology. The robot can move by curving its body into a loop and rolling forward like a wheel. Courtesy of Tokyo Institute of Technology

There are also works that consider active propulsion along the body of a snake robot, for example by equipping each link with motorised wheels (Kimura and Hirose 2002; Taal et al. 2009; Yamada and Hirose 2006b; Fjerdingen et al. 2009), or by installing tracks along the body of the snake robot (Gao et al. 2008; Granosik et al. 2006; Kamegawa et al. 2004; Masayuki et al. 2004; McKenna et al. 2008), or by installing legs on the links of the robot (Ijspeert et al. 2007), or by employing a screw drive mechanism (Hara et al. 2007). The snake robots with active propulsion presented in Granosik et al. (2006), McKenna et al. (2008), and Ijspeert et al. (2007) are shown in Figs. 1.14, 1.15, and 1.16, respectively.

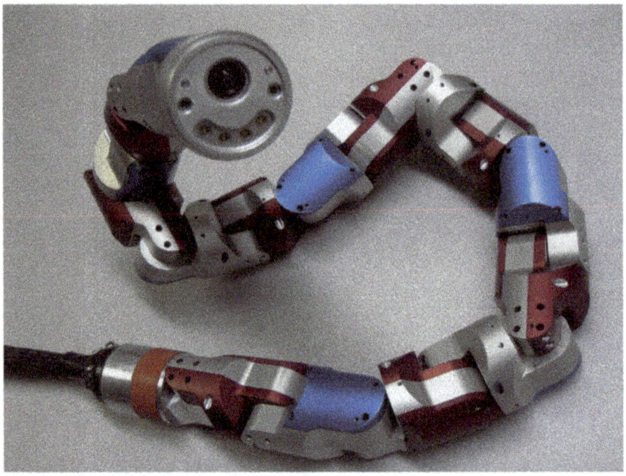

Fig. 1.12 The snake robot *Uncle Sam* developed at Carnegie Mellon University. The robot has a strong and compact joint mechanism and can climb up poles. Courtesy of Carnegie Mellon University

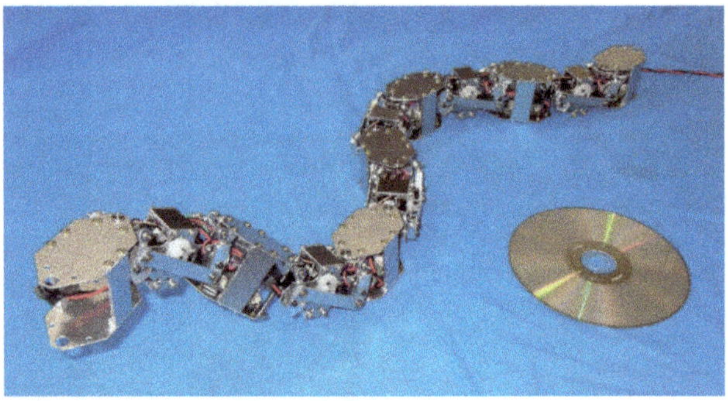

Fig. 1.13 A snake robot with a miniature joint mechanism developed at Tokyo Institute of Technology. Courtesy of Tokyo Institute of Technology

Snake Robots with Contact Force Sensors

Previous research on environment sensing for snake robots is limited. The wheeled snake robot developed by Hirose already in 1972 (Hirose 1993) was equipped with contact switches, which enabled the robot to demonstrate lateral inhibition with respect to external obstacles. Snake robots with cylindrical modules covered by force sensors are proposed in Fjerdingen et al. (2008), Liljebäck et al. (2006). The force sensor systems in these works are able to detect and, to some extent, assess the magnitude of external forces applied at certain areas of the joint modules. A snake robot with active wheels, where each wheel axis is equipped with a three-axial force sensor, is presented in Taal et al. (2009). The force sensor measures the translational forces on the wheel axis based on optical range measurements. Bayraktaroglu (2008) presents a wheel-less snake robot with contact switches and presents experimental results where the robot is propelled forward by pushing against pegs that are detected by the contact switches. A snake robot with passive wheels and strain gauge sensors is proposed in Chen et al. (2008), where the strain gauge sensors are shown to successfully measure the constraint forces on the wheels. Ideas related

Fig. 1.14 The *OmniTread* snake robot developed at the University of Michigan. The robot has pneumatic joints and is covered by motorised tracks. Courtesy of the University of Michigan

to environment sensing for snake robots are considered in Andruska and Peterson (2010), where the preliminary design of a capacitive contact sensor is proposed that can be wrapped around each module of a snake robot.

Fig. 1.15 A snake robot with a skin drive propulsion system developed at Carnegie Mellon University. A motor drives the outer skin backwards along the snake body in order to propel the robot forward. Courtesy of Carnegie Mellon University

Fig. 1.16 A salamander-like snake robot with motorised legs that propel the robot forward. The robot can operate under water. Courtesy of the Biologically Inspired Robotics Group at Ecole Polytechnique Fédérale de Lausanne (EPFL)

1.3.3 Previous Work on Control of Snake Robots

In the following, we provide an overview of previous control design efforts for snake robots. The review is structured according to Tables 1.3 and 1.4, which summarise all papers referred to in this section. The two tables separate between works that present gait patterns without explicitly controlling the position or heading of the snake robot and works that present gait patterns along with position and/or heading controllers. The review focuses on controllers based on lateral undulation, which is the most common form of snake robot locomotion and which is also most relevant to the results presented in this book.

Remark 1.1 Stability analysis of control laws for snake robots is challenging due to the complexity of existing models of these mechanisms. For this reason, applications

Table 1.3 Previous work on control of snake robot locomotion (1 of 2)

Flat surface locomotion with sideslip constraints	
Without position or heading control	Shan and Koren (1993), Kelly and Murray (1995), Ostrowski and Burdick (1998), Date and Takita (2005), Tanaka and Matsuno (2009), Ute and Ono (2002), Sato et al. (2010), Wang et al. (2010)
With position and/or heading control	Prautsch et al. (2000), Date et al. (2000, 2001a, 2001b), Yamakita et al. (2003), Matsuno and Mogi (2000), Ma et al. (2003), Matsuno and Suenaga (2003), Ye et al. (2004a), Matsuno and Sato (2005), Tanaka and Matsuno (2008a, 2008b), Wiriyacharoensunthorn and Laowattana (2002), Watanabe et al. (2008), Ishikawa (2009), Ishikawa et al. (2010), Paap et al. (1999), Linnemann et al. (1999), Murugendran et al. (2009)
Flat surface locomotion without sideslip constraints	
Without position or heading control	Dowling (1997, 1999), Ma (2001), Ma et al. (2004), Saito et al. (2002), Chernousko (2003, 2005), Transeth et al. (2007b), Burdick et al. (1995), Gonzalez-Gomez et al. (2007), Yu et al. (2008), Chirikjian and Burdick (1995), Poi et al. (1998), Yim (1994), Yim et al. (2002), Ohno and Hirose (2001), Rincon and Sotelo (2003), Hatton and Choset (2010), Yamada and Hirose (2010), Mori and Hirose (2002), Chen et al. (2004), Ohashi and Hirose (2010)
With position and/or heading control	Hicks (2003), Hicks and Ito (2005)

Table 1.4 Previous work on control of snake robot locomotion (2 of 2)

Robotic fish and eel-like mechanisms	
Without position or heading control	Morgansen et al. (2001), Melli et al. (2006), Crespi and Ijspeert (2008)
With position and/or heading control	McIsaac and Ostrowski (2003a, 2003b), Andruska and Peterson (2002, 2007), Vela et al. (2002a)
Locomotion in environments with obstacles	
Without position or heading control	Hirose (1993), Andruska and Peterson (2008), Kuwada et al. (2008), Kulali et al. (2002), Greenfield et al. (2005), Kamegawa et al. (2009), Zarrouk et al. (2010), Nilsson (1997), Chen et al. (2007), Lipkin et al. (2007), Hatton and Choset (2009b)
With position and/or heading control	Bayraktaroglu and Blazevic (2005), Bayraktaroglu (2008), Date and Takita (2007), Sfakiotakis and Tsakiris (2007)

of formal stability analysis tools in previous snake robot literature are very limited. Simulations and experimental investigations are instead the common approach in the literature for providing support of proposed control strategies.

Control of Flat Surface Locomotion with Sideslip Constraints

A majority of previous control design efforts for snake robots has focused on locomotion where the links are subjected to nonholonomic constraints, i.e. where each link is constrained from moving sideways. Shan and Koren (1993) consider a snake robot that uses solenoids for attachment to the environment and proposes gaits for forward and turning motion of this mechanism. Tools from differential geometry are employed in Kelly and Murray (1995), Ostrowski and Burdick (1998) to demonstrate that sinusoidal shape inputs to wheeled snake robots lead to propulsion.

A position and path following controller for a wheeled snake robot is proposed in Prautsch et al. (2000), where also Lyapunov analysis is employed to analyse the controller. The work also considers approaches for preventing the snake robot from attaining a straight shape, which is singular with respect to propulsion. The works Date et al. (2000, 2001a, 2001b) propose path following controllers for wheeled snake robots aimed at minimising the lateral constraint forces on the wheels during lateral undulation. The controllers are based on a measure of dynamic manipulability, which describes the ability of the robot to generate propulsive force. A similar approach is employed in Yamakita et al. (2003), which proposes a gait pattern aimed at minimising the lateral constraint forces on the wheels, and in Date and Takita (2005), which formulates and solves an optimisation problem in order to minimise the torque input. The optimisation problem is solved using a 3D continuum model of the snake robot.

In Ma et al. (2003), Matsuno and Mogi (2000), Matsuno and Sato (2005), Matsuno and Suenaga (2003), position and path following controllers are proposed for the case where some, but not all, of the snake robot links are wheeled. The wheelless links correspond to links that are lifted from the ground, which give the system more degrees of freedom that can be utilised to follow a trajectory while simultaneously maintaining a high manipulability. Similar approaches are considered in Tanaka and Matsuno (2008a, 2008b, 2009), where also strategies for sinus-lifting during lateral undulation are proposed.

Ute and Ono (2002) propose a gait based on a self-excitation principle where joint angle information determines the winding motion of a snake robot. Directional control during lateral undulation is considered in Wiriyacharoensunthorn and Laowattana (2002), Ye et al. (2004a). Watanabe et al. (2008) propose a position controller for a wheeled snake robot that takes ground friction forces into account. A similar approach is employed in Sato et al. (2010), where deviations of the joint angles from their setpoints are used to modify the oscillatory joint motion, thereby enabling the snake robot to automatically adapt its motion to variations in the ground friction conditions. The works in Ishikawa (2009), Ishikawa et al. (2010) propose position and path following controllers for three-linked and four-linked wheeled snake robots based on Lie bracket calculations and controllability analysis results. The concept of passive creeping is considered in Wang et al. (2010), which involves adjusting the motion of a snake robot based on a measure of the dissipated energy, thereby achieving adaptation of the motion to different surface conditions. Local orbital stability of state trajectories during the motion is concluded based on recurrence plots.

A snake robot with *active* wheels is considered in Linnemann et al. (1999), Paap et al. (1999), where an optimisation scheme is employed to make the robot follow the path that minimises energy dissipation due to friction forces. Active wheels are also assumed in Murugendran et al. (2009), where a path following controller for such snake robots is proposed on a kinematic level.

Remark 1.2 The works Date et al. (2000, 2001b), Ma et al. (2003), Matsuno and Mogi (2000), Matsuno and Sato (2005), Prautsch et al. (2000), which were described above, all employ a common approach for motion control in that the nonholonomic constraints on the links are used to establish an explicit connection between body shape changes and propulsion, which allows the control input to be specified directly in terms of the desired propulsion of the robot. Such approaches are, to our best knowledge, the only known approaches for motion control of wheeled snake robots which infer some formal and model-based conclusions on the propulsion of the robot.

Control of Flat Surface Locomotion Without Sideslip Constraints

The works Dowling (1997, 1999) employ Fourier series to specify periodic functions representing the gait patterns of a wheel-less snake robot. The parameters of the Fourier series are determined using certain learning techniques. In Ma (2001), computer simulations are employed to study properties of lateral undulation related to the optimality of the motion. Ma et al. (2004) propose a control strategy for sinus-lifting during lateral undulation by solving a quadratic optimisation problem.

Saito et al. (2002) consider snake robots influenced by anisotropic ground friction and optimise the gait parameters of lateral undulation based on simulations. The work also proposes a forward velocity controller for wheel-less snake robots. The works Chernousko (2003, 2005) consider several elementary motions for planar snake robots and derive conditions for the feasibility of these motions, such as required actuator strength. In Hicks and Ito (2005), Hicks (2003), methods based on numerical optimal control are considered for determining optimal gaits during positional control of snake robots influenced by anisotropic viscous ground friction. Gonzalez-Gomez et al. (2007) use *Open Dynamics Engine* (ODE) to propose and simulate various 3D gaits for translational and turning motion of snake robots, including a gait for rotation with very little displacement. Transeth et al. (2007b) propose a controller for the joints of a planar snake robot influenced by anisotropic Coulomb ground friction and prove that the resulting translational and rotational velocity of the robot is bounded.

The following works consider other gait patterns than lateral undulation, and the gaits are carried out in open loop without explicitly controlling the position and orientation of the snake robot. Gaits for sidewinding motion, which is a sideways rolling type of motion, are proposed in Burdick et al. (1995), Gonzalez-Gomez et al. (2007), Hatton and Choset (2010), Yu et al. (2008). Inchworm locomotion gaits are proposed in Chirikjian and Burdick (1995), Gonzalez-Gomez et al. (2007), Ohno and Hirose (2001), Poi et al. (1998), Rincon and Sotelo (2003), Yamada and Hirose (2010), Yim (1994), Yim et al. (2002). Lateral rolling, which is achieved by continuously forming the snake body into a vertical U-shape that tips over, is considered in Chen et al. (2004), Dowling (1997), Gonzalez-Gomez et al. (2007), Mori and Hirose (2002), Ohno and Hirose (2001). Furthermore, gaits for loop forming motion are proposed in Ohashi and Hirose (2010), Yim (1994), Yim et al. (2002), where the head and tail of the snake robot are connected to turn the robot into a rolling wheel (see Fig. 1.11).

Remark 1.3 To our best knowledge, previous literature has not presented any formal mathematical proofs regarding the propulsion of wheel-less snake robots.

Control of Robotic Fish and Eel-Like Mechanisms

A complete treatment of robotic underwater locomotion is beyond the scope of this review. However, we consider the following works to be representative of previous research related to control of such mechanisms.

Eel-like motion is considered in McIsaac and Ostrowski (2003a, 2003b), where controllers for tracking straight and curved trajectories are proposed. The works Morgansen et al. (2001, 2002, 2007), Vela et al. (2002a) consider motion control of robotic fish. Lie bracket calculations based on the dynamics of the robotic fish are used to derive gaits for forward motion and various forms of turning motion. Algorithms for closed-loop heading and depth control are also considered. Melli et al.

(2006) propose open-loop gaits for a robotic fish based on curvature plots of the mechanical connection between the shape space motion and the overall displacement of the robot. A swimming snake robot is considered in Crespi and Ijspeert (2008), where a gradient-free optimisation method is employed to adjust the gait parameters online, i.e. while the robot is moving, in order to maximise the forward velocity.

Control of Locomotion in Environments with Obstacles

Only a few works in previous literature consider control strategies for snake robots where the surface is no longer assumed to be flat (i.e. in environments with obstacles). To our best knowledge, the works in Bayraktaroglu and Blazevic (2005), Bayraktaroglu (2008), Hirose (1993) are the only works in previous literature which present control strategies for snake robots that employ explicit contact sensing in the feedback loop. Hirose (1993) proposes a strategy for lateral inhibition that modifies the shape of a snake robot based on contact force sensing along the snake body in order to avoid obstacles. Bayraktaroglu and Blazevic (2005) propose an inverse dynamics approach by formulating and numerically solving an optimisation problem in order to, for a given set of obstacle contacts, calculate the contact forces required to propel the robot in a desired direction. A strategy for calculating the actual torque inputs to the joints from the desired contacts was, however, not presented. A kinematic approach is proposed in Bayraktaroglu (2008), where a curve fitting procedure is used to determine the shape of the robot with respect to the detected obstacles. Subsequently, this shape is propagated backwards along the snake body under the assumption that this will push the robot forward.

Sensing the environment of a snake robot must not necessarily involve contact force sensing since the environment can be indirectly sensed through the joint angle measurements and/or the actuator torques. This approach is considered in Date and Takita (2007), where the joint torques of a snake robot are specified solely in terms of the measured joint angles to achieve motion through a winding corridor, in Andruska and Peterson (2008), which presents a control strategy that uses motor current measurements to adjust the shape of a snake robot moving through an elastically deformable channel, and in Kuwada et al. (2008), where the deviations of the joint angles from their setpoints are used to adapt the body shape of a snake robot moving inside pipe structures.

The remaining works presented in the following consider controllers aimed at locomotion in environments that are not flat but do not appear to involve sensing of the *interaction* between the snake robot and its environment. Kulali et al. (2002) employ a fuzzy logic controller to switch between various predefined gaits during motion in an obstacle environment. The goal of the motion controller is to avoid the obstacles. In Greenfield et al. (2005), an algorithm is presented that takes contact constraints on a snake robot into account in order to compute the joint torques that produce the desired motion. The algorithm is applied to achieve climbing motion with a snake robot. A gait for climbing motion is also proposed in Kamegawa et al. (2009). Sfakiotakis and Tsakiris (2007) use range sensor measurements to centre a crawling snake robot between the walls of a corridor. Zarrouk et al. (2010) analyse the efficiency of earthworm-like motion on compliant surfaces motivated by biomedical applications of worm robots. Moreover, various gaits aimed at motion in cluttered and uneven environments, including climbing gaits, are proposed in Chen et al. (2007), Hatton and Choset (2009b), Lipkin et al. (2007), Nilsson (1997).

1.4 The Scope of This Book

The work underlying this book has been carried out with the following scope.

1.4.1 An Analytical Approach

This book is motivated by practical applications of snake robots. However, the book is primarily a theoretical study, although experimental investigations are also considered. In our opinion, there are many aspects related to control of snake robots that have not yet been addressed. Moreover, even though research on snake robots has been conducted for several decades, our understanding of snake locomotion so far is largely based on empirical studies of biological snakes and simulation-based synthesis of properties of snake robots. In this book, we therefore take an analytical approach in an attempt to provide a basic understanding of snake robot locomotion. We hope that this approach will contribute to the mathematical foundation of the control theory of snake robots.

1.4.2 Snake Robots Without a Fixed Base

There are many works in the literature which consider the use of snake robots for manipulation purposes, where one end of the robot is fixed. In such applications, the snake robot is essentially a fixed robotic manipulator arm with many degrees of freedom (see e.g. Chirikjian 2001; Jones and Walker 2006; Transeth et al. 2008c). In this book, however, we only consider snake robots intended for locomotion purposes, i.e. snake robots without a fixed base.

1.4.3 A Planar Perspective

This book considers planar snake robot locomotion in the horizontal plane. Of course, snake locomotion is inherently a three-dimensional phenomenon, and a snake robot capable of strict planar motion will generally not be able to operate in cluttered and challenging environments. However, we believe that the essential control principles of snake robot locomotion are contained in a planar perspective. In particular, since the fully three-dimensional motion of a snake robot consists of motion components in horizontal and vertical planes, respectively, we conjecture that control laws that fulfil some control objective in a planar perspective can be extended to fulfil a similar control objective in a fully three-dimensional perspective. Moreover, we believe that the simpler case of planar locomotion should be fully understood before the more challenging problem of three-dimensional locomotion is attacked.

1.4.4 Locomotion Without Sideslip Constraints

As indicated by the literature review in Sect. 1.3, a majority of previous research has focused on snake robots where the links are subjected to nonholonomic constraints, i.e. where each link is constrained from moving sideways. Such conditions are usually obtained by installing passive wheels in the tangential direction of the snake robot links. However, it seems unrealistic to enforce a nonholonomic constraint on each link during motion in unknown and cluttered environments, which represents the long-term goal of our research. In this book, we therefore consider snake robots where the links are allowed to slip sideways, often referred to as *wheel-less* snake robot locomotion.

1.4.5 Motion Based on Lateral Undulation

This book focuses on motion analysis and control strategies where the snake robot moves according to various forms of the gait pattern lateral undulation (see Sect. 1.2.2). We have chosen this scope since lateral undulation is the fastest and most common form of snake locomotion. Moreover, we believe that this is the gait pattern which is most relevant and most efficient in a planar perspective.

1.5 An Outline of This Book

This book is structured and organised according to its title, as illustrated in Table 1.5. The book has two parts, where the first part targets snake robot locomotion on flat surfaces, while the second part targets snake robot locomotion in cluttered environments, i.e. environments containing external obstacles. In the following, we present a summary of each individual chapter of the book.

Table 1.5 The problem area treated in each chapter

		Modelling	Mechatronics	Control
Part I	Chapter 2	X		
	Chapter 3		X	
	Chapter 4			X
	Chapter 5			X
	Chapter 6	X		
	Chapter 7			X
	Chapter 8			X
Part II	Chapter 9			
	Chapter 10	X		
	Chapter 11		X	
	Chapter 12			X
	Chapter 13			X

1.5.1 Outline of Part I—Snake Robot Locomotion on Flat Surfaces

Chapter 2

Topic: We model the kinematics and dynamics of a snake robot moving on a horizontal and flat surface.

Chapter summary: Based on first principles, we develop a mathematical model of the kinematics and dynamics of a snake robot with N links moving on a horizontal and flat surface. The links of the robot are influenced by ground friction forces, which propel the motion. Due to the many degrees of freedom of the robot and the dynamical couplings between the links, the resulting model turns out to be quite complex. By partially linearising the model, we show that some of this complexity can be eliminated. This is achieved by introducing a change of coordinates which enables us to partition the model into an actuated part (the joint angles of the snake robot) and an unactuated part (the position and orientation of the snake robot). Through an input transformation, we are then able to linearise the actuated part of the model. Due to the complexity of the initial form of the model, much of the model analysis presented in this book would not have been feasible without the model transformation.

Chapter 3

Topic: We present the design of the snake robot *Wheeko*, which was developed for motion control experiments on flat surfaces.

Chapter summary: The robot consists of 10 identical joint modules covered by passive wheels in order to enable flat surface locomotion. Each module is battery-powered and has two degrees of freedom (pitch and yaw) that are actuated by electric servo motors. The internal structure of Wheeko is identical to the internal structure of the snake robot *Kulko*, which is described in Chap. 11.

Chapter 4

Topic: We employ nonlinear system analysis tools for investigating fundamental properties of snake robot dynamics.

Chapter summary: We first present a stabilisability analysis that proves that any asymptotically stabilising control law for a planar snake robot to an equilibrium point must be *time-varying*, i.e. not of pure-state feedback type. Subsequently, we present a controllability analysis of planar snake robots influenced by viscous ground friction forces that shows that a snake robot is *not* controllable when the ground friction is *isotropic*, but that a snake robot becomes *strongly accessible* when the ground friction is *anisotropic*. The analysis also shows that the snake robot does *not* satisfy sufficient conditions for *small-time local controllability*. The results from the controllability analysis are not sufficient to conclude that a snake robot with anisotropic ground friction is controllable. However, the analysis proves that propulsion of a snake robot under viscous friction conditions requires the friction to be *anisotropic* and also that the joint angles of a snake robot should be *out of phase* during snake locomotion.

In this chapter, we also derive a simple relationship between link velocities *normal* to the direction of motion and *propulsive forces* in the direction of motion that explains how snake robots influenced by anisotropic ground friction are able to locomote forward on a planar surface. Moreover, we present mathematical arguments that support the empirically derived mathematical description of lateral undulation previously proposed in Hirose (1993). We also identify an important property concerning the turning motion of a snake robot and a property related to the relative displacements of the links during lateral undulation.

Chapter 5

Topic: We consider straight line path following control of snake robots.

Chapter summary: We propose a control law that enables snake robots to track a straight path. We also employ a *Poincaré map* to show that all state variables of the snake robot, except for the position along the path, trace out an *exponentially stable* periodic orbit during path following with the proposed controller.

Chapter 6

Topic: We consider an approach for simplifying the mathematical model of the snake robot.

Chapter summary: Based on properties derived in Chap. 4, we propose a *simplified model* of planar snake robot locomotion, which is intended for control design and stability analysis purposes. Instead of considering the *rotational* link motion, the simplified model describes the body shape dynamics of a snake robot in terms of the *translational* motion of the links, which significantly simplifies the equations of motion. We provide support of the claim that the simplified model captures the essential part of the dynamics of planar snake robot locomotion by showing that the stabilisability and controllability properties of the simplified model are similar to the corresponding properties of the more complex model of the snake robot.

Chapter 7

Topic: We employ *averaging theory* to study the average effect of the joint motion that propels the snake robot during lateral undulation.

Chapter summary: We derive an *averaged* model of the velocity dynamics of a snake robot during lateral undulation. Moreover, we show that the average velocity of a snake robot during lateral undulation converges exponentially fast to a steady-state velocity and that an analytical expression for calculating the steady-state velocity is presented as a function of the gait pattern parameters. These results prove that a wheel-less snake robot with anisotropic ground friction properties achieves forward propulsion when it moves by lateral undulation. In this chapter, we also derive a set of fundamental relationships between the gait pattern parameters of lateral undulation and the resulting forward velocity of a planar snake robot. In particular, the derived properties state that the average forward velocity of a snake robot (1) is proportional to the squared amplitude of the sinusoidal motion of each joint, (2) is proportional to the angular frequency of the sinusoidal motion of each joint, (3) is proportional to a particular function of the constant phase shift between the joints, and (4) is maximised by the phase shift between the joints that also maximises the particular phase shift function. The properties are supported by experimental results based on the snake robot Wheeko.

Chapter 8

Topic: We return to the problem of straight line path following control of snake robots, but this time based on the simplified model.

Chapter summary: We propose a straight line path following controller, which, using cascaded systems theory, is proved to $\mathcal{K}$-exponentially stabilise a snake robot to any desired straight path. The proof relies on the assumption that the forward velocity of the robot is contained in some non-zero and positive interval. To our best knowledge, this is currently the only formal proof in the literature concerning path following control of snake robots without nonholonomic constraints. We present experimental results where the proposed controller successfully steers the snake robot Wheeko towards and along the desired straight path.

The chapter also describes how the straight line path following controller can be extended to path following of general curved paths by employing an approach previously proposed in the marine control literature for path following control of marine vessels. Moreover, we propose a waypoint guidance strategy for steering a snake robot along a path defined by waypoints interconnected by straight lines, and we prove that the guidance strategy is guaranteed to steer the position of the robot into the acceptance region of each waypoint.

1.5.2 Outline of Part II—Snake Robot Locomotion in Cluttered Environments

Chapter 9

Topic: This chapter gives an introduction to the second part of the book.
Chapter summary: The introductory text focuses on the importance of environment *sensing* and *adaptation* during snake robot locomotion in uneven and cluttered environments. The concept of *obstacle-aided locomotion* is also presented, which is a form of locomotion where a snake robot is propelled forward by active use of the interaction with obstacles in its environment.

Chapter 10

Topic: We extend the mathematical model of the snake robot to include contact forces from external obstacles in the environment around the robot.
Chapter summary: We present a *hybrid model* of a planar snake robot interacting with obstacles in its environment. The obstacle interaction is modelled by introducing a *unilateral velocity constraint* on each contacted link of the snake robot, which is a novel approach. In particular, the conventional approach for modelling the obstacle interaction would be to assume that the obstacle constraint force points in the direction perpendicular to the *obstacle* (see Brogliato 1999). With the approach described in this chapter, the shape of the obstacles does not have to be considered explicitly as we instead calculate constraint forces with respect to the normal direction of the *contacted links*, which simplifies the equations of motion. The existence and uniqueness properties of the hybrid model are investigated based on the theory of *linear complementarity problems*.

Chapter 11

Topic: We present the design of the snake robot *Kulko*, which was developed for the purpose of experiments related to obstacle-aided locomotion in cluttered environments.

Chapter summary: Previous snake robot design efforts have given very limited attention to the exterior gliding surface of such robots and to methods for enabling snake robots to sense their environment. In this chapter, we address these limitations by presenting the design of a ball-shaped joint mechanism for a snake robot that (1) allows the joint modules to be covered by shells, thereby giving the robot a smooth outer surface independently of how the joints are flexed and that (2) allows contact force sensors to be installed underneath the shells, thereby enabling the robot to sense its environment. Experimental results are presented that validate the function of the contact force measurement system. This chapter also proposes an alternative approach for sensing environment contact forces, which has the advantage that force measurements are only required at the locations of the joints and that the sensor system can be well protected inside the snake robot.

Chapter 12

Topic: We consider motion control of snake robots on surfaces with irregularities in the form of external obstacles.

Chapter summary: We propose a general control principle for snake robots which suggests that obstacle-aided locomotion is achieved by producing body shape changes where the links in contact with obstacles are rotated to increase the propulsive forces on the robot. Moreover, we use the control principle to propose a *hybrid controller* for obstacle-aided locomotion aimed at resolving situations where the snake robot is jammed between obstacles. We present experimental results based on the snake robot Kulko, where the hybrid controller is shown to maintain the propulsion of the snake robot in different obstacle environments. To our best knowledge, this is the first time a snake robot is propelled forward based on measurements of the contact force amplitudes along the robot body. The works in Bayraktaroglu (2008), Hirose (1993) also report experiments where a snake robot is propelled by obstacle contact forces. However, the control strategies in these experiments do not consider the amplitude of the contact forces.

Chapter 13

Topic: We consider straight line path following control of snake robots in environments containing obstacles.

Chapter summary: This chapter represents the culmination of the book and can be regarded as a fusion of the control efforts reported in the previous chapters. We first propose a general framework for motion control of snake robots, where the motion is specified in terms of a *body wave* component, an *environment adaptation* component, and a *heading control* component. Subsequently, we employ this controller framework to propose a control law for straight line path following control of snake robots in environments with obstacles. An important feature of this controller is the idea of a continuous jam resolution action that is performed in parallel with the cyclic wave motion of the robot to continuously adapt the body shape to the environment and prevent the motion from being jammed. To provide support of the proposed control law, we present experimental results where the snake robot Kulko is successfully propelled through three different obstacle environments with the proposed controller.

1.6 Publications Underlying This Book

In the following, we provide a list of journal publications and conference papers on which the majority of the material in this book is based. Note that each chapter of this book begins by describing how the chapter is related to previous literature. In these descriptions, the following list of publications is not considered to be previous literature since the results in these publications are contained in this book.

Journal Papers

- Liljebäck, P., Pettersen, K.Y., Stavdahl, Ø., Gravdahl, J.T.: Lateral undulation of snake robots: a simplified model and fundamental properties. Robotica (2012). Submitted
- Liljebäck, P., Stavdahl, Ø., Pettersen, K.Y., Gravdahl, J.T.: Two new design concepts for snake robot locomotion in unstructured environments. Paladyn. J. Behav. Robot. **1**, 154–159 (2011)
- Liljebäck, P., Pettersen, K.Y., Stavdahl, Ø., Gravdahl, J.T.: Snake robot locomotion in environments with obstacles. IEEE Trans. Mechatron. (2012). doi:10.1109/TMECH.2011.2159863
- Liljebäck, P., Pettersen, K.Y., Stavdahl, Ø., Gravdahl, J.T.: A review on modelling, implementation, and control of snake robots. Robot. Auton. Syst. **60**, 29–40 (2011). doi:10.1016/j.robot.2011.08.010
- Liljebäck, P., Haugstuen, I.U., Pettersen, K.Y.: Path following control of planar snake robots using a cascaded approach. IEEE Trans. Control Syst. Technol. **20**, 111–126 (2011). doi:10.1109/TCST.2011.2107516
- Liljebäck, P., Pettersen, K.Y., Stavdahl, Ø., Gravdahl, J.T.: Experimental investigation of obstacle-aided locomotion with a snake robot. IEEE Trans. Robot. **27**, 792–800 (2011)
- Liljebäck, P., Pettersen, K.Y., Stavdahl, Ø., Gravdahl, J.T.: Controllability and stability analysis of planar snake robot locomotion. IEEE Trans. Autom. Control **56**, 1365–1380 (2011)
- Liljebäck, P., Pettersen, K.Y., Stavdahl, Ø., Gravdahl, J.T.: Hybrid modelling and control of obstacle-aided snake robot locomotion. IEEE Trans. Robot. **26**, 781–799 (2010)
- Liljebäck, P., Stavdahl, Ø., Pettersen, K.Y.: Modular pneumatic snake robot: 3D modelling, implementation and control. Model. Identif. Control **29**, 21–28 (2008)

Conference Papers

- Liljebäck, P., Pettersen, K.Y., Stavdahl, Ø., Gravdahl, J.T.: Path following control of snake robots in unstructured environments. In: Proc. IEEE Int. Conf. Robotics and Automation, Shanghai, China, pp. 503–510 (2011)
- Liljebäck, P., Pettersen, K.Y.: Waypoint guidance control of snake robots. In: Proc. IEEE Int. Conf. Robotics and Automation, Shanghai, China, pp. 937–944 (2011)
- Liljebäck, P., Haugstuen, I.U., Pettersen, K.Y.: Experimental investigation of a path following controller for planar snake robots. In: Proc. IEEE Int. Conf. Control, Automation, Robotics, and Vision (ICARCV), Singapore, pp. 2325–2332 (2010)
- Liljebäck, P., Pettersen, K.Y., Stavdahl, Ø., Gravdahl, J.T.: Experimental investigation of fundamental properties of snake robot locomotion. In: Proc. IEEE Int. Conf. Control, Automation, Robotics, and Vision (ICARCV), Singapore, pp. 187–194 (2010). Finalist for the Best Paper Award
- Liljebäck, P., Haugstuen, I.U., Pettersen, K.Y.: Path following control of planar snake robots using a cascaded approach. In: Proc. IEEE Conf. Decision and Control, Atlanta, GA, USA, pp. 1969–1976 (2010)
- Liljebäck, P., Pettersen, K.Y., Stavdahl, Ø., Gravdahl, J.T.: Stability analysis of snake robot locomotion based on averaging theory. In: Proc. IEEE Conf. Decision and Control, Atlanta, GA, USA, pp. 1977–1984 (2010)
- Liljebäck, P., Pettersen, K.Y., Stavdahl, Ø., Gravdahl, J.T.: Fundamental properties of snake robot locomotion. In: Proc. IEEE/RSJ Int. Conf. Intelligent Robots and Systems, Taipei, Taiwan, pp. 2876–2883 (2010)
- Liljebäck, P., Pettersen, K.Y., Stavdahl, Ø., Gravdahl, J.T.: A simplified model of planar snake robot locomotion. In: Proc. IEEE/RSJ Int. Conf. Intelligent Robots and Systems, Taipei, Taiwan, pp. 2868–2875 (2010)
- Liljebäck, P., Pettersen, K.Y., Stavdahl, Ø., Gravdahl, J.T.: A hybrid model of obstacle-aided snake robot locomotion. In: Proc. IEEE Int. Conf. Robotics and Automation, Anchorage, AK, USA, pp. 675–682 (2010)
- Liljebäck, P., Pettersen, K.Y., Stavdahl, Ø.: A snake robot with a contact force measurement system for obstacle-aided locomotion. In: Proc. IEEE Int. Conf. Robotics and Automation, Anchorage, AK, USA, pp. 683–690 (2010)
- Liljebäck, P., Pettersen, K.Y., Stavdahl, Ø., Gravdahl, J.T.: Controllability analysis of planar snake robots influenced by viscous ground friction. In: Proc. IEEE/RSJ Int. Conf. Intelligent Robots and Systems, St. Louis, MO, USA, pp. 3615–3622 (2009)
- Liljebäck, P., Pettersen, K.Y., Stavdahl, Ø., Gravdahl, J.T.: Stability analysis of snake robot locomotion based on Poincaré maps. In: Proc. IEEE/RSJ Int. Conf. Intelligent Robots and Systems, St. Louis, MO, USA, pp. 3623–3630 (2009)
- Liljebäck, P., Fjerdingen, S., Pettersen, K.Y., Stavdahl, Ø.: A snake robot joint mechanism with a contact force measurement system. In: Proc. IEEE Int. Conf. Robotics and Automation, Kobe, Japan, pp. 3815–3820 (2009)

- Liljebäck, P., Pettersen, K.Y., Stavdahl, Ø.: Modelling and control of obstacle-aided snake robot locomotion based on jam resolution. In: Proc. IEEE Int. Conf. Robotics and Automation, Kobe, Japan, pp. 3807–3814 (2009)
- Liljebäck, P., Stavdahl, Ø., Beitnes, A.: SnakeFighter—development of a water hydraulic fire fighting snake robot. In: Proc. IEEE Int. Conf. Control, Automation, Robotics, and Vision (ICARCV), Singapore, December 2006
- Liljebäck, P., Stavdahl, Ø., Pettersen, K.Y.: Modular pneumatic snake robot: 3D modelling, implementation and control. In: Proc. 16th IFAC World Congress, Prague, Czech Republic, July 2005

Part I
Snake Robot Locomotion on Flat Surfaces

Chapter 2
A Complex Model of Snake Robot Locomotion on Planar Surfaces

The underlying theme of this book is analytical approaches aimed at increasing our understanding of snake robot locomotion. The mathematical model of the snake robot is the basis for these analytical studies, which means that the analysis relies heavily on the form and complexity of the model. In this chapter, we employ first principles to derive a mathematical model of the kinematics and dynamics of a snake robot with N links moving on a horizontal and flat surface. The links of the robot are influenced by ground friction forces which propel the motion. Due to the many degrees of freedom of the robot and the dynamical couplings between the links, the resulting model will turn out to be quite complex. We will eliminate some of this complexity by partially linearising the model. This is achieved by introducing a change of coordinates which enables us to partition the model into an actuated part (the joint angles of the snake robot) and an unactuated part (the position and orientation of the snake robot). Through an input transformation, we are then able to linearise the actuated part of the model. However, even the partially linearised model contains complex terms which make model-based controller design and analysis challenging. Throughout this book, we will therefore refer to the model developed in this chapter as the *complex* model of the snake robot.

In Chap. 4, the complex model will be analysed in order to deduce several fundamental properties of snake robot dynamics. Some of these properties will be instrumental in the development of a *simplified* model of the snake robot in Chap. 6, where we propose a model that captures only the 'essential' part of the dynamics of the complex model. In Part II of this book, which considers snake robot locomotion in cluttered environments, the complex model will be extended to include contact forces from external obstacles in the environment around the snake robot.

The chapter is organised as follows. The relation between this chapter and previous literature is briefly discussed in Sect. 2.1. Section 2.2 introduces some basic notation that will be used throughout the book. The parameters that characterise the snake robot are presented in Sect. 2.3. The kinematics of the snake robot is described in Sect. 2.4, while two different ground friction models are presented in Sect. 2.5. The model of the snake robot dynamics is presented in Sect. 2.6, is partitioned into an actuated and an unactuated part in Sect. 2.7, and is transformed to a simpler form

P. Liljebäck et al., *Snake Robots*, Advances in Industrial Control,
DOI 10.1007/978-1-4471-2996-7_2, © Springer-Verlag London 2013

through partial feedback linearisation in Sect. 2.8. Finally, the chapter is summarised in Sect. 2.9.

2.1 The Relation Between This Chapter and Previous Literature

Mathematical models of planar snake robot dynamics have previously been developed from first principles in Ma (2001), Saito et al. (2002), and the initial form of the model presented in this chapter is developed using the same approach as in these works. The notation and the ground friction models considered in this chapter are, however, different from the works in Ma (2001), Saito et al. (2002). Moreover, the expression for the linear velocity of individual links given in (2.13a) is novel to this work. The two main novel features of the results presented in this chapter in relation to previous literature are the change of coordinates, which enables us to partition the model into an actuated and an unactuated part (see Sect. 2.7), and the subsequent partial feedback linearisation of the model (see Sect. 2.8). Due to the complexity of the nonlinearised model of the snake robot, much of the model analysis presented in Chap. 4 would not have been feasible without the model transformation.

2.2 Basic Notation

The following notation is used throughout this book:

- The operator sgn(·) produces a vector containing the sign of each individual element of its argument.
- The operator diag(·) produces a diagonal matrix with each individual element of its argument along its diagonal.
- The sinus and cosine operators, sin(·) and cos(·), are vector operators when their argument is a vector and scalar operators when their argument is a scalar value.
- We will use subscript i to denote element i of a vector (see Table 2.1 below). When parameters of the links (joints) of the snake robot are assembled into a vector, we associate element i of this vector with link i (joint i).
- Symbols representing a vector or a matrix are indicated with a bold font.
- The matrix $\mathbf{I}_k$ represents the $k \times k$ identity matrix, and $\mathbf{0}_{i \times j}$ represents the $i \times j$ matrix of zeros.
- A vector related to link i of the snake robot is either expressed in the global coordinate system or in the local coordinate system of the link (see Fig. 2.1). This is indicated by superscript *global* or *link*, i, respectively. If not otherwise specified, a vector with no superscript is expressed in the global coordinate system.

2.3 The Parameters of the Snake Robot

The snake robot consists of N rigid links of length $2l$ interconnected by $N - 1$ motorised joints. The width of each link is not considered in the model. All N links

Table 2.1 Parameters that characterise the snake robot

Symbol	Description	Vector
N	The number of links	
l	Half the length of a link	
m	Mass of each link	
J	Moment of inertia of each link	
θ_i	Angle between link i and the global x axis	$\boldsymbol{\theta} \in \mathbb{R}^N$
ϕ_i	Angle of joint i	$\boldsymbol{\phi} \in \mathbb{R}^{N-1}$
(x_i, y_i)	Global coordinates of the CM of link i	$\mathbf{X}, \mathbf{Y} \in \mathbb{R}^N$
(p_x, p_y)	Global coordinates of the CM of the robot	$\mathbf{p} \in \mathbb{R}^2$
u_i	Actuator torque on link i from link $i+1$	$\mathbf{u} \in \mathbb{R}^{N-1}$
u_{i-1}	Actuator torque on link i from link $i-1$	$\mathbf{u} \in \mathbb{R}^{N-1}$
$(f_{R,x,i}, f_{R,y,i})$	Ground friction force on link i	$\mathbf{f}_{R,x}, \mathbf{f}_{R,y} \in \mathbb{R}^N$
$(h_{x,i}, h_{y,i})$	Joint constraint force on link i from link $i+1$	$\mathbf{h}_x, \mathbf{h}_y \in \mathbb{R}^{N-1}$
$-(h_{x,i-1}, h_{y,i-1})$	Joint constraint force on link i from link $i-1$	$\mathbf{h}_x, \mathbf{h}_y \in \mathbb{R}^{N-1}$

have the same mass m and moment of inertia $J = \frac{1}{3}ml^2$. The total mass of the snake robot is therefore Nm. The mass of each link is uniformly distributed so that the link CM (centre of mass) is located at its centre point (at length l from the joint at each side). In the following subsections, the kinematics and dynamics of the snake robot will be modelled in terms of the mathematical symbols described in Table 2.1 and illustrated in Figs. 2.1 and 2.2. We will make use of the following vectors and matrices:

$$\mathbf{A} = \begin{bmatrix} 1 & 1 & & \\ & \ddots & \ddots & \\ & & 1 & 1 \end{bmatrix} \in \mathbb{R}^{(N-1)\times N},$$

$$\mathbf{D} = \begin{bmatrix} 1 & -1 & & \\ & \ddots & \ddots & \\ & & 1 & -1 \end{bmatrix} \in \mathbb{R}^{(N-1)\times N},$$

$$\mathbf{e} = [1, \ldots, 1]^T \in \mathbb{R}^N, \qquad \mathbf{E} = \begin{bmatrix} \mathbf{e} & \mathbf{0}_{N\times 1} \\ \mathbf{0}_{N\times 1} & \mathbf{e} \end{bmatrix} \in \mathbb{R}^{2N\times 2},$$

$$\sin\boldsymbol{\theta} = [\sin\theta_1, \ldots, \sin\theta_N]^T \in \mathbb{R}^N, \qquad \mathbf{S}_\theta = \operatorname{diag}(\sin\boldsymbol{\theta}) \in \mathbb{R}^{N\times N},$$

$$\cos\boldsymbol{\theta} = [\cos\theta_1, \ldots, \cos\theta_N]^T \in \mathbb{R}^N, \qquad \mathbf{C}_\theta = \operatorname{diag}(\cos\boldsymbol{\theta}) \in \mathbb{R}^{N\times N},$$

$$\operatorname{sgn}\boldsymbol{\theta} = [\operatorname{sgn}\theta_1, \ldots, \operatorname{sgn}\theta_N]^T \in \mathbb{R}^N, \qquad \dot{\boldsymbol{\theta}}^2 = \left[\dot\theta_1^2, \ldots, \dot\theta_N^2\right]^T \in \mathbb{R}^N.$$

The matrices $\mathbf{A}$ and $\mathbf{D}$ represent, respectively, an *addition* and a *difference* matrix, which will be used, respectively, for adding and subtracting pairs of adjacent ele-

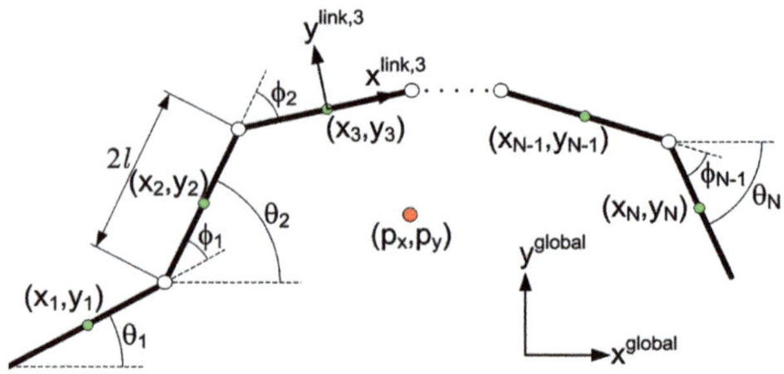

Fig. 2.1 The kinematic parameters of the snake robot

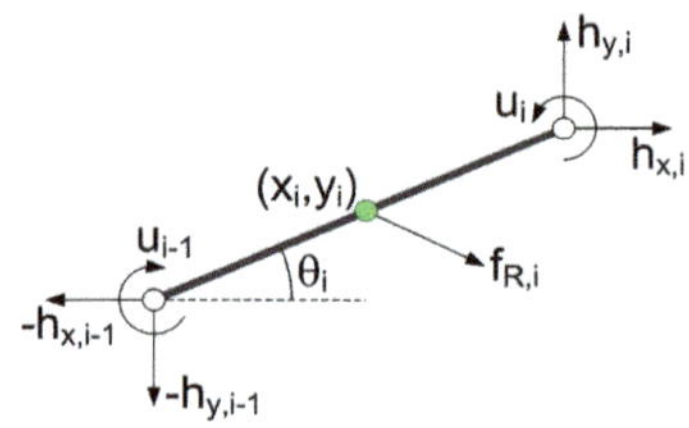

Fig. 2.2 Forces and torques acting on each link of the snake robot

ments of a vector. Furthermore, the vector $\mathbf{e}$ represents a summation vector, which will be used for adding all elements of an N-dimensional vector. The remaining vectors and matrices have been defined above since they appear several times during the development of the model.

2.4 The Kinematics of the Snake Robot

The snake robot moves on a horizontal and flat surface, and has $N+2$ degrees of freedom (N link angles and the planar position of the robot). The following definitions are illustrated in Fig. 2.1.

Definition 2.1 (Link angle) The *link angle* of link $i \in \{1, \dots, N\}$ of the snake robot is denoted by $\theta_i \in \mathbb{R}$ and is defined as the angle that the link forms with the global x axis with counterclockwise positive direction.

Definition 2.2 (Joint angle) The *joint angle* of joint $i \in \{1, \dots, N-1\}$ of the snake robot is denoted $\phi_i \in \mathbb{R}$ and is defined as

$$\phi_i = \theta_i - \theta_{i+1}. \tag{2.1}$$

Note the distinction between *link angles* and *joint angles*. A link angle is the orientation of a link with respect to the global x axis, while a joint angle is the difference between the link angles of two neighbouring links. We will quite frequently assemble the link angles and the joint angles in the vectors $\boldsymbol{\theta} = [\theta_1, \dots, \theta_N]^T \in \mathbb{R}^N$ and $\boldsymbol{\phi} = [\phi_1, \dots, \phi_{N-1}]^T \in \mathbb{R}^{N-1}$, respectively.

The snake robot has no explicitly defined orientation since there is an independent link angle associated with each link. We can still obtain a measure of the heading of the robot as follows (this approach is also considered in e.g. Hatton and Choset 2009a; Hu et al. 2009):

Definition 2.3 (Heading) The *heading* (or *orientation*) of the snake robot is denoted by $\overline{\theta} \in \mathbb{R}$ and is defined as the average of the link angles, i.e. as

$$\overline{\theta} = \frac{1}{N} \sum_{i=1}^{N} \theta_i . \tag{2.2}$$

The local coordinate system of each link is fixed in the CM of the link with x (tangential) and y (normal) axes oriented such that they are aligned with the global x and y axis, respectively, when the link angle is zero. The rotation matrix from the global frame to the frame of link i is given by

$$\mathbf{R}_{\text{link},i}^{\text{global}} = \begin{bmatrix} \cos\theta_i & -\sin\theta_i \\ \sin\theta_i & \cos\theta_i \end{bmatrix} . \tag{2.3}$$

The global frame position $\mathbf{p} \in \mathbb{R}^2$ of the CM (centre of mass) of the robot is given by

$$\mathbf{p} = \begin{bmatrix} p_x \\ p_y \end{bmatrix} = \begin{bmatrix} \frac{1}{Nm} \sum_{i=1}^{N} m x_i \\ \frac{1}{Nm} \sum_{i=1}^{N} m y_i \end{bmatrix} = \frac{1}{N} \begin{bmatrix} \mathbf{e}^T \mathbf{X} \\ \mathbf{e}^T \mathbf{Y} \end{bmatrix} , \tag{2.4}$$

where $\mathbf{e}$ was defined in Sect. 2.3, (x_i, y_i) are the global frame coordinates of the CM of link i, $\mathbf{X} = [x_1, \ldots, x_N]^T \in \mathbb{R}^N$, and $\mathbf{Y} = [y_1, \ldots, y_N]^T \in \mathbb{R}^N$. We define the velocity of the snake robot along its *forward* direction as follows:

Definition 2.4 (Forward velocity) The *forward velocity* of the snake robot is denoted by $\overline{v}_t \in \mathbb{R}$ and is defined as the component of the CM velocity $\dot{\mathbf{p}}$ along the current heading $\overline{\theta}$, i.e. as

$$\overline{v}_t = \dot{p}_x \cos\overline{\theta} + \dot{p}_y \sin\overline{\theta} . \tag{2.5}$$

Remark 2.1 Subscript t in the forward velocity $\overline{v}_t$ denotes *tangential*. The simplified model of the snake robot presented in Chap. 6 makes a clear distinction between the forward velocity v_t and the sideways velocity v_n of the robot. We have chosen to denote the forward velocity in the complex model by $\overline{v}_t$ to maintain a similar notation as in the simplified model.

The connection between link i and link $i+1$ at joint $i \in \{1, \ldots, N-1\}$ must comply with the two *holonomic* constraints

$$x_{i+1} - x_i = l\cos\theta_i + l\cos\theta_{i+1}, \tag{2.6a}$$

$$y_{i+1} - y_i = l\sin\theta_i + l\sin\theta_{i+1}. \tag{2.6b}$$

Using the notation from Sect. 2.3, we can write the joint constraints for all the links of the robot in matrix form as

$$\mathbf{DX} + l\mathbf{A}\cos\boldsymbol{\theta} = \mathbf{0}, \tag{2.7a}$$

$$\mathbf{DY} + l\mathbf{A}\sin\boldsymbol{\theta} = \mathbf{0}. \tag{2.7b}$$

We can now express the position of the individual links as a function of the CM position and the link angles of the robot by combining (2.4) and (2.7a)–(2.7b) into

$$\begin{aligned} \mathbf{TX} &= \begin{bmatrix} -l\mathbf{A}\cos\boldsymbol{\theta} \\ p_x \end{bmatrix}, \\ \mathbf{TY} &= \begin{bmatrix} -l\mathbf{A}\sin\boldsymbol{\theta} \\ p_y \end{bmatrix}, \end{aligned} \tag{2.8}$$

where

$$\mathbf{T} = \begin{bmatrix} \mathbf{D} \\ \frac{1}{N}\mathbf{e}^T \end{bmatrix} \in \mathbb{R}^{N\times N}. \tag{2.9}$$

It can be shown that

$$\mathbf{T}^{-1} = \begin{bmatrix} \mathbf{D}^T\left(\mathbf{DD}^T\right)^{-1} & \mathbf{e} \end{bmatrix}, \tag{2.10}$$

which enables us to solve (2.8) for $\mathbf{X}$ and $\mathbf{Y}$ according to

$$\mathbf{X} = \mathbf{T}^{-1}\begin{bmatrix} -l\mathbf{A}\cos\boldsymbol{\theta} \\ p_x \end{bmatrix} = -l\mathbf{K}^T\cos\boldsymbol{\theta} + \mathbf{e}p_x, \tag{2.11a}$$

$$\mathbf{Y} = \mathbf{T}^{-1}\begin{bmatrix} -l\mathbf{A}\sin\boldsymbol{\theta} \\ p_y \end{bmatrix} = -l\mathbf{K}^T\sin\boldsymbol{\theta} + \mathbf{e}p_y, \tag{2.11b}$$

where $\mathbf{K} = \mathbf{A}^T(\mathbf{DD}^T)^{-1}\mathbf{D} \in \mathbb{R}^{N\times N}$, and where $\mathbf{DD}^T$ is nonsingular and thereby invertible. The linear velocities of the links are found by differentiating (2.11a) and (2.11b) with respect to time, which gives

$$\dot{\mathbf{X}} = l\mathbf{K}^T\mathbf{S}_\theta\dot{\boldsymbol{\theta}} + \mathbf{e}\dot{p}_x, \tag{2.12a}$$

$$\dot{\mathbf{Y}} = -l\mathbf{K}^T\mathbf{C}_\theta\dot{\boldsymbol{\theta}} + \mathbf{e}\dot{p}_y. \tag{2.12b}$$

By manually investigating the structure of each row in (2.12a) and (2.12b), it can be verified that the linear velocity of the CM of link i in the global x and y directions is given by

$$\dot{x}_i = \dot{p}_x - \sigma_i\mathbf{S}_\theta\dot{\boldsymbol{\theta}}, \tag{2.13a}$$

$$\dot{y}_i = \dot{p}_y + \sigma_i\mathbf{C}_\theta\dot{\boldsymbol{\theta}}, \tag{2.13b}$$

where

$$\sigma_i = \left[a_1, a_2, \dots, a_{i-1}, \frac{a_i + b_i}{2}, b_{i+1}, b_{i+2}, \dots, b_N \right] \in \mathbb{R}^N, \tag{2.14a}$$

$$a_i = \frac{l(2i-1)}{N}, \qquad b_i = \frac{l(2i-1-2N)}{N}. \tag{2.14b}$$

2.5 The Ground Friction Models

2.5.1 The Friction Models and Their Role in This Book

As will become apparent in Chap. 4, a planar snake robot achieves forward propulsion on a flat surface by continuously changing its body shape to induce ground friction forces that propel the robot forward. The ground friction model is therefore an important part of the dynamics of the snake robot.

During planar locomotion, it is of great advantage to the propulsion of the snake robot that the ground friction forces on the links are *anisotropic*, which means that the friction coefficient describing the friction force in the tangential direction of a link (parallel to the link, i.e. along link x axis) is different from the friction coefficient describing the friction force in the direction normal to the link (perpendicular to the link, i.e. along link y axis). This friction property is exhibited by biological snakes, as was explained in the description of biological snakes in Sect. 1.2, and is also assumed in the majority of published research on snake robots, as described in the literature review of Sect. 1.3. We therefore include anisotropic friction conditions in the friction model of the snake robot. The importance of this friction property will be investigated in more detail in Chap. 4.

Note that anisotropic ground friction properties are typically implemented by mounting passive wheels along the body of the snake robot, but can also be achieved by equipping the underside of each link of the robot with edges, or grooves, that run parallel to each link (see e.g. Saito et al. 2002). If a snake robot does *not* have anisotropic ground friction properties, which is the case if the ground friction force on each link is independent of the link orientation with respect to its direction of motion, the links are said to have *isotropic* ground friction properties. This will typically be the case when the links of the snake robot are completely smooth in all directions. Note also that a snake robot with a completely smooth outer surface will still have anisotropic "friction" properties when it is swimming under water due to the higher drag forces in the direction normal to each link compared to in the tangential link direction (see e.g. Boyer et al. 2006; McIsaac and Ostrowski 2003a).

We consider two different ground friction models in this book, i.e. a *Coulomb* friction model and a *viscous* friction model. The Coulomb friction model, which assumes that the ground friction force on a link is proportional to the weight of the link, is more accurate (from a physical point of view) than the viscous friction model, which assumes that the ground friction force on a link is proportional to the velocity of the link. However, during planar locomotion, we conjecture that the anisotropic friction property of the links, which is independent of the choice of Coulomb or viscous friction, is the decisive factor of the motion. In other words, we conjecture that the motion of the snake robot is *qualitatively* (although not *quantitatively*) similar with anisotropic viscous friction as with anisotropic Coulomb friction. The viscous friction model is, however, less complex than the Coulomb friction model, which makes the viscous model more suitable for control design and analysis purposes. In this book, we will therefore mostly assume that the ground friction is viscous.

In the following, we first present the Coulomb friction model and subsequently the viscous friction model. In both models, the ground friction force on link i is assumed to act on the CM of the link only and is denoted by

$$\mathbf{f}_{R,i} = \mathbf{f}_{R,i}^{\text{global}} = \begin{bmatrix} f_{R,x,i} \\ f_{R,y,i} \end{bmatrix} \in \mathbb{R}^2. \tag{2.15}$$

The friction forces on all links are written in matrix form as

$$\mathbf{f}_R = \begin{bmatrix} \mathbf{f}_{R,x} \\ \mathbf{f}_{R,y} \end{bmatrix} \in \mathbb{R}^{2N}, \tag{2.16}$$

where $\mathbf{f}_{R,x} = [f_{R,x,1}, \dots, f_{R,x,N}]^T \in \mathbb{R}^N$ and $\mathbf{f}_{R,y} = [f_{R,y,1}, \dots, f_{R,y,N}]^T \in \mathbb{R}^N$ contain the friction forces on the links in the global x and y directions, respectively.

2.5.2 A Coulomb Friction Model

The coefficients describing the Coulomb friction force in the tangential (along link x axis) and normal (along link y axis) directions of a link, respectively, are denoted by μ_t and μ_n, respectively. We define the Coulomb friction force on link i in the local link frame, $\mathbf{f}_{R,i}^{\text{link},i} \in \mathbb{R}^2$, as

$$\mathbf{f}_{R,i}^{\text{link},i} = -mg \begin{bmatrix} \mu_t & 0 \\ 0 & \mu_n \end{bmatrix} \operatorname{sgn}\left(\mathbf{v}_i^{\text{link},i}\right), \tag{2.17}$$

where $\mathbf{v}_i^{\text{link},i} \in \mathbb{R}^2$ is the link velocity expressed in the local link frame, and g is the gravitational acceleration constant. Using (2.3), we can express the global frame Coulomb friction force on link i in the form of (2.15) as

$$\begin{aligned} \mathbf{f}_{R,i} = \mathbf{f}_{R,i}^{\text{global}} &= \mathbf{R}_{\text{link},i}^{\text{global}} \mathbf{f}_{R,i}^{\text{link},i} \\ &= -mg \mathbf{R}_{\text{link},i}^{\text{global}} \begin{bmatrix} \mu_t & 0 \\ 0 & \mu_n \end{bmatrix} \operatorname{sgn}\left(\mathbf{v}_i^{\text{link},i}\right) \\ &= -mg \mathbf{R}_{\text{link},i}^{\text{global}} \begin{bmatrix} \mu_t & 0 \\ 0 & \mu_n \end{bmatrix} \operatorname{sgn}\left(\left(\mathbf{R}_{\text{link},i}^{\text{global}}\right)^T \begin{bmatrix} \dot{x}_i \\ \dot{y}_i \end{bmatrix}\right). \end{aligned} \tag{2.18}$$

By performing the matrix multiplication in (2.18) and assembling the forces on all links in matrix form, we can rewrite the global frame Coulomb friction forces on the links in the form of (2.16) as

$$\mathbf{f}_R = \begin{bmatrix} \mathbf{f}_{R,x} \\ \mathbf{f}_{R,y} \end{bmatrix} = -mg \begin{bmatrix} \mu_t \mathbf{C}_\theta & -\mu_n \mathbf{S}_\theta \\ \mu_t \mathbf{S}_\theta & \mu_n \mathbf{C}_\theta \end{bmatrix} \operatorname{sgn}\left(\begin{bmatrix} \mathbf{C}_\theta & \mathbf{S}_\theta \\ -\mathbf{S}_\theta & \mathbf{C}_\theta \end{bmatrix} \begin{bmatrix} \dot{\mathbf{X}} \\ \dot{\mathbf{Y}} \end{bmatrix} \right) \in \mathbb{R}^{2N}. \quad (2.19)$$

2.5.3 A Viscous Friction Model

Similar to the Coulomb friction model, we assume that the viscous ground friction forces act on the CM of the links only. We present the viscous friction model for the different cases of isotropic versus anisotropic viscous friction since these two cases are analysed separately in Chap. 4.

Isotropic Viscous Friction

The isotropic viscous friction force on link i in the global x and y directions is proportional to the global frame velocity of the link given by (2.13a) and (2.13b), and is written in the form of (2.15) as

$$\mathbf{f}_{R,i} = \mathbf{f}_{R,i}^{\text{global}} = -c \begin{bmatrix} \dot{x}_i \\ \dot{y}_i \end{bmatrix} = -c \begin{bmatrix} \dot{p}_x - \sigma_i \mathbf{S}_\theta \dot{\boldsymbol{\theta}} \\ \dot{p}_y + \sigma_i \mathbf{C}_\theta \dot{\boldsymbol{\theta}} \end{bmatrix}, \quad (2.20)$$

where c is the viscous friction coefficient. The friction forces on all links are easily expressed in the form of (2.16) as

$$\mathbf{f}_R = \begin{bmatrix} \mathbf{f}_{R,x} \\ \mathbf{f}_{R,y} \end{bmatrix} = -c \begin{bmatrix} \dot{\mathbf{X}} \\ \dot{\mathbf{Y}} \end{bmatrix} = -c \begin{bmatrix} l\mathbf{K}^T \mathbf{S}_\theta \dot{\boldsymbol{\theta}} + \mathbf{e}\dot{p}_x \\ -l\mathbf{K}^T \mathbf{C}_\theta \dot{\boldsymbol{\theta}} + \mathbf{e}\dot{p}_y \end{bmatrix}, \quad (2.21)$$

where we have used the expression for the link velocities given by (2.12a) and (2.12b).

Anisotropic Viscous Friction

Under anisotropic friction conditions, a link has two viscous friction coefficients, c_t and c_n, describing the friction force in the tangential (along link x axis) and normal (along link y axis) directions of the link, respectively. We define the viscous friction force on link i in the local link frame, $\mathbf{f}_{R,i}^{\text{link},i} \in \mathbb{R}^2$, as

$$\mathbf{f}_{R,i}^{\text{link},i} = -\begin{bmatrix} c_t & 0 \\ 0 & c_n \end{bmatrix} \mathbf{v}_i^{\text{link},i}, \quad (2.22)$$

where $\mathbf{v}_i^{\text{link},i} \in \mathbb{R}^2$ is the link velocity expressed in the local link frame. Using (2.3), we can express the global frame viscous friction force on link i in the form of (2.15) as

$$\begin{aligned}\mathbf{f}_{R,i} = \mathbf{f}_{R,i}^{\text{global}} &= \mathbf{R}_{\text{link},i}^{\text{global}} \mathbf{f}_{R,i}^{\text{link},i} \\ &= -\mathbf{R}_{\text{link},i}^{\text{global}} \begin{bmatrix} c_t & 0 \\ 0 & c_n \end{bmatrix} \mathbf{v}_i^{\text{link},i} \\ &= -\mathbf{R}_{\text{link},i}^{\text{global}} \begin{bmatrix} c_t & 0 \\ 0 & c_n \end{bmatrix} \left(\mathbf{R}_{\text{link},i}^{\text{global}}\right)^T \begin{bmatrix} \dot{x}_i \\ \dot{y}_i \end{bmatrix}. \end{aligned} \tag{2.23}$$

By performing the matrix multiplication in (2.23), we get

$$\mathbf{f}_{R,i} = -\begin{bmatrix} c_t \cos^2\theta_i + c_n \sin^2\theta_i & (c_t - c_n)\sin\theta_i \cos\theta_i \\ (c_t - c_n)\sin\theta_i \cos\theta_i & c_t \sin^2\theta_i + c_n \cos^2\theta_i \end{bmatrix} \begin{bmatrix} \dot{x}_i \\ \dot{y}_i \end{bmatrix}. \tag{2.24}$$

By assembling the forces on all links in matrix form, we can rewrite the global frame viscous friction forces on the links in the form of (2.16) as

$$\mathbf{f}_R = \begin{bmatrix} \mathbf{f}_{R,x} \\ \mathbf{f}_{R,y} \end{bmatrix} = -\begin{bmatrix} c_t(\mathbf{C}_\theta)^2 + c_n(\mathbf{S}_\theta)^2 & (c_t - c_n)\mathbf{S}_\theta \mathbf{C}_\theta \\ (c_t - c_n)\mathbf{S}_\theta \mathbf{C}_\theta & c_t(\mathbf{S}_\theta)^2 + c_n(\mathbf{C}_\theta)^2 \end{bmatrix} \begin{bmatrix} \dot{\mathbf{X}} \\ \dot{\mathbf{Y}} \end{bmatrix} \in \mathbb{R}^{2N}. \tag{2.25}$$

Note that (2.25) reduces to (2.21) in the case of isotropic friction, i.e. where $c_t = c_n = c$.

2.6 The Dynamics of the Snake Robot

The $N + 2$ degrees of freedom of the snake robot are defined by the link angles $\boldsymbol{\theta} \in \mathbb{R}^N$ and the CM position $\mathbf{p} \in \mathbb{R}^2$. We now present the equations of motion of the robot in terms of the acceleration of the link angles, $\ddot{\boldsymbol{\theta}}$, and the acceleration of the CM position, $\ddot{\mathbf{p}}$.

As illustrated in Fig. 2.2, link $i \in \{1, \dots, N\}$ is influenced by the ground friction force $\mathbf{f}_{R,i} \in \mathbb{R}^2$, which acts on the CM of the link, and also the joint constraint forces $-h_{x,i-1}$, $-h_{y,i-1}$, $h_{x,i}$, and $h_{y,i}$, which keep the link connected to link $i-1$ and link $i+1$. The joint constraint forces are described in Table 2.1. Using first principles, the force balance for link i in global frame coordinates is given by

$$m\ddot{x}_i = f_{R,x,i} + h_{x,i} - h_{x,i-1}, \tag{2.26a}$$

$$m\ddot{y}_i = f_{R,y,i} + h_{y,i} - h_{y,i-1}. \tag{2.26b}$$

The force balance equations for all links may be expressed in matrix form as

$$m\ddot{\mathbf{X}} = \mathbf{f}_{R,x} + \mathbf{D}^T \mathbf{h}_x, \tag{2.27a}$$

$$m\ddot{\mathbf{Y}} = \mathbf{f}_{R,y} + \mathbf{D}^T \mathbf{h}_y, \tag{2.27b}$$

where $\mathbf{h}_x = [h_{x,1}, \ldots, h_{x,N}]^T \in \mathbb{R}^N$ and $\mathbf{h}_y = [h_{y,1}, \ldots, h_{y,N}]^T \in \mathbb{R}^N$. The link accelerations may also be expressed by differentiating (2.7a) and (2.7b) twice with respect to time, which gives

$$\mathbf{D}\ddot{\mathbf{X}} = l\mathbf{A}\big(\mathbf{C}_\theta\dot{\boldsymbol{\theta}}^2 + \mathbf{S}_\theta\ddot{\boldsymbol{\theta}}\big), \tag{2.28a}$$

$$\mathbf{D}\ddot{\mathbf{Y}} = l\mathbf{A}\big(\mathbf{S}_\theta\dot{\boldsymbol{\theta}}^2 - \mathbf{C}_\theta\ddot{\boldsymbol{\theta}}\big), \tag{2.28b}$$

where the square operator of $\dot{\boldsymbol{\theta}}^2$ means that each element of $\dot{\boldsymbol{\theta}}$ is squared ($\dot{\boldsymbol{\theta}}^2 = \mathrm{diag}(\dot{\boldsymbol{\theta}})\dot{\boldsymbol{\theta}}$). We obtain the CM acceleration by differentiating (2.4) twice with respect to time, inserting (2.27a) and (2.27b), and noting that the joint constraint forces, $\mathbf{h}_x$ and $\mathbf{h}_y$, are eliminated when the link accelerations are summed (i.e. $\mathbf{e}^T\mathbf{D}^T = \mathbf{0}$). This gives

$$\begin{bmatrix} \ddot{p}_x \\ \ddot{p}_y \end{bmatrix} = \frac{1}{N}\begin{bmatrix} \mathbf{e}^T\ddot{\mathbf{X}} \\ \mathbf{e}^T\ddot{\mathbf{Y}} \end{bmatrix} = \frac{1}{Nm}\begin{bmatrix} \mathbf{e}^T\mathbf{f}_{R,x} \\ \mathbf{e}^T\mathbf{f}_{R,y} \end{bmatrix} = \frac{1}{Nm}\mathbf{E}^T\mathbf{f}_R. \tag{2.29}$$

This equation simply states, as would be expected, that the acceleration of the CM of the snake robot equals the sum of the external forces acting on the robot divided by its mass.

The torque balance for link i is given by

$$J\ddot{\theta}_i = u_i - u_{i-1} - l\sin\theta_i(h_{x,i} + h_{x,i-1}) + l\cos\theta_i(h_{y,i} + h_{y,i-1}), \tag{2.30}$$

where u_i and u_{i-1} are the actuator torques exerted on link i from link $i+1$ and link $i-1$, respectively. Hence, the torque balance equations for all links may be expressed in matrix form as

$$J\ddot{\boldsymbol{\theta}} = \mathbf{D}^T\mathbf{u} - l\mathbf{S}_\theta\mathbf{A}^T\mathbf{h}_x + l\mathbf{C}_\theta\mathbf{A}^T\mathbf{h}_y. \tag{2.31}$$

It now remains to remove the joint constraint forces from (2.31). Premultiplying (2.27a) and (2.27b) by $\mathbf{D}$, solving for $\mathbf{h}_x$ and $\mathbf{h}_y$, and also inserting (2.28a) and (2.28b) give

$$\mathbf{h}_x = \big(\mathbf{D}\mathbf{D}^T\big)^{-1}\big(ml\mathbf{A}\big(\mathbf{C}_\theta\dot{\boldsymbol{\theta}}^2 + \mathbf{S}_\theta\ddot{\boldsymbol{\theta}}\big) - \mathbf{D}\mathbf{f}_{R,x}\big), \tag{2.32a}$$

$$\mathbf{h}_y = \big(\mathbf{D}\mathbf{D}^T\big)^{-1}\big(ml\mathbf{A}\big(\mathbf{S}_\theta\dot{\boldsymbol{\theta}}^2 - \mathbf{C}_\theta\ddot{\boldsymbol{\theta}}\big) - \mathbf{D}\mathbf{f}_{R,y}\big). \tag{2.32b}$$

By inserting (2.32a) and (2.32b) into (2.31) and solving for $\ddot{\boldsymbol{\theta}}$, we can finally rewrite the model of the snake robot as

$$\mathbf{M}_\theta\ddot{\boldsymbol{\theta}} + \mathbf{W}\dot{\boldsymbol{\theta}}^2 - l\mathbf{S}_\theta\mathbf{K}\mathbf{f}_{R,x} + l\mathbf{C}_\theta\mathbf{K}\mathbf{f}_{R,y} = \mathbf{D}^T\mathbf{u}, \tag{2.33a}$$

$$Nm\ddot{\mathbf{p}} = Nm\begin{bmatrix} \ddot{p}_x \\ \ddot{p}_y \end{bmatrix} = \begin{bmatrix} \mathbf{e}^T\mathbf{f}_{R,x} \\ \mathbf{e}^T\mathbf{f}_{R,y} \end{bmatrix} = \mathbf{E}^T\mathbf{f}_R, \tag{2.33b}$$

where $\mathbf{f}_R$ is either the Coulomb friction force given by (2.19) or the viscous friction force given by (2.25), and where

$$\mathbf{M}_\theta = J\mathbf{I}_N + ml^2\mathbf{S}_\theta\mathbf{V}\mathbf{S}_\theta + ml^2\mathbf{C}_\theta\mathbf{V}\mathbf{C}_\theta, \tag{2.34a}$$

$$\mathbf{W} = ml^2\mathbf{S}_\theta\mathbf{V}\mathbf{C}_\theta - ml^2\mathbf{C}_\theta\mathbf{V}\mathbf{S}_\theta, \tag{2.34b}$$

$$\mathbf{V} = \mathbf{A}^T\left(\mathbf{D}\mathbf{D}^T\right)^{-1}\mathbf{A}, \tag{2.34c}$$

$$\mathbf{K} = \mathbf{A}^T\left(\mathbf{D}\mathbf{D}^T\right)^{-1}\mathbf{D}. \tag{2.34d}$$

By introducing the state variable $\mathbf{x} = [\boldsymbol{\theta}^T, \mathbf{p}^T, \dot{\boldsymbol{\theta}}^T, \dot{\mathbf{p}}^T]^T \in \mathbb{R}^{2n+4}$, we can rewrite the model of the snake robot compactly in state space form as

$$\dot{\mathbf{x}} = \begin{bmatrix} \dot{\boldsymbol{\theta}} \\ \dot{\mathbf{p}} \\ \ddot{\boldsymbol{\theta}} \\ \ddot{\mathbf{p}} \end{bmatrix} = \mathbf{F}(\mathbf{x}, \mathbf{u}), \tag{2.35}$$

where the elements of $\mathbf{F}(\mathbf{x}, \mathbf{u})$ are easily found by solving (2.33a) and (2.33b) for $\ddot{\boldsymbol{\theta}}$ and $\ddot{\mathbf{p}}$, respectively.

2.7 Separating Actuated and Unactuated Dynamics

The model of the snake robot in (2.33a) and (2.33b) is rather complex for analysis and control design purposes. We therefore seek a transformation which allows us to write the model in a simpler form. Partial feedback linearisation of underactuated systems (see e.g. Gu and Xu 1993; Spong 1994) consists of linearising the dynamics corresponding to the actuated degrees of freedom of the system. We will employ this methodology in the next section. However, before partial feedback linearisation can be carried out, the model of the snake robot in (2.33a) and (2.33b) must be partitioned into two parts representing the actuated and unactuated degrees of freedom, respectively. This partitioning is now carried out.

The acceleration of the CM of the snake robot, $\ddot{\mathbf{p}}$, belongs to the unactuated part since it is not directly influenced by the input, $\mathbf{u}$. The acceleration of the link angles, $\ddot{\boldsymbol{\theta}}$, represents one unactuated degree of freedom and $N-1$ actuated degrees of freedom since there are N link accelerations ($\boldsymbol{\theta} \in \mathbb{R}^N$) and only $N-1$ control inputs ($\mathbf{u} \in \mathbb{R}^{N-1}$). However, it is not possible to partition the equation for $\ddot{\boldsymbol{\theta}}$ in (2.33a) into an actuated and an unactuated part since the matrix $\mathbf{D}^T$ in front of the control input gives a direct influence between $\mathbf{u}$ and all the link accelerations. We therefore seek a form of the model where there is a direct influence between $\mathbf{u}$ and only $N-1$ link accelerations. This is achieved by modifying the choice of generalised coordinates from absolute link angles to relative joint angles. The generalised coordinates of the model in (2.33a) and (2.33b) are given by the link angles, $\boldsymbol{\theta}$, and the CM position of the snake robot, $\mathbf{p}$. We now replace these coordinates with

$$\mathbf{q}_\phi = \begin{bmatrix} \boldsymbol{\phi} \\ \mathbf{p} \end{bmatrix} \in \mathbb{R}^{N+2}, \tag{2.36}$$

where $\overline{\boldsymbol{\phi}} = [\phi_1, \ldots, \phi_{N-1}, \theta_N]^T \in \mathbb{R}^N$ contains the $N-1$ joint angles of the snake robot and the absolute link angle, $\theta_N \in \mathbb{R}$, of the head link . The joint angles were defined in Definition 2.2. The coordinate transformation between link angles and joint angles is easily shown to be given by

$$\boldsymbol{\theta} = \mathbf{H}\overline{\boldsymbol{\phi}}, \quad \mathbf{H} = \begin{bmatrix} 1 & 1 & 1 & \cdots & 1 & 1 \\ 0 & 1 & 1 & \cdots & 1 & 1 \\ \vdots & & & & & \vdots \\ 0 & 0 & 0 & \cdots & 0 & 1 \end{bmatrix} \in \mathbb{R}^{N\times N}. \tag{2.37}$$

The model of the snake robot in the new coordinates is found by inserting (2.37) into (2.33a). This gives

$$\mathbf{M}_\theta \mathbf{H}\ddot{\overline{\boldsymbol{\phi}}} + \mathbf{W}\,\mathrm{diag}(\mathbf{H}\dot{\overline{\boldsymbol{\phi}}})\mathbf{H}\dot{\overline{\boldsymbol{\phi}}} - l\mathbf{S}_\theta \mathbf{K}\mathbf{f}_{R,x} + l\mathbf{C}_\theta \mathbf{K}\mathbf{f}_{R,y} = \mathbf{D}^T\mathbf{u}, \tag{2.38a}$$

$$Nm\ddot{\mathbf{p}} = \mathbf{E}^T\mathbf{f}_R, \tag{2.38b}$$

where we have used that $\dot{\boldsymbol{\theta}}^2 = \mathrm{diag}(\dot{\boldsymbol{\theta}})\dot{\boldsymbol{\theta}} = \mathrm{diag}(\mathbf{H}\dot{\overline{\boldsymbol{\phi}}})\mathbf{H}\dot{\overline{\boldsymbol{\phi}}}$. Finally, we premultiply (2.38a) with $\mathbf{H}^T$ in order to achieve the desired form of the input mapping matrix on the right-hand side by making the last of the N equations independent of the control input. This enables us to write the complete model of the snake robot as

$$\overline{\mathbf{M}}(\overline{\boldsymbol{\phi}})\ddot{\mathbf{q}}_\phi + \overline{\mathbf{W}}(\overline{\boldsymbol{\phi}}, \dot{\overline{\boldsymbol{\phi}}}) + \overline{\mathbf{G}}(\overline{\boldsymbol{\phi}})\mathbf{f}_R(\overline{\boldsymbol{\phi}}, \dot{\overline{\boldsymbol{\phi}}}, \dot{\mathbf{p}}) = \overline{\mathbf{B}}\mathbf{u}, \tag{2.39}$$

where

$$\mathbf{q}_\phi = \begin{bmatrix} \overline{\boldsymbol{\phi}} \\ \mathbf{p} \end{bmatrix}, \tag{2.40a}$$

$$\overline{\mathbf{M}}(\overline{\boldsymbol{\phi}}) = \begin{bmatrix} \mathbf{H}^T\mathbf{M}_\theta(\overline{\boldsymbol{\phi}})\mathbf{H} & \mathbf{0}_{N\times 2} \\ \mathbf{0}_{2\times N} & Nm\mathbf{I}_2 \end{bmatrix}, \tag{2.40b}$$

$$\overline{\mathbf{W}}(\overline{\boldsymbol{\phi}}, \dot{\overline{\boldsymbol{\phi}}}) = \begin{bmatrix} \mathbf{H}^T\mathbf{W}(\overline{\boldsymbol{\phi}})\,\mathrm{diag}(\mathbf{H}\dot{\overline{\boldsymbol{\phi}}})\mathbf{H}\dot{\overline{\boldsymbol{\phi}}} \\ \mathbf{0}_{2\times 1} \end{bmatrix}, \tag{2.40c}$$

$$\overline{\mathbf{G}}(\overline{\boldsymbol{\phi}}) = \begin{bmatrix} -l\mathbf{H}^T\mathbf{S}_{H\overline{\phi}}\mathbf{K} & l\mathbf{H}^T\mathbf{C}_{H\overline{\phi}}\mathbf{K} \\ -\mathbf{e}^T & \mathbf{0}_{1\times N} \\ \mathbf{0}_{1\times N} & -\mathbf{e}^T \end{bmatrix}, \tag{2.40d}$$

$$\overline{\mathbf{B}} = \begin{bmatrix} \mathbf{I}_{N-1} \\ \mathbf{0}_{3\times(N-1)} \end{bmatrix}, \tag{2.40e}$$

and where $\mathbf{S}_{H\overline{\phi}} = \mathbf{S}_\theta$ and $\mathbf{C}_{H\overline{\phi}} = \mathbf{C}_\theta$.

Remark 2.2 It is interesting to note that premultiplying (2.38a) with $\mathbf{H}^T$ both causes the input mapping matrix $\overline{\mathbf{B}}$ to attain a desirable form and produces a symmetrical inertia matrix $\overline{\mathbf{M}}$. Had we left the model in the form of (2.38a) and (2.38b), the inertia matrix would not have been symmetrical.

The first $N-1$ equations of (2.39) represent the dynamics of the relative joint angles of the snake robot, i.e. the *actuated* degrees of freedom of the snake robot. The last three equations represent the dynamics of the absolute orientation and position of the snake robot, i.e. the *unactuated* degrees of freedom. The model may therefore be partitioned as

$$\overline{\mathbf{M}}_{11}\ddot{\mathbf{q}}_a + \overline{\mathbf{M}}_{12}\ddot{\mathbf{q}}_u + \overline{\mathbf{W}}_1 + \overline{\mathbf{G}}_1\mathbf{f}_R = \mathbf{u}, \tag{2.41a}$$

$$\overline{\mathbf{M}}_{21}\ddot{\mathbf{q}}_a + \overline{\mathbf{M}}_{22}\ddot{\mathbf{q}}_u + \overline{\mathbf{W}}_2 + \overline{\mathbf{G}}_2\mathbf{f}_R = \mathbf{0}_{3\times 1}, \tag{2.41b}$$

where $\mathbf{q}_a = [\phi_1, \ldots, \phi_{N-1}]^T \in \mathbb{R}^{N-1}$ represents the actuated degrees of freedom, $\mathbf{q}_u = [\theta_N, p_x, p_y]^T \in \mathbb{R}^3$ represents the unactuated degrees of freedom, $\overline{\mathbf{M}}_{11} \in \mathbb{R}^{(N-1)\times(N-1)}$, $\overline{\mathbf{M}}_{12} \in \mathbb{R}^{(N-1)\times 3}$, $\overline{\mathbf{M}}_{21} \in \mathbb{R}^{3\times(N-1)}$, $\overline{\mathbf{M}}_{22} \in \mathbb{R}^{3\times 3}$, $\overline{\mathbf{W}}_1 \in \mathbb{R}^{N-1}$, $\overline{\mathbf{W}}_2 \in \mathbb{R}^3$, $\overline{\mathbf{G}}_1 \in \mathbb{R}^{(N-1)\times 2N}$, and $\overline{\mathbf{G}}_2 \in \mathbb{R}^{3\times 2N}$.

Remark 2.3 $\overline{\mathbf{M}}(\boldsymbol{\phi})$ only depends on the relative joint angles of the snake robot and not on the absolute orientation of the head link, θ_N. Formally, this is a result of the fact that θ_N is a *cyclic* coordinate (Goldstein et al. 2002). Less formally, this is quite obvious since it would not be reasonable that the inertial properties of a planar snake robot be dependent on how the snake robot is oriented in the plane. We therefore have that $\overline{\mathbf{M}} = \overline{\mathbf{M}}(\mathbf{q}_a)$.

2.8 Partial Feedback Linearisation of the Model

Based on the partitioned model in (2.41a) and (2.41b), we are now ready to transform the model of the snake robot to a simpler form through partial feedback linearisation (see Gu and Xu 1993; Spong 1994) by introducing an input transformation which linearises the dynamics of the actuated degrees of freedom in (2.41a). This conversion greatly simplifies the controllability and stabilisability analysis of the snake robot presented in Chap. 4. We will follow the approach presented in Reyhanoglu et al. (1999).

We begin by solving (2.41b) for $\ddot{\mathbf{q}}_u$ as

$$\ddot{\mathbf{q}}_u = -\overline{\mathbf{M}}_{22}^{-1}(\overline{\mathbf{M}}_{21}\ddot{\mathbf{q}}_a + \overline{\mathbf{W}}_2 + \overline{\mathbf{G}}_2\mathbf{f}_R), \tag{2.42}$$

where $\overline{\mathbf{M}}_{22}$ is an invertible 3×3 matrix as a consequence of the uniform positive definiteness of the system inertia matrix $\overline{\mathbf{M}}(\mathbf{q}_a)$. Inserting (2.42) into (2.41a) gives

$$(\overline{\mathbf{M}}_{11} - \overline{\mathbf{M}}_{12}\overline{\mathbf{M}}_{22}^{-1}\overline{\mathbf{M}}_{21})\ddot{\mathbf{q}}_a + \overline{\mathbf{W}}_1 + \overline{\mathbf{G}}_1\mathbf{f}_R - \overline{\mathbf{M}}_{12}\overline{\mathbf{M}}_{22}^{-1}(\overline{\mathbf{W}}_2 + \overline{\mathbf{G}}_2\mathbf{f}_R) = \mathbf{u}. \tag{2.43}$$

Consequently, the following linearising controller

$$\mathbf{u} = (\overline{\mathbf{M}}_{11} - \overline{\mathbf{M}}_{12}\overline{\mathbf{M}}_{22}^{-1}\overline{\mathbf{M}}_{21})\overline{\mathbf{u}} + \overline{\mathbf{W}}_1 + \overline{\mathbf{G}}_1\mathbf{f}_R - \overline{\mathbf{M}}_{12}\overline{\mathbf{M}}_{22}^{-1}(\overline{\mathbf{W}}_2 + \overline{\mathbf{G}}_2\mathbf{f}_R), \tag{2.44}$$

where $\overline{\mathbf{u}} = [\overline{u}_1, \ldots, \overline{u}_{N-1}]^T \in \mathbb{R}^{N-1}$ is a new set of control inputs, enables us to rewrite (2.41a) and (2.41b) as

$$\ddot{\mathbf{q}}_a = \overline{\mathbf{u}}, \tag{2.45a}$$

$$\ddot{\mathbf{q}}_u = \mathcal{A}(\mathbf{q}_\phi, \dot{\mathbf{q}}_\phi) + \mathcal{B}(\mathbf{q}_a)\overline{\mathbf{u}}, \tag{2.45b}$$

where

$$\mathcal{A}(\mathbf{q}_\phi, \dot{\mathbf{q}}_\phi) = -\overline{\mathbf{M}}_{22}^{-1}(\overline{\mathbf{W}}_2 + \overline{\mathbf{G}}_2\mathbf{f}_R) \in \mathbb{R}^3, \tag{2.46a}$$

$$\mathcal{B}(\mathbf{q}_a) = -\overline{\mathbf{M}}_{22}^{-1}\overline{\mathbf{M}}_{21} \in \mathbb{R}^{3\times(N-1)}. \tag{2.46b}$$

This model may be written in the standard form of a control-affine system by defining $\mathbf{x}_1 = \mathbf{q}_a$, $\mathbf{x}_2 = \mathbf{q}_u$, $\mathbf{x}_3 = \dot{\mathbf{q}}_a$, $\mathbf{x}_4 = \dot{\mathbf{q}}_u$, and $\mathbf{x} = [\mathbf{x}_1^T, \mathbf{x}_2^T, \mathbf{x}_3^T, \mathbf{x}_4^T]^T \in \mathbb{R}^{2N+4}$. This gives

$$\dot{\mathbf{x}} = \begin{bmatrix} \dot{\mathbf{x}}_1 \\ \dot{\mathbf{x}}_2 \\ \dot{\mathbf{x}}_3 \\ \dot{\mathbf{x}}_4 \end{bmatrix} = \begin{bmatrix} \mathbf{x}_3 \\ \mathbf{x}_4 \\ \overline{\mathbf{u}} \\ \mathcal{A}(\mathbf{x}) + \mathcal{B}(\mathbf{x}_1)\overline{\mathbf{u}} \end{bmatrix} = \mathbf{f}(\mathbf{x}) + \sum_{j=1}^{N-1} \big(\mathbf{g}_j(\mathbf{x}_1)\overline{u}_j\big), \tag{2.47}$$

where

$$\mathbf{f}(\mathbf{x}) = \begin{bmatrix} \mathbf{x}_3 \\ \mathbf{x}_4 \\ \mathbf{0}_{(N-1)\times 1} \\ \mathcal{A}(\mathbf{x}) \end{bmatrix}, \qquad \mathbf{g}_j(\mathbf{x}_1) = \begin{bmatrix} \mathbf{0}_{(N-1)\times 1} \\ \mathbf{0}_{3\times 1} \\ \mathbf{e}_j \\ \mathcal{B}_j(\mathbf{x}_1) \end{bmatrix}, \tag{2.48}$$

$j \in \{1, \ldots, N-1\}$, $\mathbf{e}_j$ denotes the jth standard basis vector in $\mathbb{R}^{N-1}$ (the jth column of $\mathbf{I}_{N-1}$), and $\mathcal{B}_j(\mathbf{x}_1)$ denotes the jth column of $\mathcal{B}(\mathbf{x}_1)$. In literature that considers control-affine systems, the vector $\mathbf{f}(\mathbf{x})$ is often called the drift vector field, while the vectors $\mathbf{g}_j(\mathbf{x}_1)$ are called the control vector fields.

Remark 2.4 We have used $\mathbf{x}$ to denote the state vector of the model (2.47) and also the state vector of the model (2.35) even though these two state vectors are not identical (the difference being the use of joint angles in (2.47) and link angles in (2.35)). Several models of snake robot locomotion are presented in this book, the presentation of which would be obscured by the introduction of a new symbol to represent the state vector of each model. Since the various models are treated separately, which means that it will always be clear from the context which elements that are contained in the state $\mathbf{x}$, we choose to denote the state vector in all models by $\mathbf{x}$.

Remark 2.5 The input transformation in (2.44) is nonsingular, which means that results from analysis and control design based on the partially linearised model in (2.47) are also applicable to the nonlinearised model in (2.35). This is obvious since the behaviour of the model in (2.47) with a control law for $\overline{\mathbf{u}}$ is identical to the behaviour of the model in (2.35) with the control law for $\overline{\mathbf{u}}$ transformed to $\mathbf{u}$ according to (2.44).

2.9 Chapter Summary

This chapter is summarised as follows:

- We have presented a mathematical model of a planar snake robot with N rigid links interconnected by $N-1$ motorised joints. The surface underneath the robot was assumed to be flat and horizontal.
- We have developed two different ground friction models, i.e. a Coulomb friction model given by (2.19) and a viscous friction model given by (2.25).
- The equations of motion of the snake robot in terms of the acceleration of the link angles, $\ddot{\boldsymbol{\theta}}$, and the acceleration of the CM position, $\ddot{\mathbf{p}}$, are given by (2.33a) and (2.33b).
- The equations of motion of the snake robot in terms of the acceleration of the joint angles, $\ddot{\boldsymbol{\phi}}$, the acceleration of the head link angle, $\ddot{\theta}_N$, and the acceleration of the CM position, $\ddot{\mathbf{p}}$, are given by (2.41a) and (2.41b). In this model, the *actuated* degrees of freedom of the snake robot are separated from the *unactuated* degrees of freedom.
- With the input transformation in (2.44), the model of the snake robot is partially feedback linearised to the simpler form given by (2.47). This form is more suitable for control design and analysis purposes.

Chapter 3
Development of a Mechanical Snake Robot for Motion Across Planar Surfaces

The work underlying this book includes the development of two mechanical snake robots. The first snake robot, named *Wheeko*, was developed to enable experiments related to snake robot locomotion across flat surfaces. The robot is shown in Fig. 3.1 and is described in more detail in this chapter. The second snake robot, which is called *Kulko*, is described in Chap. 11 and was developed for the purpose of experiments related to snake robot locomotion in environments containing obstacles. The robots *Wheeko* and *Kulko* are, in other words, the experimental platforms used in Part I and Part II of this book, respectively.

The internal structures of Wheeko and Kulko are identical. The difference between the two robots concern their outer structure. The joint modules of Wheeko are equipped with passive wheels to give the robot anisotropic ground friction properties, which are desirable during motion across flat surfaces. The joint modules of Kulko, on the other hand, are covered by contact force sensors and spherical shells that give the robot a smooth outer surface, thereby allowing gliding motion in uneven and cluttered environments.

This chapter is organised as follows. The relation between this chapter and previous literature is briefly discussed in Sect. 3.1. The joint actuation mechanism, the passive wheels, and the power and control system of Wheeko are presented in Sects. 3.2, 3.3, and 3.4, respectively. The setup of the experiments carried out with the robot is presented in Sect. 3.5. Finally, the chapter is summarised in Sect. 3.6.

3.1 The Relation Between This Chapter and Previous Literature

Several wheeled snake robots have been presented in previous literature over the years, such as the robots presented in Hirose (1993) (see Fig. 1.7), Endo et al. (1999), Togawa et al. (2000), Ma et al. (2001), Wiriyacharoensunthorn and Laowattana (2002), Mori and Hirose (2002) (see Fig. 1.8), Miller (2002) (see Fig. 1.9), Ye et al. (2004b, 2007), Yamada et al. (2005) (see Fig. 1.10), Crespi and Ijspeert (2008), Yu et al. (2008, 2009), and Kamegawa et al. (2009). We would argue that

P. Liljebäck et al., *Snake Robots*, Advances in Industrial Control,
DOI 10.1007/978-1-4471-2996-7_3, © Springer-Verlag London 2013

Fig. 3.1 The snake robot *Wheeko* developed for locomotion across flat surfaces

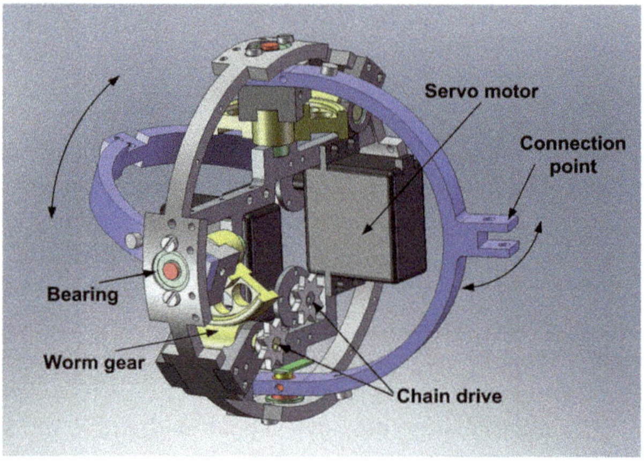

Fig. 3.2 Illustration of the articulation mechanism of the joint modules

all wheeled snake robots are qualitatively similar in the sense that they are able to move across relatively flat surfaces. In that sense, there are no features of Wheeko which are significantly novel, although the internal structure and shape of Wheeko are still quite different from previous snake robots.

3.2 The Joint Actuation Mechanism

Wheeko consists of 10 identical joint modules. The robot was developed with the same joint actuation mechanism as in Kulko to allow the same joint design to be used for both robots. As described in Chap. 11, the joint modules were designed to allow them to be covered by contact force sensors and spherical shells for the purpose of adaptive gliding motion in environments with obstacles.

As illustrated in Fig. 3.2, the articulation mechanism of each joint module has two degrees of freedom (pitch and yaw) and consists of two links supported by bearings in a steel ring. The outer diameter of the steel ring is 130 mm. Each link has a connection point at its centre that allows it to be connected to the next joint module by two screws. The axes of rotation of the two links are orthogonal and intersecting.

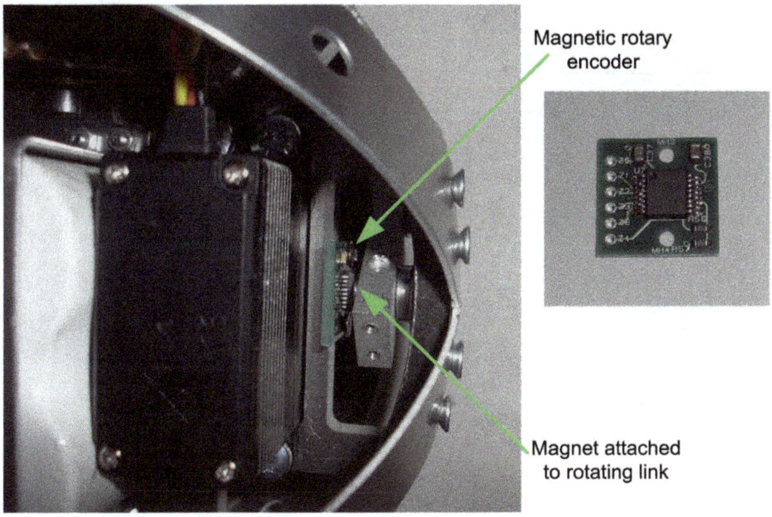

Fig. 3.3 *Magnetic rotary encoder* used for measuring the joint angle

The angle of the two moving links in the joint are measured with magnetic rotary encoders (AS5043 from austriamicrosystems). A magnet measuring 6 mm in diameter is attached to each link so that it rotates above the rotary encoder as shown in Fig. 3.3. Each encoder is attached to a custom-designed circuit board shown to the right in Fig. 3.3.

Each link is driven by a *Hitec* servo motor (HS-5955TG) by connecting the output shaft of each motor to a worm gear (gear ratio of 1 : 5.71) through a steel roller chain. The worm gear and the chain drive are shown in Figs. 3.4 and 3.5. The servo motors are manufactured to have a limited range of rotation (about ±90°). However, the gearing between the motors and the links requires the motors to rotate more than this limited range. The motors were therefore manually modified in order to enable them to rotate continuously. The process of modifying the servos is very simple and consists of disconnecting the output shaft of the servo from its internal potmeter and also removing a mechanical pin inside the servo that otherwise would prevent the servo from rotating continuously.

Worm gears have a disadvantage due to a high-friction component in the gear system. However, worm gears are advantageous in that they may essentially produce any desired gear ratio in a single gear stage. This facilitates a compact design. In addition, a worm gear is not likely to break in contrast to e.g. spur gears. This makes the joint mechanically robust. The steel roller chain between the servo motor and the worm gear is rated to handle forces significantly higher than the forces produced by the servo motor.

Experiments indicate that the servo motors produce a maximum continuous torque of about 1.6 Nm (at 6-V supply voltage with a maximum current drain of about 3 A). The rated power efficiency of the worm gears is about 75%. This should theoretically give the joint mechanism a maximum continuous torque of around

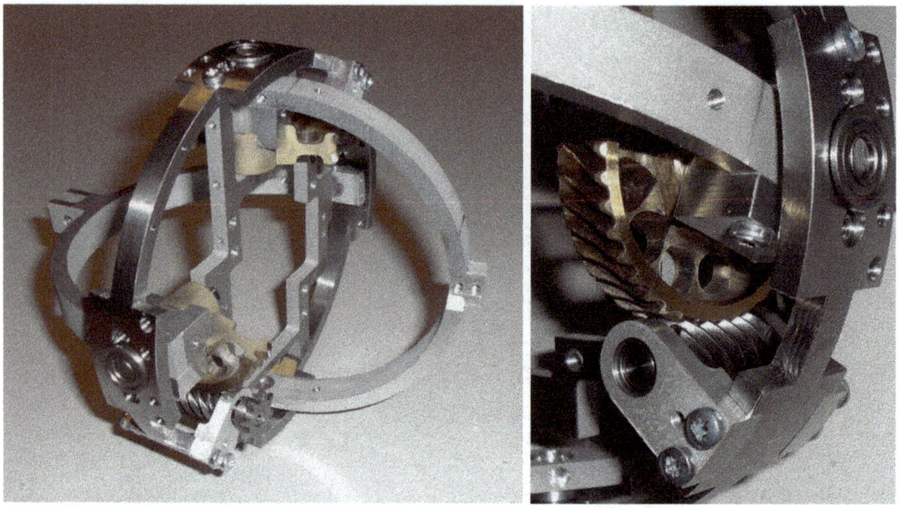

Fig. 3.4 The implemented articulation mechanism of the joint modules

Fig. 3.5 Roller chain connecting the servo motor to the worm gear

7 Nm. However, experiments with the implemented joint mechanism indicate that the maximum continuous torque lies around 4.5 Nm. This is probably due to more friction in the worm gear than expected and also some friction in the chain drive. Table 3.1 lists the parameters characterising the actuation mechanism.

3.3 The Passive Wheels

As shown in Fig. 3.6, each joint module of Wheeko is enclosed by a plastic ring mounted with 12 plastic wheels. The wheels are passive, i.e. not motorised, and ensure that the ground friction forces acting on the robot are anisotropic (see

Table 3.1 Parameters of a joint module

Parameter	Value
Total weight of a joint module	960 g
Outer diameter	130 mm
Degrees of freedom	2
Max joint travel	±45°
Max continuous joint torque	4.5 Nm
Max joint speed (no load)	70°/sec

Sect. 2.5.1), i.e. that the friction coefficient characterising the ground friction forces in the normal (sideways) direction of each joint is larger than the friction coefficient characterising the ground friction forces in the tangential (forward) direction of the joint. Note that the wheels are able to slip sideways, so they do not introduce nonholonomic constraints in the system.

3.4 The Power and Control System

The power and control system of Wheeko is described in Chap. 11 since the system is identical to the power and control system of Kulko.

3.5 The Experimental Setup of the Snake Robot

The experiments carried out with Wheeko are described in Chaps. 7 and 8. The experiments were performed on a white horizontal surface measuring about 240 cm in width and 600 cm in length. The surface is shown in Fig. 3.7. In order to measure

Fig. 3.6 A ring with 12 plastic wheels encloses each joint module in order to give the robot anisotropic ground friction properties

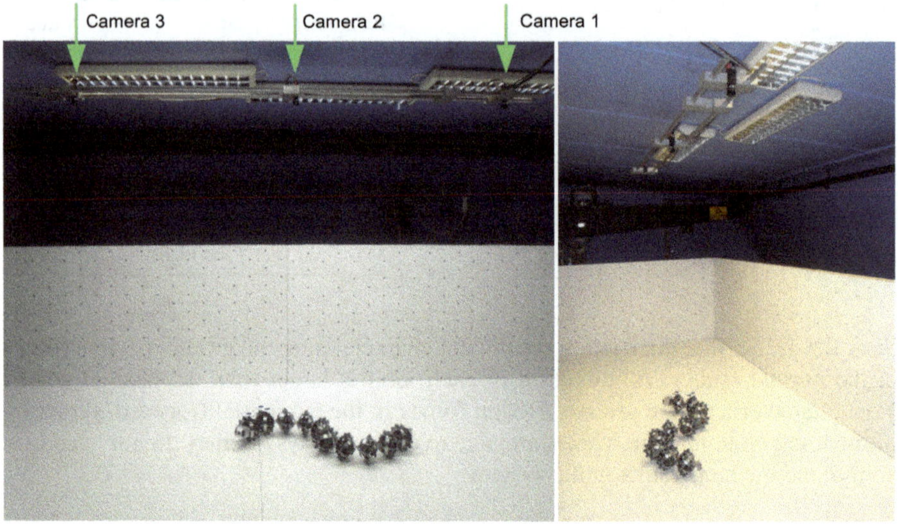

Fig. 3.7 The experimental setup. Three *cameras* mounted in the ceiling measured the position of the snake robot on a horizontal surface measuring about 240 cm in width and 600 cm in length

Fig. 3.8 *Left*: The firewire *cameras* mounted in the ceiling above the snake robot. *Right*: The *black markers* mounted on the snake robot to allow the position to be tracked by SwisTrack

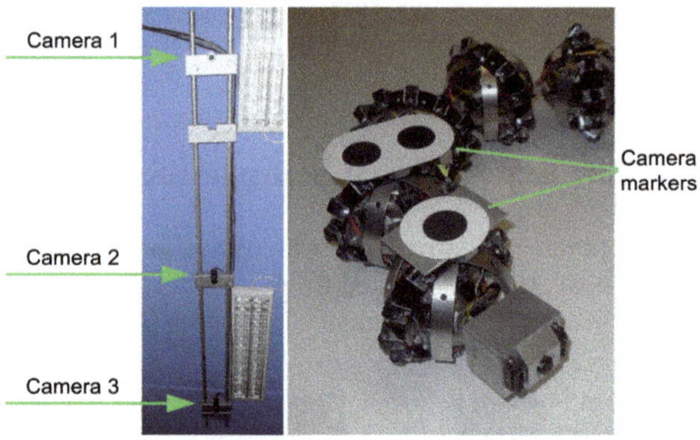

the horizontal position of the snake robot during the experiments, we employed the open source camera tracking software *SwisTrack* (Lochmatter et al. 2008). Three firewire cameras (Unibrain Fire-i 520c) were mounted in the ceiling above the snake robot as shown in Fig. 3.7 and to the left in Fig. 3.8. The use of multiple cameras allowed for position measurements over a greater distance than the area covered by a single camera. The cameras were mounted facing downwards approximately 218 cm above the floor and 132 cm apart. The distance between the cameras was chosen so that there was a slight overlap between the images from two neighbouring cameras. Each firewire camera was sampled at 15 frames per second.

SwisTrack was configured to track three black circular markers (40 mm in diameter) mounted on the snake robot as shown to the right in Fig. 3.8. The conversion from the pixel position of a marker to the real-world position (in cm) was conducted by SwisTrack based on a specific calibration method available in this software. SwisTrack estimated the maximum position error to be about 1.9 cm and the average position error to be about 0.6 cm. The global frame coordinates of the head link, (x_N, y_N), and the absolute angle of the head link, θ_N, were calculated based on the position of the three individual markers. Knowing the position and orientation of the head of the snake robot, and also the individual joint angles, the kinematic relationships presented in Sect. 2.4 enabled us to calculate the CM position, $\mathbf{p}$, and the absolute link angles, $\boldsymbol{\theta}$, of the snake robot. We ran three separate instances of SwisTrack in order to process data from all three cameras and developed our own software in order to merge the output from each SwisTrack instance into the final position measurement of the snake robot.

3.6 Chapter Summary

This chapter is summarised as follows:

- We have presented the design of the snake robot *Wheeko*, which was developed for the purpose of experiments related to snake robot locomotion across flat surfaces.
- The robot consists of 10 identical joint modules, and each joint has two degrees of freedom (pitch and yaw).
- The joint modules are equipped with passive wheels to give the robot anisotropic ground friction properties.
- The internal structure of Wheeko is identical to the internal structure of the snake robot *Kulko*, which is described in Chap. 11.

Chapter 4
Analysis and Synthesis of Snake Robot Locomotion

Research on snake robots has been conducted for several decades. However, our understanding of snake locomotion so far is for the most part based on empirical studies of biological snakes and simulation-based synthesis of relationships between parameters of the snake robot. Armed with the mathematical model of the snake robot presented in Chap. 2, we attempt in this chapter to contribute to the understanding of snake robots by employing nonlinear system analysis tools for investigating fundamental properties of their dynamics. We will also derive several interesting properties of snake robot locomotion simply by investigating the equations of motion of the robot, some of which will be instrumental in the development of a *simplified* model in Chap. 6.

In this chapter, we also investigate the motion pattern which is most common among biological snakes, namely *lateral undulation* (see Sect. 1.2.2). This motion pattern is considered in the majority of the snake robot literature and will also receive much attention throughout this book. A well-known mathematical description of the shape of a snake during lateral undulation was presented in Hirose (1993) based on empirical studies of biological snakes. In this chapter, we develop analytical results that support this mathematical description.

The chapter is organised as follows. The relation between this chapter and previous literature is elaborated in Sect. 4.1. Section 4.2 introduces some selected tools for analysing the controllability of nonlinear systems. Section 4.3 and Sect. 4.4 study, respectively, the stabilisability and controllability properties of planar snake robots. Section 4.5 explains how a snake robot is able to move forward, while Sect. 4.6 deduces how the links of the robot can be moved in order to achieve forward propulsion. The deduced link motion is identified as the gait pattern *lateral undulation* in Sect. 4.7, and two different control laws for the joints of the snake robot are presented in Sect. 4.8. Two properties related to the turning motion and the relative displacements of the links during lateral undulation are derived in Sect. 4.9 and Sect. 4.10, respectively. Finally, the chapter is summarised in Sect. 4.11.

P. Liljebäck et al., *Snake Robots*, Advances in Industrial Control,
DOI 10.1007/978-1-4471-2996-7_4, © Springer-Verlag London 2013

4.1 The Relation Between This Chapter and Previous Literature

Controllability properties of *wheeled* snake robots, i.e. snake robots modelled under the assumption that the links cannot slip sideways, have been studied in several previous works, such as in Ishikawa (2009), Kelly and Murray (1995), Krishnaprasad and Tsakiris (1994), Ostrowski and Burdick (1998), Ostrowski (1996). In this book, however, we are considering *wheel-less* snake locomotion (this scope was explained in Sect. 1.4), which differ quite significantly from wheeled locomotion due to the sideslip constraints on the links of a wheeled snake robot. Previous investigations of fundamental properties of wheel-less snake robot dynamics are very limited. The stabilisability and controllability properties of snake robots presented in this chapter are therefore novel with respect to previous literature.

Note that snake robots with anisotropic ground friction properties, which are considered in this chapter, bear some resemblance with robotic fish and eel-like mechanisms since water drag forces have a similar influence on a snake-like mechanism as anisotropic ground friction properties (this was explained in Sect. 2.5.1). Snake-like locomotion under water is analysed in e.g. McIsaac and Ostrowski (2003a), Morgansen et al. (2001, 2002, 2007), Vela et al. (2002a).

The first novel result of this chapter in relation to previous literature is a stabilisability analysis that proves that any asymptotically stabilising control law for a planar snake robot to an equilibrium point must be *time-varying*, i.e. not of a pure-state feedback type (see Theorem 4.3).

The second result is a controllability analysis of planar snake robots influenced by viscous ground friction forces. The analysis shows that a snake robot is *not* controllable when the viscous ground friction is *isotropic* (see Theorem 4.4) but that a snake robot becomes *strongly accessible* when the viscous ground friction is *anisotropic* (see Theorem 4.5). The analysis also shows that the snake robot does *not* satisfy sufficient conditions for *small-time local controllability* (see Theorem 4.6). To our best knowledge, no formal controllability analysis has previously been reported for the position and link angles of a wheel-less snake robot influenced by ground friction. The results from the controllability analysis prove that propulsion of a snake robot under viscous friction conditions requires the friction to be *anisotropic*, and also suggest that the joint angles of a snake robot should be *out of phase* during snake locomotion.

The third result is the development of a simple relationship between link velocities *normal* to the direction of motion and *propulsive forces* in the direction of motion. This relationship explains how snake robots influenced by anisotropic ground friction are able to locomote forward on a planar surface, and enables us to derive several fundamental properties of snake locomotion (see Properties 4.2, 4.3, and 4.4). In our opinion, previously published research on snake robots has not presented an explicit mathematical description that *easily* explains how a snake robot achieves forward propulsion.

As a fourth and final result, we use the derived properties of snake robot locomotion to provide support for the empirically derived mathematical description of lateral undulation proposed in Hirose (1993). We also identify an important property concerning the turning motion of a snake robot (see Property 4.7), and a property related to the relative displacements of the links during lateral undulation (see Property 4.8).

4.2 Introduction to Nonlinear Controllability Analysis

This section presents a brief summary of selected tools for analysing the controllability of nonlinear systems. The summary given below is formulated in an intuitive form that aims to be easily understandable for readers unaccustomed with nonlinear controllability analysis. For a rigorous presentation, the readers are referred to Bianchini and Stefani (1990), Nijmeijer and Schaft (1990), Sussmann (1987).

Analysing the controllability of a *linear* system is straightforward and involves checking if the system matrices satisfy the *Kalman rank condition* (see Nijmeijer and Schaft 1990). However, studying the controllability of a *nonlinear* system is far more complex and constitutes an active area of research. In the following, we summarise important controllability concepts for control-affine nonlinear systems, i.e. systems of the form

$$\dot{\mathbf{x}} = \mathbf{f}(\mathbf{x}) + \sum_{j=1}^{m} \mathbf{g}_j(\mathbf{x})\overline{u}_j, \tag{4.1}$$

where $\mathbf{x} \in \mathbb{R}^n$ is the state vector, $\overline{\mathbf{u}} \in \mathbb{R}^m$ is the control input, $\mathbf{f}(\mathbf{x}) \in \mathbb{R}^n$ is the drift vector field, and $\mathbf{g}_j(\mathbf{x}) \in \mathbb{R}^n$, $j \in \{1, \dots, m\}$, are the control vector fields of the system. Note that the model of the snake robot was written in this form in Sect. 2.8.

A nonlinear system is said to be *controllable* if there exist admissible control inputs that will move the system between two arbitrary states in finite time. However, general conditions for this kind of controllability that are both necessary and sufficient do not exist. Nonlinear controllability is instead typically analysed by investigating the local behaviour of the system near equilibrium points.

The simplest approach to studying the controllability of system (4.1) is to linearise the system about an equilibrium point $\mathbf{x}^e$ according to

$$\dot{\mathbf{z}} = \mathbf{A}\mathbf{z} + \mathbf{B}\overline{\mathbf{u}}, \tag{4.2}$$

where $\mathbf{z} = \mathbf{x} - \mathbf{x}^e$, $\mathbf{A} = \frac{\partial \mathbf{f}(\mathbf{x})}{\partial \mathbf{x}}|_{\mathbf{x}^e}$, and $\mathbf{B} = [\mathbf{g}_1, \dots, \mathbf{g}_m]$. If the linearised system satisfies the Kalman rank condition at $\mathbf{x}^e$, which requires that the controllability matrix given by $[\mathbf{B}, \mathbf{AB}, \mathbf{A}^2\mathbf{B}, \dots, \mathbf{A}^{n-1}\mathbf{B}]$ has full rank, then the nonlinear system (4.1) is controllable in the sense that the set of states that can be reached from $\mathbf{x}^e$ contains a neighborhood of $\mathbf{x}^e$. Unfortunately, many underactuated systems do not have a controllable linearisation. Moreover, a nonlinear system can be controllable even though its linearisation is not.

A necessary (but not sufficient) condition for controllability from a state $\mathbf{x}_0$ (not necessarily an equilibrium) is that the nonlinear system satisfies the *Lie algebra rank condition* (LARC), also called the *accessibility rank condition* (see Nijmeijer and

Schaft 1990). If this is the case, the system is said to be *locally accessible* from $\mathbf{x}_0$. This property means that the space that the system can reach within any time $T > 0$ is fully n-dimensional, i.e. the reachable space from $\mathbf{x}_0$ has a dimension equal to the dimension of the state space. A slightly stronger property is *strong accessibility*, which means that the space that the system can reach in *exactly* time T for any $T > 0$ is fully n-dimensional.

Accessibility of a nonlinear system is investigated by computing the *accessibility algebra*, here denoted $\boldsymbol{\Delta}$, of the system. Computation of $\boldsymbol{\Delta}$ requires knowledge of the *Lie bracket* (see Nijmeijer and Schaft 1990), which is now briefly explained. The drift and control vector fields of the nonlinear system (4.1) indicate directions in which the state $\mathbf{x}$ can move. These directions will generally only span a subset of the complete state space. However, through combined motion along two or more of these vector fields, it is possible for the system to move in directions not spanned by the original system vector fields. The Lie bracket between two vector fields $\mathbf{Y}$ and $\mathbf{Z}$ produces a new vector field defined as $[\mathbf{Y},\mathbf{Z}] = \frac{\partial \mathbf{Z}}{\partial \mathbf{x}}\mathbf{Y} - \frac{\partial \mathbf{Y}}{\partial \mathbf{x}}\mathbf{Z}$. When $\mathbf{Y}$ and $\mathbf{Z}$ are any of the system vector fields, the Lie bracket $[\mathbf{Y},\mathbf{Z}]$ *approximates* the net motion produced when the system follows these two vector fields in an alternating fashion. The classical example is parallel parking with a car, where sideways motion of the car may be achieved through an alternating turning and forward/backward motion. Note that Lie brackets can be computed from other Lie brackets, thereby producing nested Lie brackets. The *accessibility algebra*, $\boldsymbol{\Delta}$, is a set of vector fields composed of the system vector fields, $\mathbf{f}$ and $\mathbf{g}_j$, the Lie brackets between the system vector fields, and also higher-order Lie brackets generated by nested Lie brackets. The LARC is satisfied at $\mathbf{x}_0$ if the vector fields in $\boldsymbol{\Delta}(\mathbf{x}_0)$ span the entire n-dimensional state space ($\dim(\mathrm{span}(\boldsymbol{\Delta})) = n$). The following result is proved in Nijmeijer and Schaft (1990):

Theorem 4.1 *System* (4.1) *is* locally accessible *from* $\mathbf{x}_0$ *if and only if the LARC is satisfied at* $\mathbf{x}_0$. *The system is* locally strongly accessible *if the drift field* $\mathbf{f}$ *by itself (i.e. unbracketed) is not included in the* accessibility algebra.

Accessibility does *not* imply controllability since it only infers conclusions on the dimension of the reachable space from $\mathbf{x}_0$. Accessibility is, however, a necessary (although not sufficient) condition for *small-time local controllability* (STLC). STLC is desirable since it is in fact a stronger property than controllability. If a system is STLC, then the control input can steer the system in any direction in an arbitrarily small amount of time. For second-order systems, STLC is only considered from equilibrium states since it is generally not possible for a second-order system to instantly move in one direction if it already has a velocity in the opposite direction. For example, an airplane in flight is not STLC since it cannot instantly move opposite to its direction of motion.

Sufficient conditions for STLC were presented by Sussmann (1987), and later extended by Bianchini and Stefani (1990). We now summarise these conditions. For any Lie bracket term $\mathbf{b} \in \mathbb{R}^n$ generated from the system vector fields, define the

θ-degree of $\mathbf{b}$, denoted $\delta_\theta(\mathbf{b})$, and the l-degree of $\mathbf{b}$, denoted $\delta_l(\mathbf{b})$, as

$$\delta_\theta(\mathbf{b}) = \frac{1}{\theta}\delta^0(\mathbf{b}) + \sum_{j=1}^{m} \delta^j(\mathbf{b}), \tag{4.3}$$

$$\delta_l(\mathbf{b}) = \sum_{j=0}^{m} l_j \delta^j(\mathbf{b}), \tag{4.4}$$

respectively, where $\delta^0(\mathbf{b})$ is the number of times the drift vector field $\mathbf{f}$ appears in the bracket $\mathbf{b}$, $\delta^j(\mathbf{b})$ is the number of times the control vector field $\mathbf{g}_j$ appears in the bracket $\mathbf{b}$, θ is an arbitrary number satisfying $\theta \in [1, \infty)$, and l_j is an arbitrary number satisfying $l_j \geq l_0 \geq 0$ for all $j \in \{0, \dots, m\}$. The bracket $\mathbf{b}$ is said to be *bad* if $\delta^0(\mathbf{b})$ is odd and $\delta^1(\mathbf{b}), \dots, \delta^m(\mathbf{b})$ are all even. A bracket is *good* if it is not bad. As an example, we have that the bracket $[\mathbf{g}_j, [\mathbf{f}, \mathbf{g}_k]]$ is *bad* for $j = k$ and *good* for $j \neq k$. This classification is motivated by the fact that a bad bracket *may* have directional constraints. For example, the drift vector $\mathbf{f}$ is *bad* because it only allows motion in its positive direction and not in its negative direction, $-\mathbf{f}$. A bad bracket is said to be *θ-neutralised* (resp. *l-neutralised*) if it can be written as a linear combination of good brackets of lower θ-degree (resp. l-degree). The *Sussmann condition* and the *Bianchini and Stefani condition* for STLC are now combined in the following theorem:

Theorem 4.2 *System* (4.1) *is* small-time locally controllable (*STLC*) *from an equilibrium point* $\mathbf{x}^e$ ($\mathbf{f}(\mathbf{x}^e) = 0$) *if the LARC is satisfied at* $\mathbf{x}^e$ *and either all* bad *brackets are* θ-neutralised (*Sussmann* 1987) *or all* bad *brackets are* l-neutralised (*Bianchini and Stefani* 1990).

4.3 Stabilisability Properties of Planar Snake Robots

In this section, we present a fundamental theorem concerning the properties of an asymptotically stabilising control law for snake robots to any equilibrium point. The model of the snake robot is given by (2.47), and we make no assumptions regarding the ground friction forces other than that they are given on the form of (2.16).

Equation (2.47) maps the state $\mathbf{x}$ and the control input $\overline{\mathbf{u}}$ of the robot into the resulting derivative of the state vector, $\dot{\mathbf{x}}$. For any equilibrium point ($\mathbf{x}_1 = \mathbf{x}_1^e, \mathbf{x}_2 = \mathbf{x}_2^e, \mathbf{x}_3 = \mathbf{0}, \mathbf{x}_4 = \mathbf{0}$), where $(\mathbf{x}_1^e, \mathbf{x}_2^e)$ is the configuration of the system at the equilibrium point, we have that $\dot{\mathbf{x}} = \mathbf{0}$. A well-known result presented in Brockett (1983) states that a necessary condition for the existence of a *time-invariant* (i.e. not explicitly dependent on time) *continuous* state feedback law, $\overline{\mathbf{u}} = \overline{\mathbf{u}}(\mathbf{x})$, that makes $(\mathbf{x}_1^e, \mathbf{x}_2^e, \mathbf{0}, \mathbf{0})$ asymptotically stable is that the image of the mapping $(\mathbf{x}, \overline{\mathbf{u}}) \mapsto \dot{\mathbf{x}}$ contains some neighbourhood of $\dot{\mathbf{x}} = \mathbf{0}$ (this requirement is explained below). A result presented in Coron and Rosier (1994) states that a control system that can be asymptotically stabilised (in the Filippov sense) by a *time-invariant discontinuous* state feedback law can be asymptotically stabilised by a *time-varying continuous* state feedback law. If, moreover, the control system is *affine* (i.e. linear with respect to the control input), then it can be asymptotically stabilised by a *time-invariant continuous* state feedback law. We now employ these results to prove the following fundamental theorem for planar snake robots:

Theorem 4.3 *An asymptotically stabilising control law for a planar snake robot described by* (2.47) *to any equilibrium point must be time-varying, i.e. of the form* $\overline{\mathbf{u}} = \overline{\mathbf{u}}(\mathbf{x}, t)$.

Proof The result in Brockett (1983) states that the mapping $(\mathbf{x}_1, \mathbf{x}_2, \mathbf{x}_3, \mathbf{x}_4, \overline{\mathbf{u}}) \mapsto (\mathbf{x}_3, \mathbf{x}_4, \overline{\mathbf{u}}, \mathcal{A}(\mathbf{x}) + \mathcal{B}(\mathbf{x}_1)\overline{\mathbf{u}})$ must map an arbitrary neighbourhood of $(\mathbf{x}_1 = \mathbf{x}_1^e, \mathbf{x}_2 = \mathbf{x}_2^e, \mathbf{x}_3 = \mathbf{0}, \mathbf{x}_4 = \mathbf{0}, \overline{\mathbf{u}} = \mathbf{0})$ onto a neighbourhood of $\dot{\mathbf{x}} = \mathbf{0}$, i.e. a neighbourhood of $(\mathbf{x}_3 = \mathbf{0}, \mathbf{x}_4 = \mathbf{0}, \overline{\mathbf{u}} = \mathbf{0}, \mathcal{A}(\mathbf{x}) + \mathcal{B}(\mathbf{x}_1)\overline{\mathbf{u}} = \mathbf{0})$. For this to be true, points of the form $(\mathbf{x}_3 = \mathbf{0}, \mathbf{x}_4 = \mathbf{0}, \overline{\mathbf{u}} = \mathbf{0}, \mathcal{A}(\mathbf{x}) + \mathcal{B}(\mathbf{x}_1)\overline{\mathbf{u}} = \boldsymbol{\epsilon})$ must be contained in this mapping for some arbitrary $\boldsymbol{\epsilon} \neq \mathbf{0}$ because points of this form are contained in every neighbourhood of $\dot{\mathbf{x}} = \mathbf{0}$. In other words, there must exist some state and control input $(\mathbf{x}_1, \mathbf{x}_2, \mathbf{x}_3, \mathbf{x}_4, \overline{\mathbf{u}})$ that can be mapped to points of the form $(\mathbf{x}_3 = \mathbf{0}, \mathbf{x}_4 = \mathbf{0}, \overline{\mathbf{u}} = \mathbf{0}, \mathcal{A}(\mathbf{x}) + \mathcal{B}(\mathbf{x}_1)\overline{\mathbf{u}} = \boldsymbol{\epsilon} \neq \mathbf{0})$. However, these points do not exist for the system in (2.47) because $\mathbf{x}_3 = \mathbf{0}, \mathbf{x}_4 = \mathbf{0}$, and $\overline{\mathbf{u}} = \mathbf{0}$ means that $\mathcal{A}(\mathbf{x}) + \mathcal{B}(\mathbf{x}_1)\overline{\mathbf{u}} = \mathbf{0} \neq \boldsymbol{\epsilon}$. Hence, the snake robot *cannot* be asymptotically stabilised to $(\mathbf{x}_1 = \mathbf{x}_1^e, \mathbf{x}_2 = \mathbf{x}_2^e, \mathbf{x}_3 = \mathbf{0}, \mathbf{x}_4 = \mathbf{0})$ by a *time-invariant continuous* state feedback law. Moreover, since the system in (2.47) is affine and *cannot* be asymptotically stabilised by a *time-invariant continuous* state feedback law, the result in Coron and Rosier (1994) proves that the system can neither be asymptotically stabilised by a *time-invariant discontinuous* state feedback law. We can therefore conclude that an asymptotically stabilising control law for a planar snake robot to any equilibrium point must be time-varying, i.e. of the form $\overline{\mathbf{u}} = \overline{\mathbf{u}}(\mathbf{x}, t)$. □

Remark 4.1 Theorem 4.3 is independent of the choice of friction model and applies to any planar snake robot described by a friction model with the property that $\mathcal{A}(\mathbf{x}^e) = \mathbf{0}$ for any equilibrium point $\mathbf{x}^e$.

4.4 Controllability Analysis of Planar Snake Robots

This section studies the controllability of planar snake robots described by the model (2.47). As described in Sect. 2.5.1, we conjecture that the motion of a snake robot is *qualitatively* (although not *quantitatively*) similar with anisotropic viscous friction as with anisotropic Coulomb friction. In the following, we therefore assume that the ground friction is viscous since the simplicity of the viscous model compared to the Coulomb model is more suitable for controllability analysis purposes.

4.4.1 Controllability with Isotropic Viscous Friction

We begin the controllability analysis of the snake robot by first assuming that the viscous ground friction is isotropic. In this case, it turns out that the equations of motion take on a particularly simple form that enables us to study controllability through pure inspection of the equations of motion. From (2.33b), the acceleration of the CM of the snake robot is given as

$$\begin{bmatrix} \ddot{p}_x \\ \ddot{p}_y \end{bmatrix} = \frac{1}{Nm} \begin{bmatrix} \mathbf{e}^T \mathbf{f}_{R,x} \\ \mathbf{e}^T \mathbf{f}_{R,y} \end{bmatrix} = \frac{1}{Nm} \begin{bmatrix} \sum_{i=1}^N \mathbf{f}_{R,x,i} \\ \sum_{i=1}^N \mathbf{f}_{R,y,i} \end{bmatrix}. \tag{4.5}$$

By inserting (2.20) into (4.5), the CM acceleration of the robot is given as

$$\begin{bmatrix} \ddot{p}_x \\ \ddot{p}_y \end{bmatrix} = \frac{c}{Nm} \begin{bmatrix} -N\dot{p}_x + (\sum_{i=1}^N \sigma_i)\mathbf{S}_\theta \dot{\boldsymbol{\theta}} \\ -N\dot{p}_y - (\sum_{i=1}^N \sigma_i)\mathbf{C}_\theta \dot{\boldsymbol{\theta}} \end{bmatrix} = -\frac{c}{m} \begin{bmatrix} \dot{p}_x \\ \dot{p}_y \end{bmatrix} \tag{4.6}$$

because it may be shown that $\sum_{i=1}^N \sigma_i = 0$. This enables us to state the following theorem:

Theorem 4.4 *A planar snake robot influenced by* isotropic *viscous ground friction is* not controllable.

Proof In order to control the position, the snake robot must accelerate its CM. From (4.6) it is clear that the CM acceleration is proportional to the CM velocity. If the robot starts from rest ($\dot{\mathbf{p}} = \mathbf{0}$), it is therefore impossible to achieve acceleration of the CM. The position of the robot is in other words completely uncontrollable in this case, which renders the robot uncontrollable. □

4.4.2 Controllability with Anisotropic Viscous Friction

The equations of motion of the snake robot in (2.47) become far more complex under anisotropic friction conditions. We therefore employ the controllability concepts presented in Sect. 4.2 and begin by computing the Lie brackets of the system vector fields. The drift vector field $\mathbf{f}(\mathbf{x})$ and the control vector fields $\mathbf{g}_j(\mathbf{x}_1)$, $j \in \{1, \dots, m\}$, of the snake robot are defined in (2.47). As explained in Sect. 4.2, Lie bracket computation involves partial derivatives of the components of the vector fields. These computations can be carried out without dealing with the complex expressions contained in $\mathcal{A}(\mathbf{x})$ and $\mathcal{B}(\mathbf{x}_1)$ given by (2.46a) and (2.46b) since we only need to know which variables each vector field depends on. As an example, consider column j of $\mathcal{B}(\mathbf{x}_1)$. Since we know that it only depends on $\mathbf{x}_1$, we may immediately write $\frac{\partial \mathcal{B}_j(\mathbf{x}_1)}{\partial \mathbf{x}} = \left[\frac{\partial \mathcal{B}_j(\mathbf{x}_1)}{\partial \mathbf{x}_1} \; \mathbf{0}_{3\times(N+5)} \right]$. This methodology enables us to compute the following

Lie brackets of the system vector fields (evaluated at an equilibrium point):

$$[\mathbf{f}, \mathbf{g}_j]^{\dot{\mathbf{q}}_\phi=\mathbf{0}} = \begin{bmatrix} -\mathbf{e}_j \\ -\mathcal{B}_j \\ \mathbf{0}_{(N-1)\times 1} \\ -\mathcal{C}_j \end{bmatrix}, \tag{4.7a}$$

$$\big[\mathbf{f}, [\mathbf{f}, \mathbf{g}_j]\big]^{\dot{\mathbf{q}}_\phi=\mathbf{0}} = \begin{bmatrix} \mathbf{0}_{(N-1)\times 1} \\ \mathcal{C}_j \\ \mathbf{0}_{(N-1)\times 1} \\ \frac{\partial \mathcal{A}}{\partial \mathbf{x}_4}\mathcal{C}_j \end{bmatrix}, \tag{4.7b}$$

$$\big[[\mathbf{f}, \mathbf{g}_j], [\mathbf{f}, \mathbf{g}_k]\big]^{\dot{\mathbf{q}}_\phi=\mathbf{0}} = \begin{bmatrix} \mathbf{0}_{(N-1)\times 1} \\ \mathcal{D}_{jk} \\ \mathbf{0}_{(N-1)\times 1} \\ \mathcal{E}_{jk} \end{bmatrix}, \tag{4.7c}$$

where $j, k \in \{1, \dots, N-1\}$ and

$$\mathcal{C}_j = \frac{\partial \mathcal{A}}{\partial \mathbf{x}_3}\mathbf{e}_j + \frac{\partial \mathcal{A}}{\partial \mathbf{x}_4}\mathcal{B}_j, \tag{4.8}$$

$$\mathcal{D}_{jk} = \frac{\partial \mathcal{B}_k}{\partial \mathbf{x}_1}\mathbf{e}_j - \frac{\partial \mathcal{B}_j}{\partial \mathbf{x}_1}\mathbf{e}_k, \tag{4.9}$$

$$\mathcal{E}_{jk} = \frac{\partial \mathcal{C}_k}{\partial \mathbf{x}_1}\mathbf{e}_j - \frac{\partial \mathcal{C}_j}{\partial \mathbf{x}_1}\mathbf{e}_k + \frac{\partial \mathcal{C}_k}{\partial \mathbf{x}_2}\mathcal{B}_j - \frac{\partial \mathcal{C}_j}{\partial \mathbf{x}_2}\mathcal{B}_k + \frac{\partial \mathcal{C}_k}{\partial \mathbf{x}_4}\mathcal{C}_j - \frac{\partial \mathcal{C}_j}{\partial \mathbf{x}_4}\mathcal{C}_k. \tag{4.10}$$

The Lie brackets have been evaluated at zero velocity ($\dot{\mathbf{q}}_\phi = \mathbf{0}$) since we are interested in controllability from an equilibrium point. The above vector fields represent our choice of vector fields to be contained in the *accessibility algebra*, $\boldsymbol{\Delta}$, of the system. To construct $\boldsymbol{\Delta}$ of full rank, we need to find $(2N+4)$ independent vector fields since the snake robot has a $(2N+4)$-dimensional state space. Each of the four types of vector fields above represent $(N-1)$ vector fields. Solving $4(N-1) \geq 2N+4$ gives that our analysis is only valid if the snake robot has $N \geq 4$ links. This is a mild requirement, however, since a snake robot generally has more than four links. In the remainder of this controllability analysis, we assume that the robot consists of exactly $N = 4$ links (and thereby $N - 1 = 3$ active joints) and argue that the following controllability results will also be valid for snake robots with more links. In particular, a robot with $N > 4$ links can behave as a robot with $N = 4$ links by fixing $(N-4)$ joint angles at zero degrees and allowing the remaining three joint angles to move. This means that controllability of the robot with $N = 4$ is a sufficient, although not necessary, condition for controllability with $N > 4$.

With $N = 4$ links, the system has a $(2N+4) = 12$-dimensional state space. The system satisfies the *Lie algebra rank condition* (LARC) if the above vector fields span a 12-dimensional space. We therefore assemble the 12 vector fields into the following matrix, which represents the *accessibility algebra* of the system evaluated at an equilibrium point $\mathbf{x}^e$:

$$\begin{aligned}\boldsymbol{\Delta}(\mathbf{x}^e) &= \big[\mathbf{g}_1, \mathbf{g}_2, \mathbf{g}_3, [\mathbf{f}, \mathbf{g}_1], [\mathbf{f}, \mathbf{g}_2], [\mathbf{f}, \mathbf{g}_3], \\ &\quad \big[\mathbf{f}, [\mathbf{f}, \mathbf{g}_1]\big], \big[\mathbf{f}, [\mathbf{f}, \mathbf{g}_2]\big], \big[\mathbf{f}, [\mathbf{f}, \mathbf{g}_3]\big], \\ &\quad \big[[\mathbf{f}, \mathbf{g}_1], [\mathbf{f}, \mathbf{g}_2]\big], \big[[\mathbf{f}, \mathbf{g}_1], [\mathbf{f}, \mathbf{g}_3]\big], \big[[\mathbf{f}, \mathbf{g}_2], [\mathbf{f}, \mathbf{g}_3]\big]\big] \\ &= \begin{bmatrix} \mathbf{0}_{3\times3} & -\mathbf{I}_3 & \mathbf{0}_{3\times3} & \mathbf{0}_{3\times3} \\ \mathbf{0}_{3\times3} & -\mathcal{B} & \mathcal{C} & \mathcal{D} \\ \mathbf{I}_3 & \mathbf{0}_{3\times3} & \mathbf{0}_{3\times3} & \mathbf{0}_{3\times3} \\ \mathcal{B} & -\mathcal{C} & \frac{\partial \mathcal{A}}{\partial \mathbf{x}_4}\mathcal{C} & \mathcal{E} \end{bmatrix} \in \mathbb{R}^{12\times12}, \end{aligned} \tag{4.11}$$

where

$$\mathcal{C} = \frac{\partial \mathcal{A}}{\partial \mathbf{x}_3} + \frac{\partial \mathcal{A}}{\partial \mathbf{x}_4}\mathcal{B} \in \mathbb{R}^{3\times3}, \tag{4.12}$$

$$\mathcal{D} = \begin{bmatrix} \mathcal{D}_{12} & \mathcal{D}_{13} & \mathcal{D}_{23} \end{bmatrix} \in \mathbb{R}^{3\times3}, \tag{4.13}$$

$$\mathcal{E} = \begin{bmatrix} \mathcal{E}_{12} & \mathcal{E}_{13} & \mathcal{E}_{23} \end{bmatrix} \in \mathbb{R}^{3\times3}. \tag{4.14}$$

We now state the following theorem regarding the accessibility of the snake robot:

Theorem 4.5 *A planar snake robot with $N \geq 4$ links influenced by* anisotropic *viscous ground friction ($c_t \neq c_n$) is* locally strongly accessible *from any equilibrium point $\mathbf{x}^e$ ($\dot{\mathbf{q}}_\phi = \mathbf{0}$) satisfying* $\det(\mathcal{C}) \neq 0$ *and* $\det(\mathcal{E} - \frac{\partial \mathcal{A}}{\partial \mathbf{x}_4}\mathcal{D}) \neq 0$, *where* $\det(*)$ *denotes the determinant evaluated at $\mathbf{x}^e$.*

Proof By Theorem 4.1, the system is *locally strongly accessible* from $\mathbf{x}^e$ if $\boldsymbol{\Delta}(\mathbf{x}^e)$, given by (4.11), has full rank, i.e. spans a 12-dimensional space. The proof is complete if we can show that this is the case as long as $\det(\mathcal{C}) \neq 0$ and $\det(\mathcal{E} - \frac{\partial \mathcal{A}}{\partial \mathbf{x}_4}\mathcal{D}) \neq 0$ at $\mathbf{x}^e$. The matrix $\boldsymbol{\Delta}(\mathbf{x}^e)$ has full rank when all its columns are linearly independent. By investigating the particular structure of $\boldsymbol{\Delta}(\mathbf{x}^e)$, we see that the first and third rows contain an identity matrix and then zeros in the remaining elements of these rows. It is therefore impossible to write the columns containing the two identity matrices as linear combinations of other columns. We can therefore conclude that any linear dependence between the columns of $\boldsymbol{\Delta}(\mathbf{x}^e)$ must be caused by linear dependence between the columns of the following submatrix of $\boldsymbol{\Delta}(\mathbf{x}^e)$:

$$\widetilde{\boldsymbol{\Delta}}(\mathbf{x}^e) = \begin{bmatrix} \mathcal{C} & \mathcal{D} \\ \frac{\partial \mathcal{A}}{\partial \mathbf{x}_4}\mathcal{C} & \mathcal{E} \end{bmatrix} \in \mathbb{R}^{6\times6}. \tag{4.15}$$

Linear dependence between columns of a square matrix causes its determinant to become zero. We therefore calculate the determinant of $\widetilde{\boldsymbol{\Delta}}(\mathbf{x}^e)$ by employing the following well-known mathematical relationship concerning the determinant of a block matrix (see e.g. Harville 2000):

$$\det\left(\begin{bmatrix} \mathbf{A} & \mathbf{B} \\ \mathbf{C} & \mathbf{D} \end{bmatrix}\right) = \det(\mathbf{A})\det\bigl(\mathbf{D} - \mathbf{C}\mathbf{A}^{-1}\mathbf{B}\bigr), \tag{4.16}$$

where $\mathbf{A}$ and $\mathbf{D}$ are any square matrices, and $\mathbf{A}$ is non-singular. The determinant of $\widetilde{\boldsymbol{\Delta}}(\mathbf{x}^e)$ can now be calculated as

$$\det\bigl(\widetilde{\boldsymbol{\Delta}}(\mathbf{x}^e)\bigr) = \det(\mathcal{C})\det\left(\mathcal{E} - \frac{\partial \mathcal{A}}{\partial \mathbf{x}_4}\mathcal{D}\right), \tag{4.17}$$

which is zero when $\det(\mathcal{C}) = 0$ or when $\det(\mathcal{E} - \frac{\partial \mathcal{A}}{\partial \mathbf{x}_4}\mathcal{D}) = 0$. This means that $\widetilde{\boldsymbol{\Delta}}(\mathbf{x}^e)$, and thereby also $\boldsymbol{\Delta}(\mathbf{x}^e)$, has full rank whenever $\det(\mathcal{C}) \neq 0$ and $\det(\mathcal{E} - \frac{\partial \mathcal{A}}{\partial \mathbf{x}_4}\mathcal{D}) \neq 0$. This completes the proof. □

The requirement regarding the two determinants in Theorem 4.5 is not very restrictive, but it implies that the snake robot can attain configurations that are singular, i.e. certain shapes of the snake robot are obstructive from a control perspective since the dimension of the reachable space from these configurations is not full-dimensional. These singular configurations are revealed by the following property:

Property 4.1 The accessibility algebra $\boldsymbol{\Delta}(\mathbf{x}^e)$ drops rank at equilibrium points where all relative joint angles are equal ($\phi_1 = \cdots = \phi_{N-1}$).

This property can be shown to hold with a mathematical software tool such as *Matlab Symbolic Math Toolbox* since it can be verified that $\det(\mathcal{C})|_{\phi_1 = \cdots = \phi_{N-1}} = 0$, thereby violating the full-rank conditions stated in Theorem 4.5. Property 4.1 is interesting since it suggests that the joint angles of a snake robot should be *out of phase* during snake locomotion. This claim has been stated in several previous works, such as Gray (1946), Hicks (2003), Hirose (1993), Kane and Lecison (2000), but no formal mathematical proof was given.

We now show that the snake robot does *not* satisfy sufficient conditions for *small-time local controllability* (STLC).

Theorem 4.6 *At any equilibrium point* $\mathbf{x}^e$ ($\dot{\mathbf{q}}_\phi = \mathbf{0}$), *a planar snake robot with* $N \geq 4$ *links influenced by viscous ground friction does* not *satisfy the sufficient conditions for* small-time local controllability *(STLC) stated in Theorem* 4.2.

Proof The proof is complete if we can show that there are *bad* brackets of the system vector fields that cannot be neither *θ-neutralised* nor *l-neutralised* (see Theorem 4.2). The *bad* brackets with the lowest θ-degree and the lowest l-degree (except for $\mathbf{f}$, which vanishes at any equilibrium point) are $[\mathbf{g}_j, [\mathbf{f}, \mathbf{g}_j]]$, $j \in \{1, 2, 3\}$. Theorem 4.2 requires these vectors to be written as linear combinations of *good* brackets with either lower θ-degree or lower l-degree. The only such *good* brackets are $\mathbf{g}_j, [\mathbf{f}, \mathbf{g}_j], [\mathbf{f}, [\mathbf{f}, \mathbf{g}_j]], \ldots, [\mathbf{f}, [\cdots[\mathbf{f}, \mathbf{g}_j]]\cdots]$, $j \in \{1, 2, 3\}$. Brackets of the form

$[\mathbf{g}_k, \mathbf{g}_j]$ are not considered because $[\mathbf{g}_k, \mathbf{g}_j] = \mathbf{0}$, $j, k \in \{1, 2, 3\}$. For a proper choice of θ and l_j, $j \in \{0, 1, 2, 3\}$, these brackets have both lower θ-degree and lower l-degree. It is straightforward to verify that $[\mathbf{g}_j, [\mathbf{f}, \mathbf{g}_j]] \in \mathbb{R}^{2N+4=12}$ is a vector of all zeros except for element number $2N + 2 = 10$. The only way to write this vector as a linear combination of the *good* brackets listed above is if these *good* brackets span the entire 12-dimensional state space. This is not the case, however, because the vectors $[\mathbf{f}, [\mathbf{f}, \mathbf{g}_j]], \ldots, [\mathbf{f}, [\cdots [\mathbf{f}, \mathbf{g}_j]] \cdots]$, $j \in \{1, 2, 3\}$, are linearly dependent, as can be seen by assembling the matrix

$$\begin{aligned} &\big[[\mathbf{f}, [\mathbf{f}, \mathbf{g}_j]], [\mathbf{f}, [\mathbf{f}, [\mathbf{f}, \mathbf{g}_j]]], [\mathbf{f}, [\mathbf{f}, [\mathbf{f}, [\mathbf{f}, \mathbf{g}_j]]]], \ldots\big] \\ &= \begin{bmatrix} \mathbf{0}_{3\times3} & \mathbf{0}_{3\times3} & \mathbf{0}_{3\times3} & \cdots \\ \mathcal{C} & -\frac{\partial \mathcal{A}}{\partial \mathbf{x}_4}\mathcal{C} & (\frac{\partial \mathcal{A}}{\partial \mathbf{x}_4})^2\mathcal{C} & \cdots \\ \mathbf{0}_{3\times3} & \mathbf{0}_{3\times3} & \mathbf{0}_{3\times3} & \cdots \\ \frac{\partial \mathcal{A}}{\partial \mathbf{x}_4}\mathcal{C} & -(\frac{\partial \mathcal{A}}{\partial \mathbf{x}_4})^2\mathcal{C} & (\frac{\partial \mathcal{A}}{\partial \mathbf{x}_4})^3\mathcal{C} & \cdots \end{bmatrix} \end{aligned} \tag{4.18}$$

and noting that the fourth row is a multiple of the second row. It is therefore not possible to either *θ-neutralise* nor *l-neutralise* the *bad* brackets of the system in (2.47). The linear dependence in (4.18) is also present for $N > 4$ links since the six non-zero rows of (4.18) concern the position and head angle of the snake while the $2N - 2$ remaining rows will be zero regardless of N. This completes the proof. □

Note that Theorem 4.6 does not claim that the snake robot is not STLC. In other words, the snake robot may be STLC even though the sufficient conditions of Theorem 4.2 are violated. Note also that STLC is not a requirement for controllability since, as described in Sect. 4.2, it is in fact a stronger property than controllability. In summary, the above results do not enable us to conclude that a wheel-less snake robot influenced by anisotropic viscous ground friction is controllable. However, the above results are hopefully an important step towards formally proving that such mechanisms are controllable, which we consider highly likely to be the case.

We end this section with a note on Theorem 4.5. This theorem clearly shows that *anisotropic* friction is an important property for a snake robot. In the snake robot literature, it is common for snake robots to exhibit the property $c_n \gg c_t$. The extreme case of this property is realised by installing passive wheels (that cannot slip sideways) along the snake body since this ideally means that $c_t = 0$ and $c_n = \infty$. However, from Theorem 4.5 it is clear that the only requirement for *strong accessibility* is that the friction coefficients are *not equal*. The property $c_t > c_n$ is therefore also valid. This means that the passive wheels commonly mounted tangential to the snake body may equally well be mounted transversal to the snake body. The resulting motion will of course be different, but the *strong accessibility* property is still preserved.

4.5 Analysis of Propulsive Forces During Snake Locomotion

Having derived some fundamental properties of snake robot locomotion in Sects. 4.3 and 4.4, we now turn our attention to the following question: How and why does anisotropic viscous ground friction enable snake robots to locomote forward on a planar surface? We will answer this question simply by investigating the equations of motion of the snake robot.

We begin by deriving an expression for the total force propelling the CM of the snake robot forward as a function of the linear link velocities. We call this the *propulsive force* on the robot and denote it by $F_{\text{prop}} \in \mathbb{R}$. The forward direction of motion is assumed to be along the global positive x axis, which means that the propulsive force is simply the sum of all external forces on the snake robot in the global x direction. From (2.33b) and (2.25) we can calculate F_{prop} as

$$\begin{aligned} F_{\text{prop}} &= Nm\ddot{p}_x \\ &= \mathbf{e}^T\mathbf{f}_{R,x} = -\mathbf{e}^T\big(\big(c_t(\mathbf{C}_\theta)^2 + c_n(\mathbf{S}_\theta)^2\big)\dot{\mathbf{X}} + (c_t - c_n)\mathbf{S}_\theta\mathbf{C}_\theta\dot{\mathbf{Y}}\big). \end{aligned} \tag{4.19}$$

The result of multiplying the vector $\mathbf{e}^T = [1, \ldots, 1] \in \mathbb{R}^N$ with $\mathbf{f}_{R,x}$ in (4.19) is an addition of all elements in $\mathbf{f}_{R,x}$, which means that (4.19) can be written as

$$F_{\text{prop}} = -\sum_{i=1}^{N}\big(\big(c_t\cos^2\theta_i + c_n\sin^2\theta_i\big)\dot{x}_i + (c_t - c_n)\sin\theta_i\cos\theta_i\,\dot{y}_i\big). \tag{4.20}$$

The propulsive force contribution from a single link, $F_{\text{prop},i} \in \mathbb{R}$, is in other words given by

$$F_{\text{prop},i} = -F_x(\theta_i)\dot{x}_i - F_y(\theta_i)\dot{y}_i, \tag{4.21}$$

where

$$F_x(\theta_i) = c_t\cos^2\theta_i + c_n\sin^2\theta_i, \tag{4.22}$$

$$F_y(\theta_i) = (c_t - c_n)\sin\theta_i\cos\theta_i, \tag{4.23}$$

and where we recall from Sect. 2.4 that the angle θ_i of link i is expressed with respect to the global x axis with counterclockwise positive direction. We see from (4.21) that $F_{\text{prop},i}$ consists of two components, i.e. one involving the linear velocity of the link in the *forward* direction of motion, $F_x(\theta_i)\dot{x}_i$, and one involving the linear velocity *normal* to the direction of motion, $F_y(\theta_i)\dot{y}_i$. Due to the minus signs in (4.21), the products $F_x(\theta_i)\dot{x}_i$ and $F_y(\theta_i)\dot{y}_i$ provide a positive contribution to the propulsive force only if they are *negative*. Since the viscous friction coefficients, c_t and c_n, are always positive, the expression $F_x(\theta_i)$ is obviously always positive. We assume that the snake robot is not generating waves that involve x direction velocities of any of the links opposite to the direction of motion. When the snake robot is moving in the forward direction ($\dot{p}_x > 0$), we therefore have that $\dot{x}_i > 0$, which means that the product $F_x(\theta_i)\dot{x}_i$ of the propulsive force is always positive.

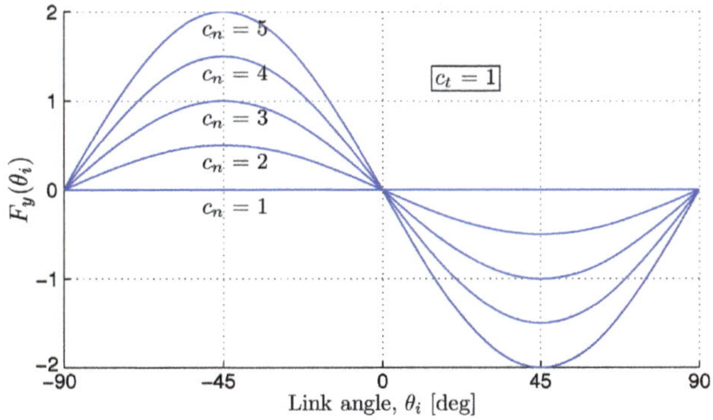

Fig. 4.1 The mapping from sideways link motion to forward propulsion for different viscous friction coefficients

This product is therefore *not* contributing to the forward propulsion of the robot, but rather opposing it. This is also expected since the snake robot must naturally be subjected to a friction force opposite to the direction of motion.

Any maintained propulsive force in the forward direction of motion must therefore be produced by the *sideways* motion of the links, i.e. the product $F_y(\theta_i)\dot{y}_i$. A plot of $F_y(\theta_i)$ for different values of the normal friction coefficient c_n, while keeping the tangential friction coefficient c_t fixed, is shown in Fig. 4.1. For each plot, the angle between the link and the forward direction, θ_i, is varied from $-90°$ to $90°$. The sideways motion of a link has no effect on the propulsive force on the snake robot when the friction coefficients are equal since this gives $F_y(\theta_i) = 0$. However, when $c_n > c_t$, Fig. 4.1 reveals that $F_y(\theta_i)$ is negative as long as θ_i is positive, and vice versa. This means that the product $F_y(\theta_i)\dot{y}_i$ is negative (the sideways motion of link i contributes to the propulsion) as long as $\text{sgn}(\theta_i) = \text{sgn}(\dot{y}_i)$. The sideways motion of link i is in other words contributing to the propulsion of the snake robot as long as θ_i is *positive during leftward motion of the link* (left with respect to the direction of motion) and *negative during rightward motion of the link* (right with respect to the direction of motion). This fundamental relationship may be written $\text{sgn}(F_{\text{prop},i}) = \text{sgn}(\text{sgn}(\theta_i)\,\text{sgn}(\dot{y}_i))$.

It is straightforward to calculate that the extrema of $F_y(\theta_i)$ occur at $\theta_i = \pm 45°$. This is also seen from Fig. 4.1. This means that, for a given $\dot{y}_i$, a link produces its highest propulsive force when it forms an angle of $\pm 45°$ with the forward direction of motion. It is also evident from (4.21) that the magnitude of $F_y(\theta_i)\dot{y}_i$, and thereby the magnitude of the propulsive force, $|F_{\text{prop},i}|$, is increased by increasing c_n with respect to c_t, or by increasing the magnitude of the sideways link velocity, $|\dot{y}_i|$.

It should now be clear that the function $F_y(\theta_i)$ maps the link velocities normal to the direction of motion into force components in the direction of motion. The following simple analogy may help understand this result. Imagine a small, handheld, wheeled wagon of some sort. The direction of the wheels corresponds to the tangential direction of a snake robot link. Obviously, the friction coefficient of the wagon in the direction of the wheels is smaller than the friction coefficient normal to the wheels. Now assume that you push the wagon across a table in the direction of the wheels. While maintaining constant direction of motion, assume that you slowly rotate the wagon about the vertical axis, thereby forcing the wheels to slip.

The hands that push and rotate the wagon will now feel a tendency of the wagon to move sideways in the same direction towards which the wagon was rotated. This is in accordance with the results presented above.

The above analysis is summarised by the following properties of planar snake robot locomotion under viscous friction conditions:

Property 4.2 For a snake robot described by (2.35) with $c_n > c_t$, forward propulsion is produced by the link velocity components that are transversal to the forward direction. The function $F_y(\theta_i)$ maps the link velocities transversal to the direction of motion into force components in the direction of motion.

Property 4.3 For a snake robot described by (2.35) with $c_n > c_t$, the propulsive force generated by the transversal motion of link i is positive as long as $\operatorname{sgn}(\theta_i) = \operatorname{sgn}(\dot{y}_i)$.

Property 4.4 For a snake robot described by (2.35) with $c_n > c_t$, the magnitude of the propulsive force produced by link i, $|F_{\text{prop},i}|$, is increased by increasing c_n with respect to c_t, or by increasing the magnitude of the sideways link velocity, $|\dot{y}_i|$, or by increasing $|\theta_i|$ as long as $|\theta_i| < 45°$.

Property 4.5 For a given $\dot{y}_i$, a link produces its highest propulsive force when it forms an angle of $\theta_i = \pm 45°$ with the forward direction of motion.

4.6 Synthesis of Propulsive Motion for the Snake Robot

The results presented so far in this chapter are general in the sense that no assumptions have yet been made regarding the actual motion pattern displayed by the snake robot. In this chapter, we use these results as a basis for deducing how the links of a snake robot can be moved in order to propel the robot forward along the global x axis. In order to deduce the propulsive link motion, we focus on manipulating the magnitude, $|F_{\text{prop},i}|$, and direction, $\operatorname{sgn}(F_{\text{prop},i})$, of the propulsive force on each link.

According to Theorem 4.3 in Sect. 4.3, asymptotic stabilisation of the snake robot to an equilibrium point requires a time-varying control law. If we assume that the reference angles of the links are included in the control law of the snake robot, we can satisfy the requirement in Theorem 4.3 regarding a time-varying control law by choosing the reference angles of the links to be explicitly time-varying. Furthermore, Property 4.2 shows that propulsive forces are generated by moving the links *transversally* to the desired direction of motion. We therefore conclude that the links must have a *periodic* velocity component normal to the direction of motion, which implies that each link should be moved alternatingly to the left and right with respect to the direction of motion. The above discussion strongly suggests that the trajectory of each link angle should be of the form

$$\theta = \alpha \sin(\omega t), \tag{4.24}$$

where θ is the angle of a link with respect to the direction of motion, $\alpha > 0$ is the amplitude of the link angle during the locomotion, $\omega > 0$ is the angular frequency of the periodic motion, and t denotes time. Link reference angles of the form in (4.24) are a natural choice based on the above discussion since the reference is time-varying and because periodic link motion from side to side must naturally also involve a periodic link angle.

The question now is how the link angles should be related to each other. Property 4.1 in Sect. 4.4.2 states that the accessibility algebra of the snake robot drops rank at equilibrium points where all the relative joint angles are equal. This property implies that we should prevent the joint angles of the snake robot from ever being all equal since this improves the controllability properties of the robot. A simple approach for satisfying this requirement is to offset the phase of each link angle with respect to the angle of its adjacent link by some phase shift δ since this ensures that the joint angles are always *out of phase* and never all equal. In other words, each of the N links of the snake robot should be moved according to

$$\theta_i = \alpha \sin\big(\omega t + (i-1)\delta\big), \tag{4.25}$$

where $i \in \{1, \dots, N\}$, and δ is the phase shift between adjacent links. For simplicity, we assume that α, ω, and δ are constant and identical for all links.

Let us now investigate how α, ω, and δ affect $|F_{\mathrm{prop},i}|$ and $\mathrm{sgn}(F_{\mathrm{prop},i})$ as the snake robot moves along the global x axis. To simplify the analysis, we assume that the snake robot consists of only $N = 3$ links. This is the minimum number of links required to achieve propulsion since phase shift between joints requires at least two joints. The link angle trajectories are given from (4.25) as

$$\theta_1 = \alpha \sin(\omega t), \tag{4.26a}$$

$$\theta_2 = \alpha \sin(\omega t + \delta), \tag{4.26b}$$

$$\theta_3 = \alpha \sin(\omega t + 2\delta), \tag{4.26c}$$

which, when differentiated with respect to time, gives the angular link velocities

$$\dot{\theta}_1 = \alpha\omega \cos(\omega t), \tag{4.27a}$$

$$\dot{\theta}_2 = \alpha\omega \cos(\omega t + \delta), \tag{4.27b}$$

$$\dot{\theta}_3 = \alpha\omega \cos(\omega t + 2\delta). \tag{4.27c}$$

The normal direction velocity of each link is given by (2.13b). We disregard the normal direction velocity of the snake robot by setting $\dot{p}_y \approx 0$. This approximation is a fairly accurate during motion along the global x axis, which is the case for this analysis. Inserting (4.26a)–(4.26c) and (4.27a)–(4.27c) into (2.13b) gives

$$\begin{aligned}
\dot{y}_1 = &-\frac{\alpha\omega l}{3}\Big(2\cos(\omega t)\cos\big(\alpha \sin(\omega t)\big)\Big) \\
&-\frac{\alpha\omega l}{3}\Big(3\cos(\omega t + \delta)\cos\big(\alpha \sin(\omega t + \delta)\big)\Big)
\end{aligned}$$

$$- \frac{\alpha\omega l}{3}\big(\cos(\omega t + 2\delta)\cos\big(\alpha\sin(\omega t + 2\delta)\big)\big), \tag{4.28a}$$

$$\dot{y}_2 = \frac{\alpha\omega l}{3}\big(\cos(\omega t)\cos\big(\alpha\sin(\omega t)\big)\big) - \frac{\alpha\omega l}{3}\big(\cos(\omega t + 2\delta)\cos\big(\alpha\sin(\omega t + 2\delta)\big)\big), \tag{4.28b}$$

$$\dot{y}_3 = \frac{\alpha\omega l}{3}\big(\cos(\omega t)\cos\big(\alpha\sin(\omega t)\big)\big) + \frac{\alpha\omega l}{3}\big(3\cos(\omega t + \delta)\cos\big(\alpha\sin(\omega t + \delta)\big)\big) + \frac{\alpha\omega l}{3}\big(2\cos(\omega t + 2\delta)\cos\big(\alpha\sin(\omega t + 2\delta)\big)\big). \tag{4.28c}$$

Property 4.4 tells us that $|F_{\text{prop},i}|$ is increased by increasing $|\dot{y}_i|$. From (4.28a)–(4.28c) it is therefore clear that the propulsive force $|F_{\text{prop},i}|$ generated by link i is increased by increasing α and/or ω.

We now determine if δ should be positive or negative in order to achieve $\text{sgn}(F_{\text{prop},i}) = 1$, which is necessary to propel the snake robot forward along the global x axis. From Property 4.3 we know that $\text{sgn}(F_{\text{prop},i}) = 1$ requires $\text{sgn}(\theta_i) = \text{sgn}(\dot{y}_i)$. Considering $\dot{y}_2$ in (4.28b) (since this expression is easy to analyse), it is seen through pure inspection that $\dot{y}_2 = 0$ when $\omega t = -\delta$. When $\omega t = -\delta$, we see from (4.26b) and (4.27b) that $\theta_2 = 0$ and $\dot{\theta}_2 = \alpha\omega > 0$. θ_2 is in other words about to become positive, which means that we also require $\dot{y}_2$ to become positive. This is the case if $\ddot{y}_2 > 0$ when $\omega t = -\delta$. Differentiating (4.28b) with respect to time gives

$$\ddot{y}_2|_{\omega t = -\delta} = \frac{2\alpha\omega^2 l}{3}\big(\alpha\cos^2(\delta)\sin\big(\alpha\sin(\delta)\big) + \sin(\delta)\cos\big(\alpha\sin(\delta)\big)\big), \tag{4.29}$$

from which it is easily seen that $\ddot{y}_2 > 0$ when $\delta > 0$, i.e. $\text{sgn}(F_{\text{prop},i}) = 1$ when $\delta > 0$. This indicates that the links generate positive propulsive forces if $\delta > 0$.

In order to verify that forward propulsion requires $\delta > 0$, we have plotted (4.26a)–(4.26c) and (4.28a)–(4.28c) together in Fig. 4.2 for $\alpha = 70°$, $\omega = 70°$, $l = 0.07$ m, and for different positive values of δ over a period of ωt from 0 to 2π. The figures show that $\text{sgn}(F_{\text{prop},i}) = \text{sgn}(\text{sgn}(\theta_i)\,\text{sgn}(\dot{y}_i)) = 1$ is always satisfied for link 2, but only satisfied over about half the period for link 1 and 3 when δ is small. As δ is increased, $\text{sgn}(F_{\text{prop},i}) = 1$ is satisfied over a larger portion of the period. The optimal choice of δ will be derived in Chap. 7. For now, however, we simply conclude that achieving positive propulsive forces requires $\delta > 0$.

Although the above analysis assumes that the snake robot consists of only $N = 3$ links, we conjecture that the results also apply to robots with $N > 3$ links. In the following property, we therefore summarise the above analysis for snake robots consisting of N links.

Property 4.6 A snake robot described by (2.35) with anisotropic viscous ground friction properties ($c_n > c_t$) achieves forward propulsion by moving its links ac-

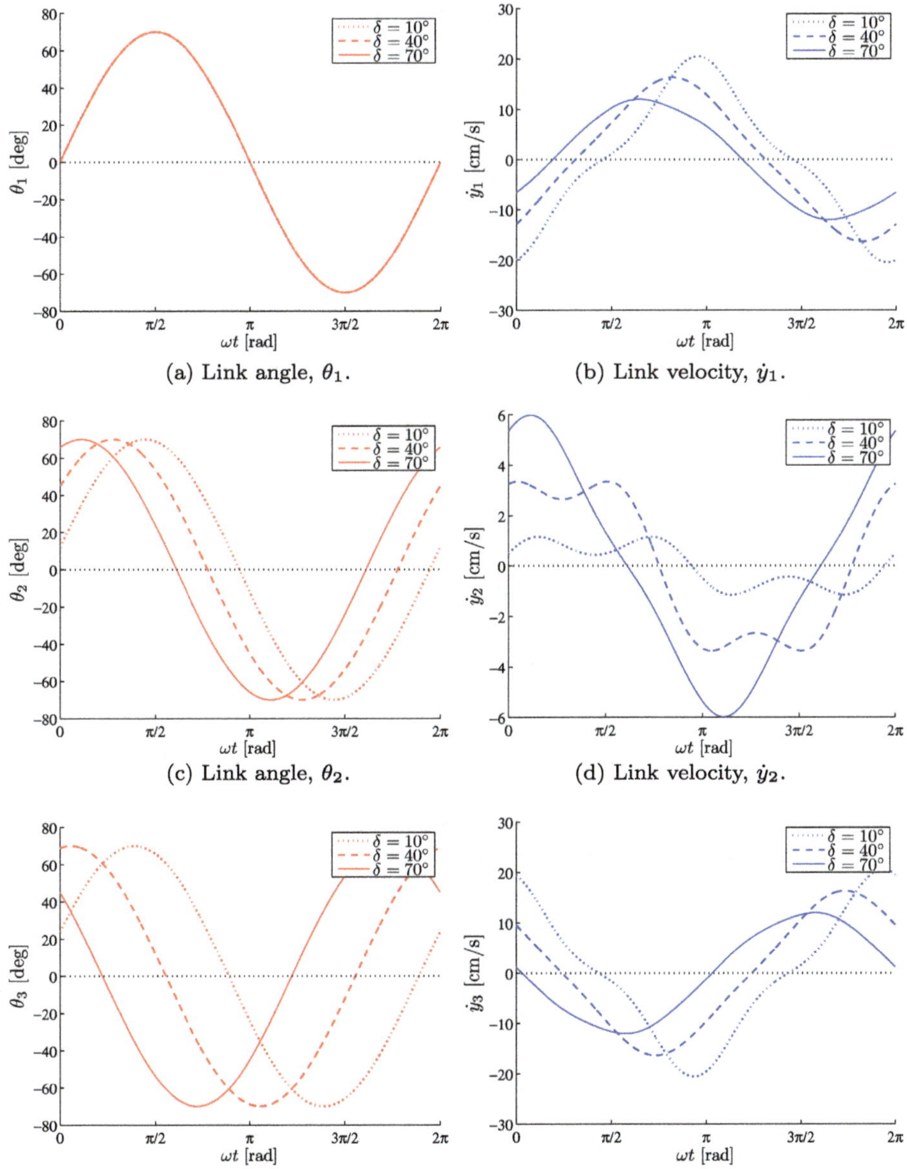

(a) Link angle, θ_1. (b) Link velocity, $\dot{y}_1$.

(c) Link angle, θ_2. (d) Link velocity, $\dot{y}_2$.

(e) Link angle, θ_3. (f) Link velocity, $\dot{y}_3$.

Fig. 4.2 The relation between θ_i and $\dot{y}_i$, $i \in \{1, 2, 3\}$, for $\alpha = 70°$, $\omega = 70°$, and $\delta = 10°$ (*dotted*), $40°$ (*dashed*), and $70°$ (*solid*), respectively

cording to $\theta_i = \alpha \sin(\omega t + (i-1)\delta)$ where $i \in \{1, \dots, N\}$, $\alpha > 0$, $\omega > 0$, and $\delta > 0$. Increasing α and/or ω increases the magnitude of the propulsive forces generated by the links.

4.7 The Gait Pattern Lateral Undulation

During locomotion according to the gait pattern presented in Property 4.6, the snake robot produces continuous body waves that are propagated backwards from the head to the tail. This form of motion is called *lateral undulation* and is the most common form of locomotion displayed by biological snakes (see Sect. 1.2.2). As described in Sect. 1.4, motion by lateral undulation has a central role in the control design efforts described in this book since we consider this motion pattern to be most relevant and most efficient for planar snake robot locomotion.

A well-known mathematical description of lateral undulation was presented in Hirose (1993) based on empirical studies of biological snakes. Hirose discovered that a close approximation to the shape of a biological snake during lateral undulation is given by a planar curve whose curvature varies sinusoidally. Hirose named this curve the *serpenoid curve* and described it by

$$\begin{aligned} x(s) &= \int_0^s \cos\bigl(a\cos(b\sigma) + c\sigma\bigr)\,\mathrm{d}\sigma, \\ y(s) &= \int_0^s \sin\bigl(a\cos(b\sigma) + c\sigma\bigr)\,\mathrm{d}\sigma, \end{aligned} \tag{4.30}$$

where $(x(s), y(s))$ are the coordinates of the point along the curve at arc length s from the origin, and where a, b, and c are positive scalars. The curvature κ of the serpenoid curve varies sinusoidally according to $\kappa(s) = |ab\sin(bs) - c|$. Note that the above curve is not dependent on time and describes the shape of a snake at some time instant. Note also that we have used the notation from Saito et al. (2002), which considers motion control of snake robots based on the serpenoid curve. It is shown in Saito et al. (2002) that a serpenoid curve of arc length 1 can be approximated by N identical discrete segments by calculating the angle θ_i of segment $i \in \{1, \dots, N\}$ with respect to the x axis according to

$$\theta_i = a\cos\left(\frac{ib}{N}\right). \tag{4.31}$$

This implies that a snake robot with N identical discrete links attains a discrete approximation to the serpenoid curve by moving its link angles sinusoidally with a constant phase shift between the links, which means that the motion pattern that we derived in Property 4.6 is, in every time instant, a discrete approximation to the serpenoid curve. The analysis leading up to Property 4.6 therefore supports the serpenoid curve motion proposed by Hirose. However, while Hirose derived the serpenoid curve based on empirical studies of biological snakes, we have in this chapter based our arguments on the mathematical properties derived from the equations of motion of the snake robot. The choices made in the analysis leading up to Property 4.6 are obviously inspired by the serpenoid curve. However, it is still interesting to see how logical arguments that support an empirically derived result can be developed through a mathematical analysis.

Since there are N link angles, but only $N-1$ control inputs (i.e. the snake robot is underactuated), it is more suitable to specify the serpenoid curve motion by the $N-1$ joint angles. Since the angle of joint $i \in \{1, \dots, N-1\}$ is given from (2.1) as $\phi_i = \theta_i - \theta_{i+1}$, it is easy to verify that the reference motion of the joint angles will have the same form as the reference motion of the link angles specified in Property 4.6. In accordance with Hirose (1993), we also introduce a joint angle offset ϕ_o in the reference motion of the joints. It will be shown in Sect. 4.9 that this offset can be used to control the direction of the locomotion since the offset makes the link motion asymmetrical with respect to the current heading of the robot. We now summarise the above discussion.

Definition 4.1 (Lateral undulation) The gait pattern *lateral undulation* is achieved by moving the joints of a planar snake robot according to

$$\phi_{i,\mathrm{ref}} = \alpha \sin\bigl(\omega t + (i-1)\delta\bigr) + \phi_o, \tag{4.32}$$

where $i \in \{1, \dots, N-1\}$, α and ω are the amplitude and angular frequency, respectively, of the sinusoidal joint motion, δ determines the phase shift between the joints, and ϕ_o is a joint offset, which we assume to be identical for all joints.

4.8 The Control System of the Joints

In this book, we consider two different control laws for making the joint angles $\boldsymbol{\phi} = [\phi_1, \dots, \phi_{N-1}]^T \in \mathbb{R}^{N-1}$ track the joint reference angles given by $\boldsymbol{\phi}_{\mathrm{ref}} = [\phi_{1,\mathrm{ref}}, \dots, \phi_{N-1,\mathrm{ref}}]^T \in \mathbb{R}^{N-1}$. We assume that the control input $\mathbf{u} \in \mathbb{R}^{N-1}$ of the model (2.35) is set according to the linearising control law (2.44) so that the joint dynamics of the snake robot is given by $\ddot{\boldsymbol{\phi}} = \overline{\mathbf{u}}$. Both control laws are therefore defined in terms of the linearised control input $\overline{\mathbf{u}} \in \mathbb{R}^{N-1}$.

4.8.1 A Simple Joint Controller

The first control law, which we refer to as the *simple joint controller*, combines proportional action with velocity damping as follows:

$$\overline{\mathbf{u}} = k_p(\boldsymbol{\phi}_{\mathrm{ref}} - \boldsymbol{\phi}) - k_d\dot{\boldsymbol{\phi}}, \tag{4.33}$$

where $k_p > 0$ and $k_d > 0$ are controller gains, and where $\dot{\boldsymbol{\phi}}_{\mathrm{ref}} = \mathbf{0}$ since the purpose of the derivative part is simply to damp the joint motion if the joint velocities become large. The advantage of this controller is that it does not require calculation of the derivative of $\boldsymbol{\phi}_{\mathrm{ref}}$ with respect to time, which e.g. allows $\boldsymbol{\phi}_{\mathrm{ref}}$ to be discontinuous. The controller in (4.33) is, in other words, easy to implement. The disadvantage of the controller, however, is that it is unable to track time-varying joint reference angles perfectly, which is not the case for the control law presented next.

4.8.2 An Exponentially Stable Joint Controller

The second control law, which we refer to as the *exponentially stable joint controller*, is defined as

$$\overline{\mathbf{u}} = \ddot{\boldsymbol{\phi}}_{\text{ref}} + k_d(\dot{\boldsymbol{\phi}}_{\text{ref}} - \dot{\boldsymbol{\phi}}) + k_p(\boldsymbol{\phi}_{\text{ref}} - \boldsymbol{\phi}), \tag{4.34}$$

where $k_p > 0$ and $k_d > 0$ are scalar controller gains. Since the joint dynamics is given by $\ddot{\boldsymbol{\phi}} = \overline{\mathbf{u}}$, the resulting error dynamics of the joints is given by

$$(\ddot{\boldsymbol{\phi}}_{\text{ref}} - \ddot{\boldsymbol{\phi}}) + k_d(\dot{\boldsymbol{\phi}}_{\text{ref}} - \dot{\boldsymbol{\phi}}) + k_p(\boldsymbol{\phi}_{\text{ref}} - \boldsymbol{\phi}) = \mathbf{0}, \tag{4.35}$$

which is clearly *exponentially stable* (see Khalil 2002). During lateral undulation according to (4.32), we can easily calculate $\dot{\boldsymbol{\phi}}_{\text{ref}}$ and $\ddot{\boldsymbol{\phi}}_{\text{ref}}$ if ϕ_o is assumed to be a constant offset since this gives

$$\dot{\phi}_{i,\text{ref}} = \alpha\omega\cos\big(\omega t + (i-1)\delta\big), \tag{4.36a}$$

$$\ddot{\phi}_{i,\text{ref}} = -\alpha\omega^2\sin\big(\omega t + (i-1)\delta\big). \tag{4.36b}$$

If, on the other hand, ϕ_o is a complex function of time so that $\dot{\boldsymbol{\phi}}_{\text{ref}}$ and $\ddot{\boldsymbol{\phi}}_{\text{ref}}$ cannot easily be calculated analytically, then $\dot{\boldsymbol{\phi}}_{\text{ref}}$ and $\ddot{\boldsymbol{\phi}}_{\text{ref}}$ can be obtained by passing the commanded reference angles, $\boldsymbol{\phi}_{\text{ref}}$, through a 3rd-order low-pass filtering reference model, as described in Appendix C.2.

4.9 Analysis of Turning Motion During Lateral Undulation

Having determined in the previous subsection how propulsion is generally achieved with a snake robot, we now investigate how a snake robot achieves turning motion. We assume that the robot moves according to the gait pattern lateral undulation.

As described in Sect. 4.7, lateral undulation is achieved by controlling the joints of the snake robot according to (4.32). The simplest and most common approach for changing the direction of the motion during this gait pattern is to use the parameter ϕ_o in (4.32), which represents a joint angle offset. The reason why a non-zero ϕ_o induces rotational motion of a snake robot is illustrated in Fig. 4.3. In particular, the analysis presented in the previous subsection implies that the net propulsive ground friction forces produced by a link points in the direction about which the link angle oscillates. When $\phi_o = 0$, we see from the top of Fig. 4.3 that each link angle will oscillate about the forward direction of the robot. These oscillations produce propulsive forces which propel the robot forward along a straight line. When ϕ_o is non-zero, on the other hand, we see from the bottom of Fig. 4.3 that the direction of the net propulsive forces on the links will point along an arc, thereby inducing rotational motion of the snake robot.

To further investigate how a joint angle offset influences the turning motion of a snake robot, we have simulated two trials of lateral undulation of a snake robot

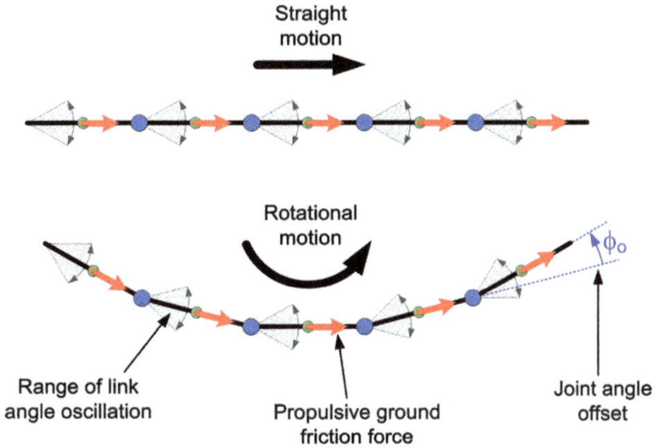

Fig. 4.3 The net propulsive ground friction forces produced by a link points in the direction about which the link angle oscillates. By offsetting the joint angles, the direction of the net propulsive forces on the links will point along an arc, thereby inducing rotational motion of the snake robot

described by (2.47). The model in (2.47) was implemented in *Matlab R2008b*, where the dynamics of the model was calculated by use of the *ode45* solver with a relative and absolute error tolerance of 10^{-6}. We considered a snake robot with $N = 10$ links of length $2l = 0.14$ m, mass $m = 1$ kg, and moment of inertia $J = 0.0016$ kgm^2. The ground friction forces were assumed to be of the viscous type defined in (2.25) with friction coefficients $c_t = 1$ and $c_n = 10$. The joint angles were controlled according to the simple joint controller in (4.33).

In the first trial, the snake robot was controlled according to (4.32) with $\alpha = 30°$, $\omega = 50°/\text{s}$, and $\delta = 40°$. In order to study the effect of ϕ_o on the motion, the offset angle was set to $\phi_o = 5°$ in the time interval $t \in [20, 30]$ and $\phi_o = -10°$ in the time interval $t \in [50, 60]$. The offset angle was $\phi_o = 0°$ outside these two time intervals. The parameters of the second trial were identical to the first trial except that we increased the frequency of the joint motion to $\omega = 80°/\text{s}$ in order to increase the forward velocity. The trace of the head during the motion from the first and the second trial is shown in Fig. 4.4(a) and Fig. 4.4(b), respectively, while Fig. 4.4(c) shows the average joint angle during both trials, which is defined as $\overline{\phi} = \frac{1}{N-1}\sum_{i=1}^{N-1} \phi_i$. The triangles pointing up and down in Fig. 4.4(a)–(b) indicate, respectively, the beginning and end of the two time intervals where ϕ_o is non-zero.

Figure 4.4 shows that the snake robot crawls forward without turning as long as the average joint angle, $\overline{\phi}$, is zero. However, when the average joint angle is non-zero, the direction of the motion changes. We see from the figure that a positive (resp. negative) average joint angle produces a counterclockwise (resp. clockwise) rotation of the snake robot. We also see that the rate of directional change is increased by increasing the amplitude of the average joint angle. This result is supported by the directional controllers for snake locomotion considered in e.g. Saito et al. (2002), Sfakiotakis and Tsakiris (2007). Moreover, Fig. 4.4 shows that the rate of directional change is larger in the second trial. Since the only difference between the two trials is that the forward velocity is larger in the second trial, we can conclude that the rate of directional change for some fixed joint angle offset is also increased by increasing the forward velocity of the snake robot.

The following property summarises the above analysis:

Fig. 4.4 Two simulated trials of lateral undulation of a snake robot with $N = 10$ links. In both trials, a joint angle offset of $\phi_o = 5°$ and $\phi_o = -10°$ is introduced at $t = 20$ s and $t = 50$ s, respectively

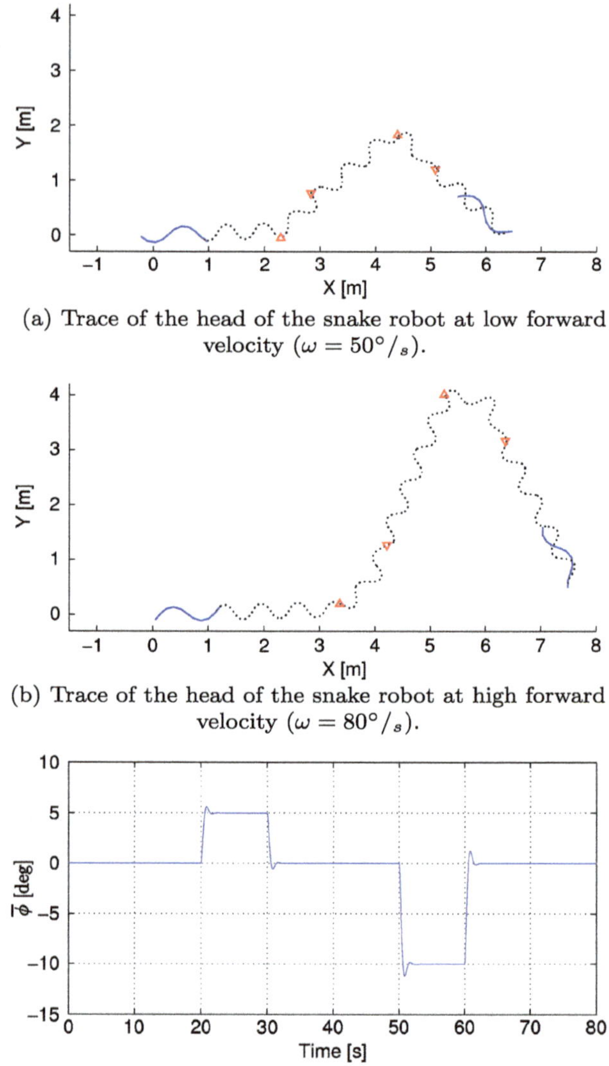

(a) Trace of the head of the snake robot at low forward velocity ($\omega = 50°/_s$).

(b) Trace of the head of the snake robot at high forward velocity ($\omega = 80°/_s$).

(c) The average joint angle during both trials.

Property 4.7 During *lateral undulation* with a snake robot described by (2.47) with $c_n > c_t$, the overall direction of the locomotion remains constant as long as the average joint angle is zero but will change in the counterclockwise (resp. clockwise) direction when the average joint angle is positive (resp. negative). The rate of directional change of the locomotion is increased by increasing the amplitude of the average joint angle and/or by increasing the forward velocity (assuming that the average joint angle is non-zero).

Remark 4.2 Although Property 4.7 was developed under the assumption that the snake robot moves according to lateral undulation, we claim that the property also applies to other gait patterns. In particular, we claim that any gait pattern which propels a snake robot forward along a straight line will induce rotational motion of the robot if a non-zero joint angle offset is introduced at each joint.

Remark 4.3 Due to the complexity of the mathematical model of the snake robot in (2.47), we are currently unable to provide a formal proof that ϕ_o affects the motion direction of the snake robot. One possible approach is to prove this property by investigating Lie brackets of the system vector fields (see Sect. 4.2). Such an approach is employed in e.g. Morgansen et al. (2001) to study the motion of robotic fish. A challenge associated with snake robots, however, is that a snake robot with revolute joints has no explicitly defined orientation since there is an independent link angle associated with each link. An estimate of the heading of the robot was given in Definition 2.3 as the average of the link angles. The summed dynamics of all the links is, however, very complex, which makes it difficult to analyse the Lie bracket motion of the heading.

4.10 Analysis of Relative Motion Between Consecutive Links During Lateral Undulation

From the results presented so far in this chapter, it should be clear that planar snake robot locomotion consists of periodic body shape changes that generate external forces that propel the robot forward. Since body shape changes are equivalent to displacing the links relative to each other, we can always partition the body shape changes into relative link displacements that are, respectively, *tangential* and *transversal* to the forward direction of the robot. We know from Property 4.2 that the transversal link displacements are what propel the snake robot forward.

With respect to the gait pattern *lateral undulation* presented in Definition 4.1, it seems natural to ask how large part of the body shape changes during lateral undulation that constitute tangential and transversal link displacements, respectively. To answer this question, we employed the simulation setup presented in Sect. 4.9 to simulate a snake robot with $N = 10$ links during lateral undulation along the global x axis with $\alpha = 30°$, $\omega = 30°/\text{s}$, $\delta = 40°$, and $\phi_o = 0°$. The trace of the head of the snake robot is plotted in the top of Fig. 4.5, while the two bottom plots in Fig. 4.5 show the relative displacements between the CM of two arbitrarily chosen links (link 4 and link 5) in the global x and y directions, respectively. The plots indicate that, during lateral undulation, the relative displacements between the CM of two adjacent links along the forward direction of motion are approximately constant, while the relative displacements normal to the direction of motion oscillate around zero. This observation is formalised as follows:

Property 4.8 The change in body shape during lateral undulation consists mainly of relative displacements of the CM of the links *normal* to the forward direction

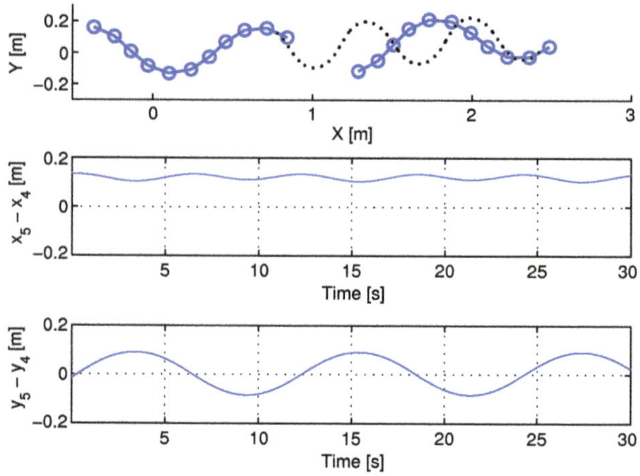

Fig. 4.5 *Top*: Simulated motion of a snake robot with $N = 10$ links. *Middle*: Relative displacements between the CM of link 4 and link 5 in the global x direction. *Bottom*: Relative displacements between the CM of link 4 and link 5 in the global y direction

of motion. The relative displacements of the CM of the links along the forward direction are approximately constant.

Remark 4.4 The relative link displacements *normal* to the direction of motion will *not* dominate over the relative link displacements *tangential* to the direction of motion when the amplitudes of the link angles become large. Property 4.8 is therefore valid only as long as the link angles are limited.

Remark 4.5 Properties 4.8 and 4.2 constitute the basis for a simplified model of the snake robot presented in Chap. 6. The properties tell us two things, namely (1) that lateral undulation mainly consists of link displacements that are transversal to the direction of motion and (2) that the transversal link displacements are what propel the robot forward. These results are appealing since they suggest that snake robot locomotion can be described in terms of the transversal displacements of the links instead of the more complex rotational link motion. We elaborate this approach in Chap. 6.

4.11 Chapter Summary

This chapter is summarised as follows:

- We have presented a stabilisability analysis that proves that any asymptotically stabilising control law for a planar snake robot to an equilibrium point must be *time-varying*, i.e. not of a pure-state feedback type (see Theorem 4.3).
- We have presented a controllability analysis of planar snake robots influenced by viscous ground friction forces that proves that:
  - A snake robot is *not* controllable when the viscous ground friction is *isotropic* (see Theorem 4.4).

  - A snake robot is *strongly accessible* from any equilibrium point (except for certain singular configurations) when the viscous ground friction is *anisotropic* (see Theorem 4.5).
  - A snake robot does *not* satisfy sufficient conditions for *small-time local controllability* (see Theorem 4.6).
  - The joint angles of a snake robot should be *out of phase* during snake locomotion (suggested by Property 4.1).
- We have identified a simple relationship between link velocities *normal* to the direction of motion and the resulting *propulsive forces* in the direction of motion that:
  - Revealed several fundamental properties of snake robot locomotion (see Properties 4.2, 4.3, and 4.4).
  - Explains how anisotropic ground friction enables snake robots to locomote forward on a planar surface.
- We have employed the derived properties to synthesise how the links of a snake robot can be moved in order to propel the robot forward (see Property 4.6).
- We have identified the synthesised motion as the gait pattern *lateral undulation* (see Definition 4.1), which is the most common form of biological snake locomotion.
- We have explained how offsetting the joint angles during lateral undulation enables directional control of the motion (see Property 4.7).
- We have shown that the body shape changes during lateral undulation (assuming that the link angles are limited) mainly consist of relative displacements of the links *normal* to the forward direction of motion, while the relative displacements of the links along the forward direction are approximately constant (see Property 4.8).

Chapter 5
Path Following Control and Analysis of Snake Robots Based on the Poincaré Map

In Chap. 4, we derived the gait pattern *lateral undulation*, which enables planar snake robots with anisotropic ground friction properties to locomote forward on a planar surface. We now turn to the problem of controlling the heading and position of the snake robot, and in particular, we consider the problem of enabling the robot to track a straight path. Straight line path following capabilities are important since they enable a snake robot to follow a desired path given by waypoints interconnected by straight lines. Straight line path following is therefore relevant for many future applications of snake robots, such as automated inspection rounds in inaccessible areas of industrial process facilities or mapping of confined spaces by moving along prescribed paths.

Control design for snake robots is challenging since these mechanisms are underactuated. In particular, the model of the snake robot in (2.47) contains $N+2$ degrees of freedom, but only $N-1$ control inputs. The underactuated degrees of freedom, i.e. the heading and position of the robot, make it impossible to independently control all degrees of freedom of the robot. During path following control of snake robots, there is additionally the challenge that the CM position $\mathbf{p}$ does not trace out a straight path during forward locomotion, but rather oscillates periodically about the straight line pointing in the forward direction of the robot. Moreover, we also expect the heading $\overline{\theta}$ of the robot, which was defined in (2.2), to oscillate periodically during forward locomotion since we cannot generally assume that the average of the link angles always is zero. As long as the heading and position of the robot display such oscillating behaviour, it makes no sense to attempt to control these states to stationary values with respect to the desired straight path (which we normally would do during path following control of more conventional propulsion mechanisms, such as wheeled robots and marine vessels).

With these challenges in mind, it becomes clear that we need a mathematical tool which allows us to study the periodically oscillating behaviour of the system states. We find such a tool in the theory of *Poincaré maps*. The Poincaré map represents a widely used tool for analysing the existence and stability of periodic orbits of dynamical systems. In this chapter, we first propose a path following controller for planar snake robots, and subsequently we analyse the stability of the locomotion

P. Liljebäck et al., *Snake Robots*, Advances in Industrial Control,
DOI 10.1007/978-1-4471-2996-7_5, © Springer-Verlag London 2013

along the path by use of a Poincaré map. In particular, we show that all state variables of the snake robot, except for the position along the path, trace out an *exponentially stable* periodic orbit during path following with the proposed controller. We also present simulation results that illustrate the performance of the controller. Note that the path following controller considered in this chapter is extended in Chap. 8, where we employ cascaded systems theory to investigate the convergence of the snake robot to the desired path based on the simplified model presented in Chap. 6. The path following controller in Chap. 8 circumvents the oscillating behaviour of the snake robot through a coordinate transformation combined with feedback cancellation of the oscillating dynamics.

This chapter is organised as follows. The relation between this chapter and previous literature is discussed in Sect. 5.1. Section 5.2 gives a brief presentation of the Poincaré map. The path following controller is presented in Sect. 5.3, and the stability of this controller is investigated in Sect. 5.4 by use of a Poincaré map. Simulation results are presented in Sect. 5.5. Finally, the chapter is summarised in Sect. 5.6.

5.1 The Relation Between This Chapter and Previous Literature

Position and path following control of snake robots has been considered in several previous works. Most of these previous works consider *wheeled* snake robots, such as the works in Date et al. (2000), Ishikawa (2009), Ma et al. (2003), Matsuno and Mogi (2000), Matsuno and Sato (2005), Prautsch et al. (2000), Tanaka and Matsuno (2008a). Previous research on position and path following control of *wheel-less* snake robots is very limited but is considered in e.g. Hicks and Ito (2005), Hicks (2003), McIsaac and Ostrowski (2003a, 2003b).

The various approaches for position and path following control proposed in previous literature are quite different. For instance, many of the previous works related to wheeled snake locomotion are able to make conclusions regarding the motion of the snake robot due to the sideslip constraints on the links. In particular, the sideslip constraints are used to establish an explicit connection between body shape changes and propulsion, which allows the control input to be specified directly in terms of the desired propulsion of the robot. This approach enables the position and heading of wheeled snake robots to be controlled based on various computed torque controllers in e.g. Date et al. (2000), Ma et al. (2003), Matsuno and Mogi (2000), Matsuno and Sato (2005), Prautsch et al. (2000), Tanaka and Matsuno (2008a) and based on Lie bracket calculations and controllability analysis results in Ishikawa (2009), Ishikawa et al. (2010). Position and heading control of swimming snake robots is achieved in McIsaac and Ostrowski (2003a, 2003b) by moving the joints according to a predetermined gait pattern, where an angular offset is introduced in each joint to steer the robot to some desired path. Note that a similar approach is employed in this chapter to steer the heading of the snake robot. Yet another approach is proposed in Hicks and Ito (2005), Hicks (2003), where methods based on numerical optimal control are considered for determining optimal gaits during positional control of snake robots. To our best knowledge, no formal stability analysis of a path following controller for a snake robot has been presented in previous literature.

The novelty of this chapter in relation to previous literature is a control strategy based on a Line-of-Sight (LOS) guidance law, which enables snake robots to track a straight path. The LOS guidance law has previously been used for e.g. path following control of marine surface vessels (see e.g. Fossen 2002; Fredriksen and Pettersen 2006) but has, to our best knowledge, not been employed previously for directional control of snake robots.

The stability of the locomotion along the path is analysed in this chapter by use of a Poincaré map, which represents a widely used tool for analysing the existence and stability of periodic orbits of dynamical systems (see e.g. Westervelt et al. 2007). A Poincaré map has, to our best knowledge, never before been used to study the stability properties of snake robot locomotion. We therefore consider the methodology of this chapter to be a contribution to the snake robot literature.

5.2 Introduction to Poincaré Maps

In this section, we give an introduction to the Poincaré map since this is used as a stability analysis tool in Sect. 5.4. For further details on the topic, the reader is referred to Parker and Chua (1989) and Westervelt et al. (2007).

5.2.1 General Description of Poincaré Maps

The Poincaré map represents a widely used tool for analysing the existence and stability of periodic orbits of dynamical systems. Consider an autonomous (not explicitly dependent on time) n-dimensional dynamical system of the form

$$\dot{\mathbf{x}} = \mathbf{f}(\mathbf{x}), \quad \mathbf{x} \in \mathbb{R}^n, \tag{5.1}$$

where $\mathbf{f}(\mathbf{x})$ is assumed to be *continuously differentiable.* Assume that the solution of this differential equation for a particular initial condition is a *limit cycle*. Then the flow of $\mathbf{x}$ in the n-dimensional state space will return to the initial condition after a time T, corresponding to the period of the limit cycle.

We now define an $(n-1)$-dimensional hyperplane $\mathcal{S}$ (called a *Poincaré section*) such that the limit cycle intersects and passes through $\mathcal{S}$ at some instant in time. We denote by $\overline{\mathbf{x}} \in \mathbb{R}^{n-1}$ the $(n-1)$-dimensional state vector when $\mathbf{x}$ is constrained to $\mathcal{S}$. The point on $\mathcal{S}$ where the limit cycle intersects $\mathcal{S}$ is denoted $\overline{\mathbf{x}}^* \in \mathbb{R}^{n-1}$. Assume now that we initialise (5.1) on the hyperplane $\mathcal{S}$ somewhere close to $\overline{\mathbf{x}}^*$. Due to the continuity of the solutions of (5.1) with respect to the initial condition, the flow of $\mathbf{x}$ will, in approximately T seconds, return to and intersect S somewhere close to $\overline{\mathbf{x}}^*$. This is illustrated in Fig. 5.1. The mapping from an initial point $\overline{\mathbf{x}}$ on $\mathcal{S}$ to the next

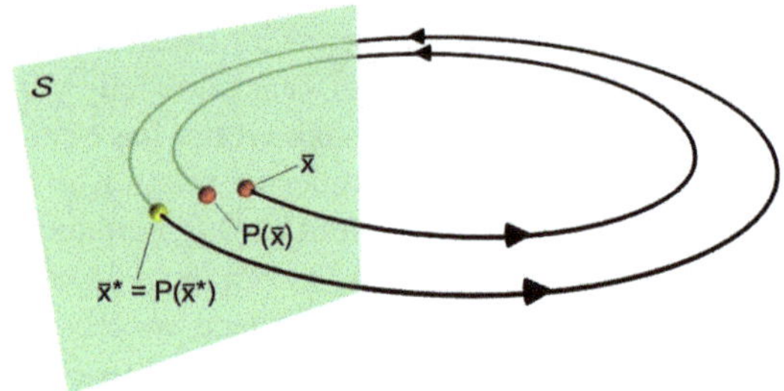

Fig. 5.1 Illustration of the Poincaré map corresponding to a Poincaré section $\mathcal{S}$

point where the flow of $\mathbf{x}$ intersects $\mathcal{S}$ is called the *Poincaré map* and is denoted by $\mathbf{P}(\overline{\mathbf{x}}) \in \mathbb{R}^{n-1}$. The Poincaré map is in other words a function that accepts an initial point on a Poincaré section as input and outputs where the Poincaré section will be intersected next by the flow of $\mathbf{x}$. This is written more formally as $\mathbf{P} : \mathcal{S} \to \mathcal{S}$. The point $\overline{\mathbf{x}}^*$ is called a *fixed point* of the Poincaré map since the Poincaré map maps $\overline{\mathbf{x}}^*$ back to itself. This is also illustrated in Fig. 5.1. We only consider *one-sided* Poincaré maps, i.e. we only consider crossings of $\mathcal{S}$ in directions corresponding to the direction of $\dot{\mathbf{x}}$ when $\mathbf{x}$ initially left $\mathcal{S}$.

The Poincaré map can be interpreted as a *discrete-time* system with an $(n-1)$-dimensional state space that evolves on the Poincaré section. This is seen by denoting by $\overline{\mathbf{x}}[k] \in \mathcal{S}$ the point of the kth intersection with $\mathcal{S}$ by the flow of $\mathbf{x}$. The Poincaré map may then be written as

$$\overline{\mathbf{x}}[k+1] = \mathbf{P}(\overline{\mathbf{x}}[k]), \quad \overline{\mathbf{x}}[0] \in \mathcal{S}. \tag{5.2}$$

The usefulness of the Poincaré map for stability analysis lies in the fact that *local exponential stability* of the fixed point $\overline{\mathbf{x}}^*$ on the Poincaré section is equivalent to *local exponential stability* of the underlying periodic orbit (see e.g. Westervelt et al. 2007), i.e. nearby orbits converge exponentially to the periodic orbit. Note that the stability is only *asymptotic* (i.e. not *exponential*) if $\mathbf{f}(\mathbf{x})$ in (5.1) is *continuous* but not *continuously differentiable* (Westervelt et al. 2007). The problem of determining if a periodic orbit of system (5.1) is exponentially stable is, in other words, reduced to determining if $\overline{\mathbf{x}}^*$ is an exponentially stable equilibrium point of the *discrete-time* system in (5.2), which is a much simpler problem to solve. A significant drawback of Poincaré maps is that they provide little insight into properties of the system dynamics.

Note that the method of Poincaré maps may also be applied to *non-autonomous* periodic systems, i.e. systems of the form $\dot{\mathbf{x}} = \mathbf{f}(\mathbf{x}, t)$, by incapsulating the time t in an augmented periodic state variable $\beta = 2\pi t/T$. This is performed for the snake robot in Sect. 5.4.1.

5.2.2 Practical Application of Poincaré Maps

This section provides an informal description of the practical use of Poincaré maps. The aim is to show how this method can be employed in practice in order to investigate the stability properties of a time-periodic dynamical system.

Calculating the Poincaré Map

It is difficult to determine the Poincaré map analytically since it requires the solution of the differential equation (5.1). However, the Poincaré map of (5.1) is simply the forward integration of this differential equation. It is therefore possible to compute the Poincaré map $\mathbf{P}(\overline{\mathbf{x}}_0)$ numerically by initialising (5.1) on $\mathcal{S}$ at $\overline{\mathbf{x}}_0$ and simulating (5.1) until $\mathcal{S}$ is intersected. The state corresponding to this intersection is the Poincaré map $\mathbf{P}(\overline{\mathbf{x}}_0)$.

Locating Fixed Points of the Poincaré Map

The easiest way of locating a fixed point $\overline{\mathbf{x}}^*$ of the Poincaré map is to simply let the simulation of (5.1) run until it reaches the steady state. This is called the *brute-force approach* and has three serious disadvantages. First of all, convergence to the fixed point can be exceedingly slow. Second, the method can only locate stable fixed points. Third, it may be difficult to tell when the steady state has been reached.

A more sophisticated method is to exploit the fact that locating $\overline{\mathbf{x}}^*$ is equivalent to locating zeros of the error function

$$\mathbf{E}(\overline{\mathbf{x}}) = \mathbf{P}(\overline{\mathbf{x}}) - \overline{\mathbf{x}}, \quad \mathbf{E}(\overline{\mathbf{x}}) \in \mathbb{R}^{n-1}, \tag{5.3}$$

since we have that $\overline{\mathbf{x}}^* = \mathbf{P}(\overline{\mathbf{x}}^*)$. The *Newton–Raphson algorithm* (Parker and Chua 1989) is a general algorithm for locating zeros of a differentiable function, and it may therefore be employed for locating $\overline{\mathbf{x}}^*$. By starting from an initial guess, $\overline{\mathbf{x}}^k$, of the fixed point, the Newton–Raphson algorithm calculates a more accurate estimate of $\overline{\mathbf{x}}^*$ through the formula

$$\overline{\mathbf{x}}^{k+1} = \overline{\mathbf{x}}^k - \mathbf{J}_E\left(\overline{\mathbf{x}}^k\right)^{-1}\mathbf{E}\left(\overline{\mathbf{x}}^k\right), \tag{5.4}$$

where $\mathbf{J}_E(\overline{\mathbf{x}}^k)$, which is the Jacobian of the error function $\mathbf{E}(\overline{\mathbf{x}})$, is defined by

$$\mathbf{J}_E = \frac{\partial \mathbf{E}}{\partial \overline{\mathbf{x}}} = \begin{bmatrix} \frac{\partial E_1}{\partial \overline{x}_1} & \cdots & \frac{\partial E_1}{\partial \overline{x}_{n-1}} \\ \vdots & \ddots & \vdots \\ \frac{\partial E_{n-1}}{\partial \overline{x}_1} & \cdots & \frac{\partial E_{n-1}}{\partial \overline{x}_{n-1}} \end{bmatrix} \in \mathbb{R}^{(n-1)\times(n-1)}. \tag{5.5}$$

The Jacobian $\mathbf{J}_E(\overline{\mathbf{x}}^k)$ can be calculated numerically by defining

$$\mathrm{d}\overline{\mathbf{x}}_i = [0, \ldots, 0, \Delta_i, 0, \ldots, 0]^T \in \mathbb{R}^{n-1}, \tag{5.6}$$

where the ith element is non-zero, and Δ_i is a small perturbation of $\overline{x}_i$ along $\mathcal{S}$. Column i of $\mathbf{J}_E(\overline{\mathbf{x}}^k)$ may then be approximated numerically as

$$\frac{\partial \mathbf{E}}{\partial \overline{x}_i}(\overline{\mathbf{x}}^k) \approx \frac{\mathbf{E}(\overline{\mathbf{x}}^k + \mathrm{d}\overline{\mathbf{x}}_i) - \mathbf{E}(\overline{\mathbf{x}}^k)}{\Delta_i}. \tag{5.7}$$

This enables a column-wise construction of $\mathbf{J}_E(\overline{\mathbf{x}}^k)$. If the initial condition is within the basin of attraction of a periodic orbit, the Newton–Raphson algorithm will converge rapidly towards the fixed point $\overline{\mathbf{x}}^*$.

Analysing Stability of a Periodic Orbit

As explained in Sect. 5.2.1, a fixed point $\overline{\mathbf{x}}^*$ of the Poincaré map corresponds to a periodic orbit of the underlying dynamical system. Once the fixed point has been found using e.g. the Newton–Raphson algorithm, the stability of the periodic orbit may be tested by investigating if the fixed point is a stable equilibrium point of the Poincaré map. This is done by calculating the Jacobian linearisation of the Poincaré map about the fixed point, i.e. by calculating the Jacobian $\mathbf{J}_P(\overline{\mathbf{x}}^*) = \frac{\partial \mathbf{P}}{\partial \overline{\mathbf{x}}}|_{\overline{\mathbf{x}}=\overline{\mathbf{x}}^*} \in \mathbb{R}^{(n-1)\times(n-1)}$. $\mathbf{J}_P(\overline{\mathbf{x}}^*)$ is calculated by following the same procedure as for calculating $\mathbf{J}_E(\overline{\mathbf{x}}^k)$ in (5.5). The Poincaré map linearised about the fixed point is thereby given as $\overline{\mathbf{x}}[k+1] = \mathbf{J}_P(\overline{\mathbf{x}}^*)\overline{\mathbf{x}}[k]$. This is a linear discrete-time system which is *exponentially stable* if the magnitude of all the eigenvalues of $\mathbf{J}_P(\overline{\mathbf{x}}^*)$ are strictly less than one. The fixed point $\overline{\mathbf{x}}^*$ of the Poincaré map, and thereby also the periodic orbit of the underlying dynamical system, is therefore *locally exponentially stable* if the magnitude of all the eigenvalues of $\mathbf{J}_P(\overline{\mathbf{x}}^*)$ are strictly less than one.

5.3 Straight Line Path Following Control of Snake Robots

In this section, we propose a control law that enables a snake robot to track a straight path. We assume that the snake robot is described by the model in (2.35) and that the robot is influenced by anisotropic viscous ground friction forces described by (2.25).

5.3.1 Control Objective

In order to track the desired straight path, we define the global coordinate system so that the global x axis is aligned with the desired path. The position of the snake robot along the global y axis, p_y, is then the shortest distance from the robot to the desired path (i.e. the cross-track error), and the heading $\overline{\theta}$ of the robot, which was defined in (2.2), is the angle that the robot forms with the desired path. The control objective is thereby to regulate p_y and $\overline{\theta}$ so that they oscillate about zero, i.e. so that their trajectories trace out a limit cycle containing $(p_y = 0, \overline{\theta} = 0)$ in its interior. As explained in the introduction of this chapter, we do not attempt to regulate p_y and $\overline{\theta}$ to zero since we expect the heading and position of the robot to display oscillating behaviour during locomotion along the desired path.

Since snake robot locomotion is a slow form of robotic mobility, which is generally employed for traversability purposes, we consider it less important to accurately control the forward velocity of the robot. During path following with a snake robot, it therefore makes sense to focus all the control efforts on converging to the path and subsequently progressing along the path at some non-zero forward velocity $\overline{v}_t(t) > 0$, where $\overline{v}_t(t)$ is the forward velocity of the robot defined in (2.5).

From the above discussion, the control problem is to design a feedback control law such that for all $t > t_c \geq 0$, there exists a $\tau \in [t, t + T]$ satisfying

$$p_y(\tau) = 0, \tag{5.8}$$

$$\overline{\theta}(\tau) = 0, \tag{5.9}$$

$$\overline{v}_t(t) > 0, \tag{5.10}$$

where t_c is some (unknown) finite time duration corresponding to the time it takes the snake robot to converge to the desired straight path, and $T > 0$ is some constant that characterises the time period of the cyclic gait pattern of the snake robot. In other words, we require that p_y and $\overline{\theta}$ are zero at least once *within* each cycle of the locomotion since this means that p_y and $\overline{\theta}$ oscillate about zero. Note that we require $\overline{v}_t(t) > 0$ for all $t > t_c$.

5.3.2 The Straight Line Path Following Controller

We choose to propel the snake robot forward according to the gait pattern lateral undulation defined in (4.32), which is achieved by controlling joint $i \in \{1, \dots, N-1\}$ according to

$$\phi_{i,\text{ref}} = \alpha \sin\bigl(\omega t + (i-1)\delta\bigr) + \phi_o, \tag{5.11}$$

where α and ω are the amplitude and frequency, respectively, of the sinusoidal joint motion, δ determines the phase shift between the joints, and ϕ_o is a joint offset, which we assume to be identical for all joints. Note that with the gait pattern in (5.11), the period of the cyclic locomotion considered in control objectives (5.8) and (5.9) will be $T = 2\pi/\omega$.

The heading $\overline{\theta}$ of the robot was defined in (2.2) as

$$\overline{\theta} = \frac{1}{N}\sum_{i=1}^{N}\theta_i. \tag{5.12}$$

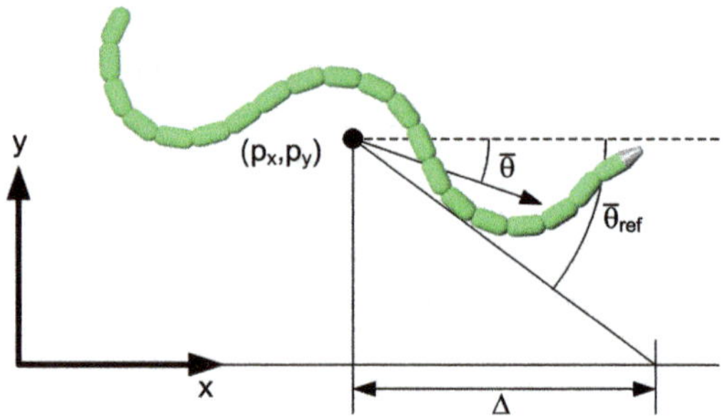

Fig. 5.2 The Line-of-Sight (LOS) guidance law

In order to steer the snake robot towards the desired straight path (i.e. the global x axis), we define the heading reference angle according to the Line-of-Sight (LOS) guidance law

$$\overline{\theta}_{\text{ref}} = -\arctan\left(\frac{p_y}{\Delta}\right), \tag{5.13}$$

where p_y is the cross-track error, and $\Delta > 0$ is a design parameter referred to as the *look-ahead distance* that influences the rate of convergence to the desired path. This LOS guidance law is commonly used during e.g. path following control of marine surface vessels (see e.g. Fossen 2002; Fredriksen and Pettersen 2006). As illustrated in Fig. 5.2, the LOS angle $\overline{\theta}_{\text{ref}}$ corresponds to the orientation of the snake robot when it is headed towards the point located a distance Δ ahead of itself along the desired path.

We know from Property 4.7 that ϕ_o can be used to control the direction of the locomotion, and we therefore conjecture that we can steer the heading $\overline{\theta}$ according to the LOS angle in (5.13) by defining this joint offset angle according to

$$\phi_o = k_\theta(\overline{\theta} - \overline{\theta}_{\text{ref}}), \tag{5.14}$$

where $k_\theta > 0$ is a controller gain. To make the joints track the reference angles given by (5.11), we set the control input $\mathbf{u} \in \mathbb{R}^{N-1}$ of the model (2.35) according to the linearising control law (2.44), and we set the new control input $\overline{\mathbf{u}} \in \mathbb{R}^{N-1}$ according to the simple joint controller defined in (4.33) as

$$\overline{u}_i = k_p(\phi_{i,\text{ref}} - \phi_i) - k_d\dot{\phi}_i, \tag{5.15}$$

where $k_p > 0$ and $k_d > 0$ are controller gains.

5.4 Stability Analysis of the Path Following Controller Based on the Poincaré Map

In this section, we employ the theory of *Poincaré maps* (see Sect. 5.2) to prove that the path following controller proposed in Sect. 5.3 generates a *locally exponentially stable periodic orbit* in the state space of the snake robot as the robot locomotes along the desired straight path, i.e. the global x axis.

5.4.1 Converting the Snake Robot Model to a Time-Periodic Autonomous System

Stability analysis of the time-periodic state variables of the snake robot by use of Poincaré maps requires that the model of the snake robot represents an *autonomous* system, i.e. a system not explicitly dependent on time. More specifically, the stability conclusions described in Sect. 5.2.2 are not valid if the Jacobian matrix $\mathbf{J}_P(\overline{\mathbf{x}}^*)$ is a function of time t. However, since the path following controller proposed in Sect. 5.3 depends explicitly on time, i.e. $\mathbf{u} = \mathbf{u}(\mathbf{x}, t)$, the model of the snake robot (2.35) with the path following controller can be written as

$$\dot{\mathbf{x}} = \begin{bmatrix} \dot{\boldsymbol{\theta}} \\ \dot{\mathbf{p}} \\ \ddot{\boldsymbol{\theta}} \\ \ddot{\mathbf{p}} \end{bmatrix} = \mathbf{F}(\mathbf{x}, t), \tag{5.16}$$

which is a *non-autonomous* system since time t is explicitly present in the system equations. We therefore follow the approach described in Parker and Chua (1989) in order to convert the snake robot model to an autonomous system by simply augmenting the state vector $\mathbf{x}$ with an extra state $\beta = 2\pi t/T$, where $T = 2\pi/\omega$ is the period of the cyclic locomotion generated by the gait pattern in (5.11). We make β periodic by enforcing that $0 \le \beta < 2\pi$, i.e. we set β to zero each time $\beta = 2\pi$. The model (2.35) with the path following controller can thereby be written as the following *autonomous* system:

$$\begin{aligned} \dot{\mathbf{x}} &= \mathbf{F}\left(\mathbf{x}, \frac{T}{2\pi}\beta\right), \qquad && \mathbf{x}(t_0) = \mathbf{x}_0, \\ \dot{\beta} &= \frac{2\pi}{T}, && \beta(t_0) = \frac{2\pi t_0}{T}. \end{aligned} \tag{5.17}$$

We have, in other words, encapsulated time t in the new state variable β, which is periodic since $0 \le \beta < 2\pi$.

5.4.2 Specification of the Poincaré Section for the Snake Robot

During locomotion along the global positive x axis, our goal is that the x axis position of the snake robot, p_x, increases, while all other states of the snake robot in (2.35) trace out a stable limit cycle in the state space. We therefore exclude p_x from the Poincaré map of the snake robot, which produces a *partial Poincaré map* (Westervelt et al. 2007). Exclusion of p_x has no effect on the other state variables since p_x is not present in any of their derivatives in (2.35). The synthesis and analysis of lateral undulation in Chap. 4, which is the basis of the path following controller proposed in Sect. 5.3, enable us to argue that forward motion along the x axis (increase of p_x) is achieved as long as the remaining state variables trace out a stable periodic orbit.

We choose the global x axis as the Poincaré section $\mathcal{S}$ of the system in (5.17). Since p_x is not included in the Poincaré map, we write $\mathcal{S} = \{(\boldsymbol{\theta}, p_y, \dot{\boldsymbol{\theta}}, \dot{\mathbf{p}}, \beta) | p_y = 0\}$. Following the notation in Sect. 5.2, the vector of *independent* time-periodic states constrained to $\mathcal{S}$ is given by

$$\overline{x} = \begin{bmatrix} \boldsymbol{\theta} \\ \dot{\boldsymbol{\theta}} \\ \dot{\mathbf{p}} \\ \beta \end{bmatrix} \in \mathbb{R}^{2N+3}. \tag{5.18}$$

Note that the considered Poincaré map is *one-sided*, which means that the Poincaré section is crossed when p_y switches from a positive to a negative value, i.e. when the CM position of the robot crosses the x axis from above.

5.4.3 Stability Analysis of the Poincaré Map

In order to investigate the stability of the path following controller, we considered a three-linked snake robot where $N = 3$, $l = 0.07$ m, $m = 1$ kg, and $J = 0.0016$ kgm^2. The ground friction coefficients were chosen as $c_t = 1$ and $c_n = 10$, and the parameters of the path following controller in Sect. 5.3 were $\alpha = 70°$, $\omega = 70°/\text{s}$, $\delta = 70°$, $k_p = 20$, $k_d = 5$, $k_\theta = 0.3$, and $\Delta = 0.42$ m.

The Poincaré map of the snake robot model in (5.17) was calculated as described in Sect. 5.2.2 using *Matlab R2008b* on a laptop running *Windows XP*. The *ode45* solver in Matlab was used with a relative and absolute error tolerance of 10^{-6}. The Newton–Raphson algorithm described in Sect. 5.2.2 calculated the fixed point, $\overline{\mathbf{x}}^* \in \mathbb{R}^9$, of the Poincaré map as

$$\begin{aligned} \overline{\mathbf{x}}^* = \big[&-15.0°, -32.6°, 27.6°, -72.4°/\text{s}, 13.7°/\text{s}, 66.7°/\text{s}, \\ &4.6 \text{ cm/s}, -1.2 \text{ cm/s}, 182.5°\big]^T. \end{aligned} \tag{5.19}$$

A plot of the cyclic locomotion of the snake robot over one period is shown in Fig. 5.3. The initial state of the snake robot was given by $\overline{\mathbf{x}}^*$ and the initial position was $\mathbf{p} = \mathbf{0}$. After one period of the motion, the state variables returned to their initial value, $\overline{\mathbf{x}}^*$. At this point, however, the position of the snake robot along the x axis had increased, which was also our goal. To clearly illustrate the limit cycle behaviour of the periodic state variables in (5.18), a 3D plot of the three absolute link angles over one period is given to the left in Fig. 5.4.

The Jacobian linearisation of the Poincaré map about the fixed point (5.19) was calculated as described in Sect. 5.2.2. The magnitude of the eigenvalues of $\mathbf{J}_P(\overline{\mathbf{x}}^*) \in \mathbb{R}^{9\times 9}$ was

$$\begin{aligned} \big|\text{eig}(\mathbf{J}_P(\overline{\mathbf{x}}^*))\big| = \big[&0.78, 0.78, 0.0022, 2.1 \times 10^{-4}, 4.9 \times 10^{-5}, \\ &4.1 \times 10^{-5}, 9.6 \times 10^{-6}, 2.9 \times 10^{-6}, 1.6 \times 10^{-6}\big]^T. \end{aligned} \tag{5.20}$$

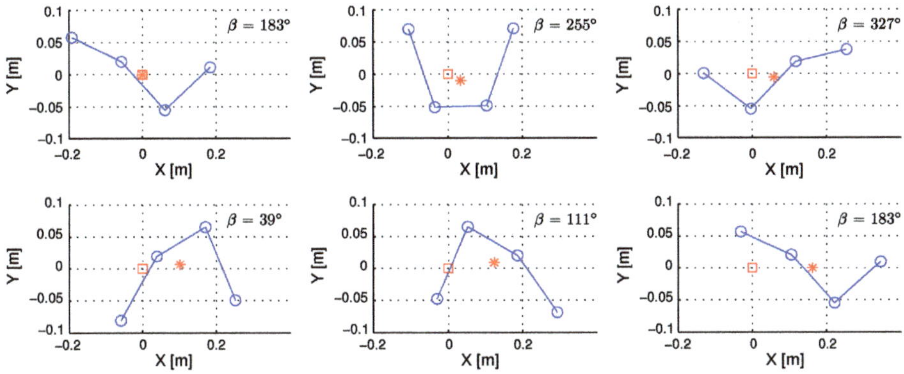

Fig. 5.3 The motion of the snake robot over one period of the cyclic locomotion

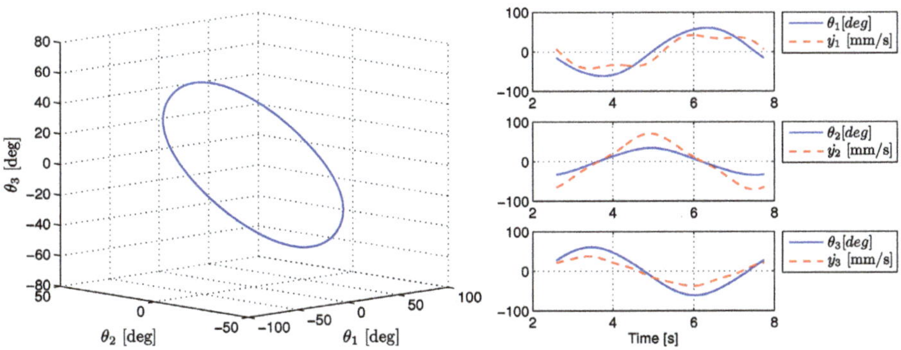

Fig. 5.4 *Left*: The limit cycle traced out by the link angles of the snake robot. *Right*: Plot of θ_i and $\dot{y}_i$ $(i = 1, 2, 3)$ over one period of the cyclic locomotion

The magnitude of all the eigenvalues are strictly less than one, which means that the periodic orbit traced out by the variables in (5.18) is *locally exponentially stable* for the given choice of controller parameters. All initial states inside the basin of attraction of this periodic orbit will converge *exponentially* to this periodic orbit.

We now summarise the above analysis with regards to the control objectives in (5.8), (5.9), and (5.10). Since the path following controller generates an exponentially stable periodic orbit, and since $p_y = 0$ is the Poincaré section of this periodic orbit (i.e. the system returns to $p_y = 0$ with time period T), we can conclude that control objective (5.8) is achieved from all initial states inside the basin of attraction of this periodic orbit. Since the snake robot is conducting lateral undulation under anisotropic ground friction conditions, the synthesis and analysis of this gait pattern in Chap. 4 implies that the robot is locomoting forward, i.e. control objective (5.10) is satisfied. In particular, Property 4.3 in Sect. 4.5 states that the propulsive force generated by the transversal motion of link $i \in \{1, 2, 3\}$ is positive as long as $\text{sgn}(\theta_i) = \text{sgn}(\dot{y}_i)$. A plot of θ_i and $\dot{y}_i$ over one period is given to the right in Fig. 5.4, which clearly shows that $\text{sgn}(\theta_i) = \text{sgn}(\dot{y}_i)$ over the majority of the pe-

Fig. 5.5 Simulation of the straight line path following controller

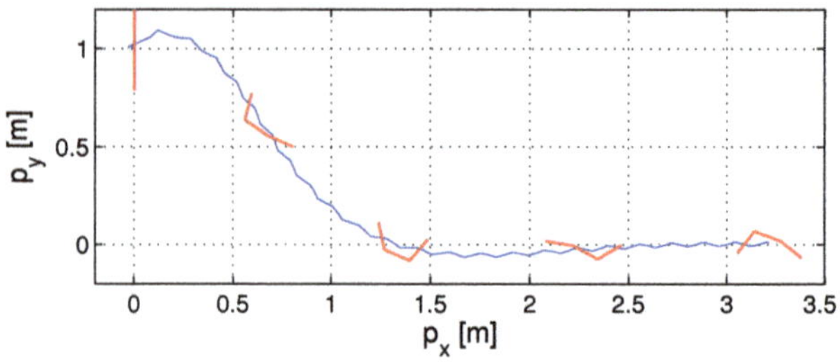

(a) The path of the CM of the snake robot.

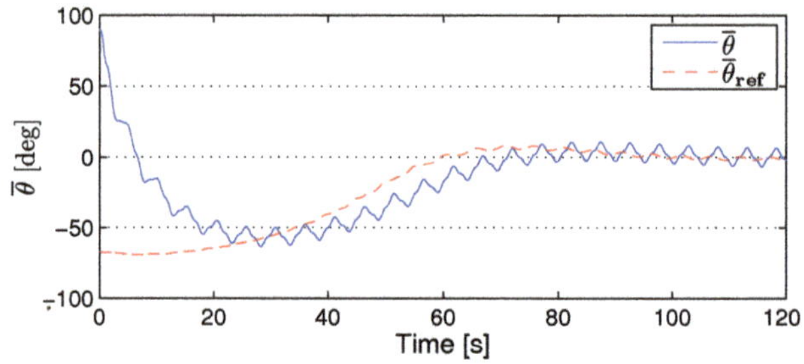

(b) The heading angle, $\overline{\theta}$.

riod. This means that the net propulsive force on the robot is positive. Finally, since control objectives (5.8) and (5.10) are both satisfied, control objective (5.9) must also be satisfied. In particular, if the heading did not oscillate about zero, then the snake robot would drift away from the x axis, which contradicts the achievement of objective (5.8).

Remark 5.1 Note that since the Poincaré map and its Jacobian linearisation are found basically by simulating the model of the snake robot, we have only proved that the periodic orbit of the state variables is locally exponentially stable for the given choice of numerical parameters described in the beginning of this section. This is a drawback of stability analysis based on the Poincaré map.

5.5 Simulation Study: The Performance of the Path Following Controller

In order to illustrate the performance of the path following controller, we provide in this section a simulation result where the three-linked snake robot starts from rest with initial configuration given by $\boldsymbol{\theta} = \mathbf{0}^\circ$, $p_x = 0$ m, and $p_y = 1$ m, i.e. the snake robot is initially headed *away* from the desired path (the x axis) and the initial distance from the CM to the desired path is 1 m. We employ the model and controller parameters described in the beginning of the previous subsection.

The path traced out by the CM of the snake robot is shown in Fig. 5.5(a), where the shape and position of the robot are shown in red at $t = 0$ s, $t = 30$ s, $t = 60$ s,

$t = 90$ s, and $t = 120$ s, respectively. We see that the position of the snake robot converges nicely to the desired path, i.e. the x axis. From Fig. 5.5(b), which shows the heading of the robot (solid blue) and the reference angle of the heading (dashed red), we see that the heading also converges nicely to zero, i.e. to the direction of the desired path. During the motion along the x axis near the end of the simulation, Fig. 5.5 clearly shows the oscillating behaviour of the heading and the position, which was predicted in the introduction of this chapter.

5.6 Chapter Summary

This chapter is summarised as follows:

- We have proposed a control law that enables snake robots to track straight paths.
- We have analysed the stability of the path following controller by use of a Poincaré map, and in particular, we have shown that all state variables of the snake robot, except for the position along the path, trace out an *exponentially stable* periodic orbit during path following with the proposed controller.
- We have presented simulation results that illustrate the performance of the path following controller.

Chapter 6
A Simplified Model of Snake Robot Locomotion on Planar Surfaces

Faced with the problem of *proving* that a control strategy for the snake robot satisfies some control objective, an attractive idea is to base the controller analysis on a *simplified* model of the snake robot that avoids the complex expressions contained in the model given by (2.47). The hypothesis behind this idea is of course that the complex model in (2.47) contains nonlinear dynamics that is not essential to the overall locomotion of the snake robot. We have already seen support of this claim in the simulation results of the path following controller proposed in the previous chapter. In particular, Fig. 5.5 clearly shows that the heading and position of a snake robot display an oscillating behaviour during locomotion. However, for control design and analysis purposes, we are not particularly interested in this oscillatory dynamics as we are primarily concerned with the *overall* motion of the heading and position of the robot.

In this chapter, we therefore propose a simplified model of a planar snake robot aimed at simplifying *analytical* investigations of the equations of motion. The basic idea behind the model is to capture only the essential properties of snake robot dynamics that we derived in Chap. 4, i.e. the features that determine the overall behaviour of the snake. In particular, we propose to describe the body shape dynamics of the snake robot in terms of the translational motion of the links by assuming that the snake robot consists of a serial connection of *prismatic* (translational) joints instead of *revolute* joints. This approach significantly simplifies the equations of motion since translational motion is generally less complex to model than rotational motion. Note that there are several limitations of this modelling approach, which are elaborated in this chapter. The modelling approach is for instance limited to snake locomotion where the link angles are limited.

In order to verify that the essential features of snake robot locomotion are contained in the simplified model, we will in this chapter repeat the stabilisability and controllability analysis presented in Chap. 4, but this time based on the simplified model. It will be shown that the stabilisability and controllability properties of the simplified model are indeed similar to the properties of the complex model. Simulation results that compare the complex and the simplified model are presented to illustrate the validity of the simplified modelling approach.

P. Liljebäck et al., *Snake Robots*, Advances in Industrial Control,
DOI 10.1007/978-1-4471-2996-7_6, © Springer-Verlag London 2013

This chapter is organised as follows. The relation between this chapter and previous literature is briefly discussed in Sect. 6.1. An overview of the simplified modelling approach is given in Sect. 6.2. The models of the snake robot kinematics, the ground friction forces, and the snake robot dynamics are presented in Sects. 6.3, 6.4, and 6.5, respectively. Subsequently, the complete simplified model is summarised in Sect. 6.6. The stabilisability and controllability properties of the model are studied in Sect. 6.8 and Sect. 6.9, respectively, and simulation results that compare the simplified and the complex model are presented in Sect. 6.10. Finally, the chapter is summarised in Sect. 6.11.

6.1 The Relation Between This Chapter and Previous Literature

As elaborated in the literature review of Sect. 1.3, several mathematical models of snake robot kinematics and dynamics have been proposed over the years. All previous approaches either assume that the snake robot consists of a finite number of *revolute* joints (this modelling approach was employed in Chap. 2) or employ a continuum approach by modelling the snake robot as a *continuous curve* (see e.g. Chirikjian and Burdick 1995; Chirikjian 1992; Date and Takita 2005; Hirose 1993; Yamada and Hirose 2006a). These approaches generally result in complex equations of motion which are challenging to analyse from a motion control perspective. In this chapter, on the other hand, we employ a completely different modelling approach by assuming that the snake robot consists of a serial connection of *prismatic* (translational) joints. The approach is, in other words, to describe the body shape dynamics of the snake robot in terms of the translational motion of the links, which significantly simplifies the equations of motion. This approach has, to our best knowledge, not been considered previously. The model of snake robot dynamics presented in this chapter is therefore novel in relation to previous literature.

6.2 Overview of the Modelling Approach

Property 4.8 from Sect. 4.10 tells us that lateral undulation mainly consists of link displacements that are transversal to the direction of motion. At the same time, Property 4.2 from Sect. 4.5 tells us that the transversal link displacements are what propel the robot forward. From these results sprung the idea of describing the mapping from body shape changes to propulsion in terms of the translational displacements of the links instead of the rotational joint motion. The motivation behind this idea is that translational motion is generally less complex to model than rotational motion. In particular, the model given by (2.47), which describes the mapping from the rotational joint motion to the propulsion of a snake robot, is quite complex.

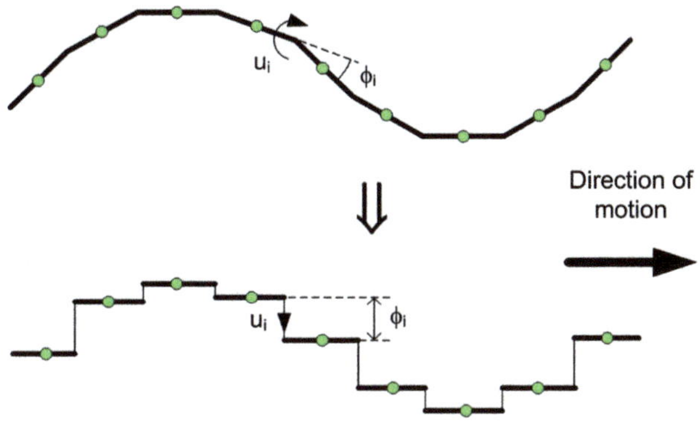

Fig. 6.1 The snake robot is modelled as a series of prismatic joints that displace the CM of each link transversal to the *direction of motion*

The simplified modelling approach is illustrated in Fig. 6.1, where we see that the body shape changes of the snake robot are described, not as rotational link displacements, but as linear link displacements transversal to the forward direction of motion. This essentially means that we will model the revolute joints of a snake robot as prismatic (translational) joints. The rotational motion of the links during body shape changes will in other words be disregarded. However, the model will still capture the *effect* of the rotational link motion during body shape changes, which we know from Property 4.8 to be primarily a linear displacement of the CM of the links normal to the forward direction of motion.

Remark 6.1 Property 4.2 makes no assumption regarding the gait pattern of the snake robot. Property 4.8, however, is only valid for gait patterns where the relative link displacements transversal to the direction of motion dominate over the relative link displacements tangential to the direction of motion. The simplified model should therefore only be used to study gait patterns that comply with this condition, i.e. gait patterns with limited link angles with respect to the forward direction. So far, the authors have only used the simplified model to study lateral undulation with limited link angles. However, it is conceivable that the model can be used to study other gait patterns as well. A discussion of the validity of the simplified model is provided in Sect. 6.7.

Remark 6.2 We emphasise that the simplified model presented in this chapter is *not* intended as an accurate simulation model of snake robot locomotion. The model is intentionally based on several simplifying assumptions in order to arrive at equations of motion that are manageable for control design and stability analysis purposes. To this end, the model only needs to be *qualitatively* similar to the complex model in (2.47).

The kinematics and dynamics of the snake robot will be detailed in the following subsections in terms of the mathematical symbols described in Table 6.1 and illustrated in Figs. 6.2 and 6.3.

Table 6.1 Parameters that characterise the snake robot

Symbol	Description
N	Number of links
l	Length of a link
m	Mass of each link
ϕ_i	Normal direction distance between links i and $i+1$
$v_{\phi,i}$	Relative velocity between links i and $i+1$
θ	Orientation of the snake robot
v_θ	Angular velocity of the snake robot
(t_i, n_i)	Coordinates of the CM of link i in the t–n frame
(p_t, p_n)	Coordinates of the CM of the robot in the t–n frame
(p_x, p_y)	Coordinates of the CM of the robot in the global frame
(v_t, v_n)	Forward and normal direction velocity of the robot
u_i	Actuator force at joint i
$(f_{R,x,i}, f_{R,y,i})$	Friction force on link i in the global frame
$(f_{t,i}, f_{n,i})$	Friction force on link i in the t–n frame

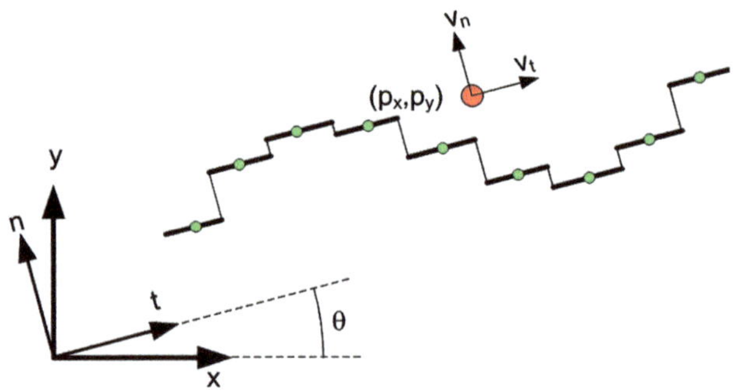

Fig. 6.2 Illustration of the two coordinate frames employed in the simplified model. The global x–y frame is fixed. The t–n frame is always aligned with the snake robot

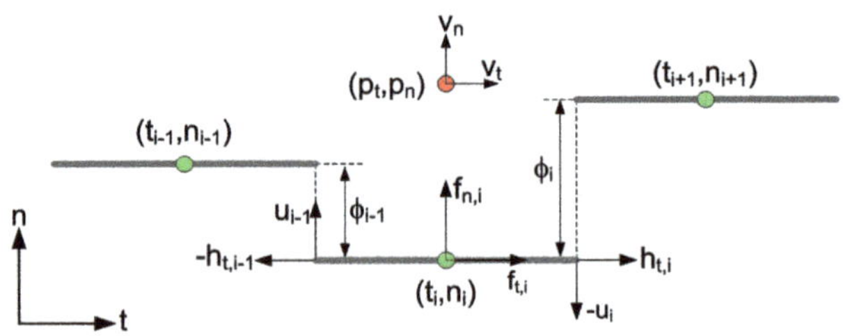

Fig. 6.3 Parameters characterising the kinematics and dynamics of the snake robot

The following vectors and matrices are used in the development of the model:

$$\mathbf{A} = \begin{bmatrix} 1 & 1 & & \\ & \ddots & \ddots & \\ & & 1 & 1 \end{bmatrix} \in \mathbb{R}^{(N-1)\times N},$$

$$\mathbf{D} = \begin{bmatrix} 1 & -1 & & \\ & \ddots & \ddots & \\ & & 1 & -1 \end{bmatrix} \in \mathbb{R}^{(N-1)\times N},$$

$$\mathbf{e} = [1, \ldots, 1]^T \in \mathbb{R}^N, \qquad \bar{\mathbf{e}} = [1, \ldots, 1]^T \in \mathbb{R}^{N-1},$$

$$\overline{\mathbf{D}} = \mathbf{D}^T \left(\mathbf{D}\mathbf{D}^T\right)^{-1} \in \mathbb{R}^{N\times(N-1)}.$$

The matrices **A** and **D** represent, respectively, an *addition* and a *difference* matrix, which will be used, respectively, for adding and subtracting pairs of adjacent elements of a vector. Furthermore, the vectors **e** and $\bar{\mathbf{e}}$ represent summation vectors, which will be used for adding all elements of N-dimensional and $(N-1)$-dimensional vectors, respectively. The matrix $\overline{\mathbf{D}}$ has been defined above since it appears several times during the development of the model.

6.3 The Kinematics of the Snake Robot

We consider a planar snake robot with N links of length l interconnected by $N-1$ motorised *prismatic* (translational) joints. All N links have the same mass m, and the total mass of the snake robot is thus Nm. Note that we denote the total link length in the simplified model by l, whereas the total link length in the complex model is $2l$ for notational convenience.

The snake robot moves on a horizontal and flat surface, and has $N+2$ degrees of freedom. We define the motion of the robot with respect to the two coordinate frames illustrated in Fig. 6.2. The x–y frame is the fixed global frame. The t–n frame is always aligned with the snake robot, i.e. the t and n axes always point in the *tangential* and *normal* direction of the robot, respectively. The origin of both frames are fixed and coincide. We will denote the direction of the t axis as the *tangential* or *forward* direction of the robot, and the direction of the n axis as the *normal* direction. Note that we do not refer to the t–n frame as the *body* frame of the snake robot since the t–n frame is not fixed to the robot. However, if a body frame fixed to the robot had been defined, the orientation of this frame would be identical to the orientation of the t–n frame.

The position of the snake robot is described through the coordinates of its CM (centre of mass). As seen in Figs. 6.2 and 6.3, the global frame position of the robot is denoted by $(p_x, p_y) \in \mathbb{R}^2$, while the t–n frame position is denoted by $(p_t, p_n) \in \mathbb{R}^2$. The global frame orientation of the robot is denoted by $\theta \in \mathbb{R}$ and is expressed with respect to the global x axis with counterclockwise positive direction. The angle between the global x axis and the t axis is also θ since the t–n frame is always aligned with the robot. Describing the position in a frame which is always aligned with the snake robot is inspired by and similar to a coordinate transformation proposed in Pettersen and Egeland (1996).

Remark 6.3 There are many possible choices of orientation for a snake robot with revolute joints. In Definition 2.3 in Sect. 2.4, we estimated the orientation of the snake robot in the complex model (2.47) as the mean of the absolute link angles. On the other hand, the simplified model proposed in this section provides an explicit representation of the orientation of the snake robot in terms of the scalar variable θ. This explicit representation of the orientation is a significant advantage of the simplified model since it simplifies analytical investigations which involve the orientation of the snake robot.

The relationship between the t–n frame position and the global frame position is given by

$$p_t = p_x \cos\theta + p_y \sin\theta, \tag{6.1a}$$

$$p_n = -p_x \sin\theta + p_y \cos\theta. \tag{6.1b}$$

As illustrated in Fig. 6.2, the forward and normal direction velocities of the CM of the snake robot are denoted by $v_t \in \mathbb{R}$ and $v_n \in \mathbb{R}$, respectively. Using (2.3), the relationship between the global frame velocity of the robot and the t–n frame velocity is given by

$$\dot{p}_x = v_t \cos\theta - v_n \sin\theta, \tag{6.2a}$$

$$\dot{p}_y = v_t \sin\theta + v_n \cos\theta, \tag{6.2b}$$

and the inverse relationship is given by

$$v_t = \dot{p}_x \cos\theta + \dot{p}_y \sin\theta, \tag{6.3a}$$

$$v_n = -\dot{p}_x \sin\theta + \dot{p}_y \cos\theta. \tag{6.3b}$$

Differentiating (6.1a) and (6.1b) with respect to time and inserting (6.3a) and (6.3b) gives

$$\dot{p}_t = v_t + p_n\dot{\theta}, \tag{6.4a}$$

$$\dot{p}_n = v_n - p_t\dot{\theta}. \tag{6.4b}$$

We denote the t–n frame position of the CM of link i by $(t_i, n_i) \in \mathbb{R}^2$. The $N-1$ prismatic joints of the snake robot control the normal direction distance between the links. As seen in Fig. 6.3, the normal direction distance between link i and link $i+1$ is given by

$$\phi_i = n_{i+1} - n_i \tag{6.5}$$

and represents the coordinate of joint i. The controlled distance ϕ_i replaces the controlled joint angle in the original model given by (2.47).

Remark 6.4 The state ϕ_i of joint i in the simplified model is a translational *distance*, while the state ϕ_i of joint i in the complex model is a joint *angle*. In the simplified model, we therefore refer to ϕ_i as a joint coordinate instead of a joint angle.

The link positions are constrained by the prismatic joints according to

$$t_i - t_{i+1} + l = 0, \tag{6.6a}$$

$$n_i - n_{i+1} + \phi_i = 0. \tag{6.6b}$$

These holonomic constraints may be expressed in matrix form for all links as

$$\mathbf{Dt} + l\bar{\mathbf{e}} = \mathbf{0}, \tag{6.7a}$$

$$\mathbf{Dn} + \boldsymbol{\phi} = \mathbf{0}, \tag{6.7b}$$

where $\mathbf{D}$ and $\bar{\mathbf{e}}$ are defined in Sect. 6.2, $\mathbf{t} = [t_1, \ldots, t_N]^T \in \mathbb{R}^N$, $\mathbf{n} = [n_1, \ldots, n_N]^T \in \mathbb{R}^N$, and $\boldsymbol{\phi} = [\phi_1, \ldots, \phi_{N-1}]^T \in \mathbb{R}^{N-1}$. The t–n frame position of the CM of the snake robot can be written in terms of the link positions as

$$p_t = \frac{1}{N}\mathbf{e}^T\mathbf{t}, \tag{6.8a}$$

$$p_n = \frac{1}{N}\mathbf{e}^T\mathbf{n}, \tag{6.8b}$$

where $\mathbf{e}$ is defined in Sect. 6.2. Combining (6.7a)–(6.7b) and (6.8a)–(6.8b) gives

$$\begin{gathered} \begin{bmatrix} \mathbf{D} \\ \frac{1}{N}\mathbf{e}^T \end{bmatrix} \mathbf{t} = \begin{bmatrix} -l\bar{\mathbf{e}} \\ p_t \end{bmatrix}, \\ \begin{bmatrix} \mathbf{D} \\ \frac{1}{N}\mathbf{e}^T \end{bmatrix} \mathbf{n} = \begin{bmatrix} -\boldsymbol{\phi} \\ p_n \end{bmatrix}. \end{gathered} \tag{6.9}$$

Similar to the approach in the complex model, we can use (2.10) to solve (6.9) for the link positions as

$$\mathbf{t} = p_t\mathbf{e} - l\overline{\mathbf{D}}\bar{\mathbf{e}}, \tag{6.10a}$$

$$\mathbf{n} = p_n\mathbf{e} - \overline{\mathbf{D}}\boldsymbol{\phi}, \tag{6.10b}$$

where $\overline{\mathbf{D}} = \mathbf{D}^T(\mathbf{D}\mathbf{D}^T)^{-1} \in \mathbb{R}^{N\times(N-1)}$. By differentiating (6.10a) and (6.10b) with respect to time and inserting (6.4a) and (6.4b), the individual link velocities are given as

$$\dot{\mathbf{t}} = (v_t + p_n\dot{\theta})\mathbf{e}, \tag{6.11a}$$

$$\dot{\mathbf{n}} = (v_n - p_t\dot{\theta})\mathbf{e} - \overline{\mathbf{D}}\dot{\boldsymbol{\phi}}. \tag{6.11b}$$

6.4 The Ground Friction Model

Following the argumentation in Sect. 2.5.1, we assume that the snake robot is influenced by anisotropic viscous ground friction forces. In order to describe these forces

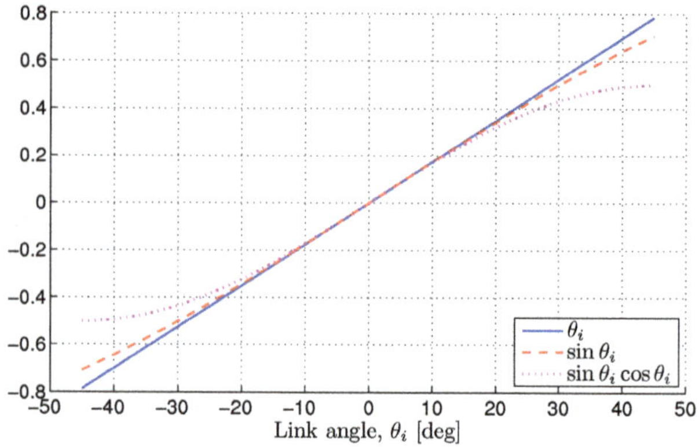

Fig. 6.4 Plots which show that the approximations $\sin\theta_i \approx \theta_i$ and $\sin\theta_i\cos\theta_i \approx \theta_i$ are valid for limited link angles

in the simplified model, we use the viscous ground friction model defined in (2.24) as a starting point. In particular, since the simplified model targets snake locomotion with limited link angles with respect to the forward direction (see Remark 6.1), we approximate the friction model in (2.24) under the assumption that the link angles are small.

Let us begin by considering a snake robot with revolute joints. When the robot is headed along the global positive x axis, the angle of link i with respect to the forward direction is given by θ_i. Moreover, when θ_i is small, the following approximations are valid:

$$\sin^2\theta_i \approx 0, \tag{6.12}$$

$$\cos^2\theta_i \approx 1, \tag{6.13}$$

$$\sin\theta_i\cos\theta_i \approx \theta_i. \tag{6.14}$$

The approximation in (6.14) is illustrated in Fig. 6.4, where we see that $\sin\theta_i\cos\theta_i$ and θ_i are very close for $|\theta_i| < 20°$ and start to deviate as $|\theta_i|$ approaches $30°$. Inserting (6.12), (6.13), and (6.14) into the ground friction model (2.24) gives

$$\begin{bmatrix} f_{R,x,i} \\ f_{R,y,i} \end{bmatrix} = -\begin{bmatrix} c_t & (c_t - c_n)\theta_i \\ (c_t - c_n)\theta_i & c_n \end{bmatrix} \begin{bmatrix} \dot{x}_i \\ \dot{y}_i \end{bmatrix}. \tag{6.15}$$

Since the snake robot is headed along the global x axis, the velocities $\dot{x}_i$ and $\dot{y}_i$ correspond to the velocity of link i in the tangential and normal directions of the snake robot, respectively, while the forces $f_{R,x,i}$ and $f_{R,y,i}$ correspond to the friction force on link i in the tangential and normal directions, respectively. By denoting the friction force components on link i in the t–n frame of the simplified model by $f_{t,i}$ and $f_{n,i}$, respectively, and recalling from Sect. 6.3 that the t–n frame velocity components of link i are given by $\dot{t}_i$ and $\dot{n}_i$, respectively, we therefore have that

$$\dot{x}_i = \dot{t}_i, \qquad \dot{y}_i = \dot{n}_i, \tag{6.16}$$

$$f_{R,x,i} = f_{t,i}, \qquad f_{R,y,i} = f_{n,i}. \tag{6.17}$$

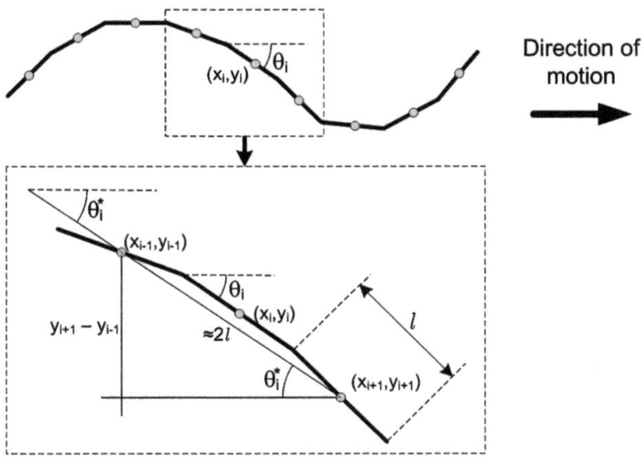

Fig. 6.5 The body angle of link i is θ_i for a snake robot with revolute joints. For a snake robot with prismatic joints, we approximate the body angle of link i based on the relative transversal distance to its two neighbouring links

Before we can use the friction model in (6.15) in the simplified model of the snake robot, it remains to write the link angle θ_i in terms of the simplified joint coordinates defined in (6.5). This is achieved with the approximation illustrated in Fig. 6.5. In particular, we estimate θ_i as the angle of the straight line from the CM of link $i-1$ to the CM of link $i+1$ with respect to the forward direction, which is denoted by θ_i^* in Fig. 6.5. We also see from Fig. 6.5 that the distance from the CM of link $i-1$ to the CM of link $i+1$ can be approximated as $2l$ when the link angles are limited. We can therefore estimate the angle of link i as

$$\theta_i \approx \sin\theta_i \approx \sin\theta_i^* \approx \frac{y_{i+1} - y_{i-1}}{2l}. \tag{6.18}$$

The fact that $\sin\theta_i \approx \theta_i$ for limited link angles is shown in Fig. 6.4. Furthermore, since the joint coordinate ϕ_i of the simplified model corresponds to the normal direction distance between link i and link $i+1$, we have that

$$y_{i+1} - y_{i-1} = \phi_{i-1} + \phi_i. \tag{6.19}$$

By inserting the identities in (6.16), (6.17), (6.18), and (6.19) into (6.15), we can write the viscous ground friction force on link i in the simplified model as

$$\begin{bmatrix} f_{t,i} \\ f_{n,i} \end{bmatrix} = \begin{bmatrix} -c_t & c_p(\phi_{i-1} + \phi_i) \\ c_p(\phi_{i-1} + \phi_i) & -c_n \end{bmatrix} \begin{bmatrix} \dot{t}_i \\ \dot{n}_i \end{bmatrix}_{\dot{\theta}=0}, \tag{6.20}$$

where

$$c_p = \frac{c_n - c_t}{2l}. \tag{6.21}$$

The parameter c_p is a propulsion coefficient (p is short for *propulsion*) which maps the normal direction link velocities and the joint coordinates into propulsive friction forces in the forward (tangential) direction of the snake robot. The subscript $\dot{\theta} = 0$ after the link velocity in (6.20) means that we choose to disregard the link velocity components due to the angular velocity $\dot{\theta}$ of the snake robot (we are now considering the simplified model of the snake robot, where θ denotes the orientation of the robot and not a link angle). Calculating $\dot{t}_i$ and $\dot{n}_i$ under the assumption that $\dot{\theta} = 0$ is a reasonable approximation since the dynamics of the angular rotation of the snake robot will generally be much slower than the body shape dynamics. This assumption also simplifies the ground friction model significantly.

The friction forces in (6.20) on all N links can now be written in matrix form as

$$\begin{bmatrix} \mathbf{f}_t \\ \mathbf{f}_n \end{bmatrix} = \begin{bmatrix} -c_t \mathbf{I}_N & c_p \operatorname{diag}(\mathbf{A}^T \boldsymbol{\phi}) \\ c_p \operatorname{diag}(\mathbf{A}^T \boldsymbol{\phi}) & -c_n \mathbf{I}_N \end{bmatrix} \begin{bmatrix} \dot{\mathbf{t}} \\ \dot{\mathbf{n}} \end{bmatrix}_{\dot{\theta}=0}, \tag{6.22}$$

where $\mathbf{f}_t \in \mathbb{R}^N$ and $\mathbf{f}_n \in \mathbb{R}^N$ contain, respectively, the tangential and normal direction friction forces on the links, $\mathbf{I}_N$ is the $N \times N$ identity matrix, $\mathbf{A}$ is defined in Sect. 6.2, and the operator $\operatorname{diag}(\cdot)$ produces a diagonal matrix with the elements of its argument along its diagonal. Inserting (6.11a) and (6.11b) into (6.22) with $\dot{\theta} = 0$ finally gives

$$\mathbf{f}_t = -c_t v_t \mathbf{e} + c_p \operatorname{diag}\big(\mathbf{A}^T \boldsymbol{\phi}\big)\big(v_n \mathbf{e} - \overline{\mathbf{D}}\dot{\boldsymbol{\phi}}\big), \tag{6.23a}$$

$$\mathbf{f}_n = -c_n v_n \mathbf{e} + c_n \overline{\mathbf{D}}\dot{\boldsymbol{\phi}} + c_p v_t \operatorname{diag}\big(\mathbf{A}^T \boldsymbol{\phi}\big)\mathbf{e}. \tag{6.23b}$$

Note that a discussion related to the validity of the derived ground friction model is given in Sect. 6.7.

6.5 The Dynamics of the Snake Robot

In this section, we derive the model of the translational and rotational accelerations of the snake robot.

6.5.1 The Translational Dynamics of the Snake Robot

We use first principles to describe the translational dynamics. In particular, we can see from Fig. 6.3 that the force balance for link i is given by

$$m\ddot{t}_i = f_{t,i} + h_{t,i} - h_{t,i-1}, \tag{6.24a}$$

$$m\ddot{n}_i = f_{n,i} - u_i + u_{i-1}, \tag{6.24b}$$

where $f_{t,i}$ and $f_{n,i}$ are the ground friction forces defined in (6.20), $h_{t,i}$ and $-h_{t,i-1}$ are the joint constraint forces on link i from link $i+1$ and link $i-1$, respectively, and $-u_i$ and u_{i-1} are the actuator forces exerted on link i from link $i+1$ and link $i-1$, respectively. The joint constraint forces, $h_{t,i}$ and $-h_{t,i-1}$, *prevent* relative motion between the links in the tangential direction, while the actuator forces, $-u_i$

and u_{i-1}, *produce* relative motion between the links in the normal direction. The force balance for all links can be written in matrix form as

$$m\ddot{\mathbf{t}} = \mathbf{f}_t + \mathbf{D}^T \mathbf{h}_t, \tag{6.25a}$$

$$m\ddot{\mathbf{n}} = \mathbf{f}_n - \mathbf{D}^T \mathbf{u}, \tag{6.25b}$$

where $\mathbf{D}$ is defined in Sect. 6.2, $\mathbf{h}_t = [h_{t,1}, \ldots, h_{t,N-1}]^T \in \mathbb{R}^{N-1}$, and $\mathbf{u} = [u_1, \ldots, u_{N-1}]^T \in \mathbb{R}^{N-1}$. Premultiplying (6.25b) by $\frac{1}{m}\mathbf{D}$ gives

$$\mathbf{D}\ddot{\mathbf{n}} = \frac{1}{m}\mathbf{D}\mathbf{f}_n - \frac{1}{m}\mathbf{D}\mathbf{D}^T \mathbf{u}. \tag{6.26}$$

By differentiating (6.7b) twice with respect to time, it is easily seen that $\mathbf{D}\ddot{\mathbf{n}} = -\ddot{\boldsymbol{\phi}}$. We can therefore write the body shape dynamics of the snake robot as

$$\ddot{\boldsymbol{\phi}} = -\frac{1}{m}\mathbf{D}\mathbf{f}_n + \frac{1}{m}\mathbf{D}\mathbf{D}^T \mathbf{u}. \tag{6.27}$$

Inserting (6.23b) into (6.27) and using the easily verifiable relations $\mathbf{D}\mathbf{e} = \mathbf{0}$, $\mathbf{D}\overline{\mathbf{D}} = \mathbf{I}_{N-1}$, and $\mathbf{D}\,\mathrm{diag}(\mathbf{A}^T\boldsymbol{\phi})\mathbf{e} = -\mathbf{A}\mathbf{D}^T\boldsymbol{\phi}$, we get

$$\ddot{\boldsymbol{\phi}} = -\frac{c_n}{m}\dot{\boldsymbol{\phi}} + \frac{c_p}{m}v_t\mathbf{A}\mathbf{D}^T\boldsymbol{\phi} + \frac{1}{m}\mathbf{D}\mathbf{D}^T\mathbf{u}. \tag{6.28}$$

The tangential and normal direction accelerations of the CM of the snake robot, denoted by $\dot{v}_t$ and $\dot{v}_n$, respectively, are given as the sum of all tangential and normal direction forces on the links divided by the mass of the snake robot, Nm. This is written

$$\dot{v}_t = \frac{1}{Nm}(\mathbf{e}^T m\ddot{\mathbf{t}}) = \frac{1}{Nm}\mathbf{e}^T\mathbf{f}_t, \tag{6.29a}$$

$$\dot{v}_n = \frac{1}{Nm}(\mathbf{e}^T m\ddot{\mathbf{n}}) = \frac{1}{Nm}\mathbf{e}^T\mathbf{f}_n, \tag{6.29b}$$

where we note that the joint constraint forces, $\mathbf{h}_t$, and the actuator forces, $\mathbf{u}$, are eliminated when the link accelerations are summed (i.e. $\mathbf{e}^T\mathbf{D}^T = \mathbf{0}$). Inserting (6.23a)–(6.23b) into (6.29a)–(6.29b) and using the easily verifiable relations $\mathbf{e}^T\,\mathrm{diag}(\mathbf{A}^T\boldsymbol{\phi})\mathbf{e} = 2\bar{\mathbf{e}}^T\boldsymbol{\phi}$, $\mathbf{e}^T\overline{\mathbf{D}} = \mathbf{0}$, and $\mathbf{e}^T\,\mathrm{diag}(\mathbf{A}^T\boldsymbol{\phi})\overline{\mathbf{D}} = \boldsymbol{\phi}^T\mathbf{A}\overline{\mathbf{D}}$, we get

$$\dot{v}_t = -\frac{c_t}{m}v_t + \frac{2c_p}{Nm}v_n\bar{\mathbf{e}}^T\boldsymbol{\phi} - \frac{c_p}{Nm}\boldsymbol{\phi}^T\mathbf{A}\overline{\mathbf{D}}\dot{\boldsymbol{\phi}}, \tag{6.30a}$$

$$\dot{v}_n = -\frac{c_n}{m}v_n + \frac{2c_p}{Nm}v_t\bar{\mathbf{e}}^T\boldsymbol{\phi}. \tag{6.30b}$$

6.5.2 The Rotational Dynamics of the Snake Robot

The translational motion of a snake robot with revolute joints is produced by the translational displacements of the links (see Property 4.2). The essence of the approach underlying the simplified model of the snake robot is therefore to disregard the rotational link motion and instead only consider the translational displacements of the links. The *rotational* motion of a snake robot, on the other hand, is determined by the orientation of the links. A drawback of disregarding the rotational link motion is therefore that we remove the primary cause of the rotational motion of the robot. For this reason, a pure first principles perspective on the rotational dynamics of a snake robot with *translational* joints (as considered in the simplified model) will *not* produce a model which resembles the rotational dynamics of a snake robot with *revolute* joints.

Instead of using first principles, we therefore choose to develop a simplified model of the rotational dynamics of the snake robot based on our *qualitative* understanding of how the rotational motion is produced. In particular, we will base the model on Property 4.7, which was derived in Sect. 4.9. This property states that the direction of the forward motion changes when the average of the joint angles is non-zero and that the rate of directional change is increased by increasing the average of the joint angles and/or by increasing the forward velocity (assuming that the average of the joint angles is non-zero). In the simplified model, the direction of the forward motion is given by the orientation θ, the forward velocity is given by v_t, and the average of the joint angles corresponds to the average of the joint coordinates, $\overline{\mathbf{e}}^T \boldsymbol{\phi}/(N-1)$. Property 4.7 therefore suggests that we can describe the overall torque that induces the rotational motion of a snake robot as

$$\ddot{\theta}_{\text{rotation}} = \lambda_2 v_t \frac{\overline{\mathbf{e}}^T \boldsymbol{\phi}}{N-1}, \tag{6.31}$$

where λ_2 is some constant parameter which determines the scaling of the mapping from average joint coordinate and forward velocity to rotational acceleration. Furthermore, there must necessarily act ground friction forces on the snake robot which induce a ground friction torque that opposes the rotational motion. Since the ground friction forces are of viscous type, we also assume that the rotational friction torque is viscous. We choose to model this viscous friction torque as

$$\ddot{\theta}_{\text{resistance}} = -\lambda_1 \dot{\theta}, \tag{6.32}$$

where λ_1 is some constant parameter which determines the scaling of the mapping from rotational velocity to rotational acceleration. By combining (6.31) and (6.32), we can write the simplified model of the rotational dynamics of the snake robot as

$$\ddot{\theta} = -\lambda_1 \dot{\theta} + \frac{\lambda_2}{N-1} v_t \overline{\mathbf{e}}^T \boldsymbol{\phi}. \tag{6.33}$$

Although the model of $\ddot{\theta}$ is not based on first principles, we conjecture that the behaviour of this model will be *qualitatively* similar to the behaviour of a snake robot with revolute joints, and also *quantitatively* similar when the rotation parameters λ_1 and λ_2 are properly chosen. This claim is supported by the stabilisability and controllability analysis of the simplified model presented in Sects. 6.8 and 6.9, respectively, and by the simulation results presented in Sect. 6.10.

Remark 6.5 Note that the model in (6.33) is very general and does not specify how to set the parameters λ_1 and λ_2 as functions of the remaining parameters of the snake robot. We conjecture that λ_1 is influenced by the friction coefficient c_n since the forces that induce the rotational friction torque are likely to act normally to the snake robot. Moreover, since the induced rotation of the robot is tightly coupled with the induced forward propulsion, we conjecture that λ_2 is influenced by the propulsion coefficient c_p.

6.6 The Complete Simplified Model of the Snake Robot

We now summarise the complete simplified model of the snake robot. Since the robot has $N+2$ degrees of freedom, a state vector containing the generalised coordinates and velocities of the robot will have dimension $2N+4$. We choose the state vector of the system as

$$\mathbf{x} = \left[\boldsymbol{\phi}^T, \theta, p_x, p_y, \mathbf{v}_\phi^T, v_\theta, v_t, v_n\right]^T \in \mathbb{R}^{2N+4}, \tag{6.34}$$

where $\boldsymbol{\phi} \in \mathbb{R}^{N-1}$ are the joint coordinates, $\theta \in \mathbb{R}$ is the absolute orientation, $(p_x, p_y) \in \mathbb{R}^2$ is the global frame position of the CM, $\mathbf{v}_\phi = \dot{\boldsymbol{\phi}} \in \mathbb{R}^{N-1}$ are the joint velocities, $v_\theta = \dot{\theta} \in \mathbb{R}$ is the angular velocity, and $(v_t, v_n) \in \mathbb{R}^2$ are the tangential and normal direction velocities of the snake robot. From (6.2a)–(6.2b), (6.28), (6.30a)–(6.30b), and (6.33), we can write the complete model of the snake robot as

$$\dot{\boldsymbol{\phi}} = \mathbf{v}_\phi, \tag{6.35a}$$

$$\dot{\theta} = v_\theta, \tag{6.35b}$$

$$\dot{p}_x = v_t \cos\theta - v_n \sin\theta, \tag{6.35c}$$

$$\dot{p}_y = v_t \sin\theta + v_n \cos\theta, \tag{6.35d}$$

$$\dot{\mathbf{v}}_\phi = -\frac{c_n}{m}\mathbf{v}_\phi + \frac{c_p}{m} v_t \mathbf{A}\mathbf{D}^T \boldsymbol{\phi} + \frac{1}{m}\mathbf{D}\mathbf{D}^T \mathbf{u}, \tag{6.35e}$$

$$\dot{v}_\theta = -\lambda_1 v_\theta + \frac{\lambda_2}{N-1} v_t \bar{\mathbf{e}}^T \boldsymbol{\phi}, \tag{6.35f}$$

$$\dot{v}_t = -\frac{c_t}{m} v_t + \frac{2c_p}{Nm} v_n \bar{\mathbf{e}}^T \boldsymbol{\phi} - \frac{c_p}{Nm} \boldsymbol{\phi}^T \mathbf{A}\bar{\mathbf{D}} \mathbf{v}_\phi, \tag{6.35g}$$

$$\dot{v}_n = -\frac{c_n}{m} v_n + \frac{2c_p}{Nm} v_t \bar{\mathbf{e}}^T \boldsymbol{\phi}, \tag{6.35h}$$

where $\mathbf{u} \in \mathbb{R}^{N-1}$ are the actuator forces at the joints, $\mathbf{A}$, $\mathbf{D}$, $\overline{\mathbf{D}}$, and $\overline{\mathbf{e}}$ are defined in Sect. 6.2, c_t and c_n correspond, respectively, to the tangential and normal direction friction coefficients of the links in the complex model of the snake robot, c_p is the propulsion coefficient defined in (6.21), and λ_1 and λ_2 are positive scalar constants which characterise the rotational motion of the snake robot.

Similar to the partial feedback linearisation performed for the complex model in Sect. 2.8, we will usually assume that the actuator forces of the simplified model are set according to the linearising control law

$$\mathbf{u} = m\big(\mathbf{D}\mathbf{D}^T\big)^{-1}\left(\overline{\mathbf{u}} + \frac{c_n}{m}\dot{\boldsymbol{\phi}} - \frac{c_p}{m}v_t\mathbf{A}\mathbf{D}^T\boldsymbol{\phi}\right), \tag{6.36}$$

where $\overline{\mathbf{u}} = [\overline{u}_1, \ldots, \overline{u}_{N-1}]^T \in \mathbb{R}^{N-1}$ is a new set of control inputs. This control law transforms the joint dynamics (6.35e) into

$$\dot{\boldsymbol{v}}_\phi = \overline{\mathbf{u}}. \tag{6.37}$$

6.7 Discussion of the Simplified Model

In this section, we provided a discussion of limitations and accuracy issues related to the simplified model of the snake robot.

6.7.1 Applications of the Simplified Model

The simplified model in (6.35a)–(6.35h) is *not* intended as an accurate simulation model of snake robot locomotion. The model is intentionally based on several simplifying assumptions in order to arrive at equations of motion which are manageable for control design and stability analysis purposes. To this end, the model only needs to be *qualitatively* similar to the complex model presented in (2.47). This *qualitative* similarity is supported by the stabilisability and controllability analysis of the simplified model presented in Sects. 6.8 and 6.9, respectively. There is also a *quantitative* similarity between the complex and the simplified model under conditions which are discussed in the following.

6.7.2 Accuracy Issues of the Simplified Kinematics

The essence of the simplified modelling approach is to describe the body shape changes of a snake robot as linear link displacements normal to the forward direction of motion (i.e. to disregard the rotational link motion). This approach is only valid for gait patterns where the relative link displacements *normal* to the direction of motion dominate over the relative link displacements *tangential* to the direction of motion. Since the relative tangential link displacements increase with increasing link angles, the simplified model should only be used to study gait patterns with limited link angles with respect to the forward direction.

There is also an extension issue inherent in the simplified kinematics of the snake robot. In particular, as seen in (6.6a), the simplified model assumes that the relative distance between two links in the tangential direction of the snake robot is constant. This assumption causes the total length of the snake robot to increase when the relative normal direction distances between the links are increased. However, we conjecture that the effect of this extension issue will be minimal as long as the link angles are limited.

6.7.3 Accuracy Issues of the Ground Friction Model

The assumptions made during the development of the ground friction model in (6.20) are all valid for limited link angles. Unfortunately, it is difficult to derive a precise bound for the range of link angles where the friction model is valid. However, Fig. 6.4 suggests that (6.20) relies on approximations which are valid for link angles limited by approximately $|\theta_i| < 20°$. It therefore seems reasonable to expect the ground friction model in (6.20) to be *quantitatively* similar to the friction model in (2.24) for link angles satisfying this approximate bound.

Although the *quantitative* discrepancy between the models in (6.20) and (2.24) increases with increasing link angles, the properties derived in Sect. 4.5 regarding the qualitative behaviour of a snake robot allow us to argue that the friction model in (6.20) is *qualitatively* similar to the friction model in (2.24) also for large link angles. In particular, we can easily see from (6.20) that the propulsive ground friction forces on link i that propel the robot forward are produced by the normal direction link velocity, $\dot{n}_i$. This is in direct accordance with Property 4.2. Furthermore, we see from (6.20) that the magnitude of the propulsive ground friction forces produced by link i is increased by increasing $|\phi_{i-1} + \phi_i|$, which, from (6.18) and (6.19), corresponds to increasing $|\theta_i|$. This is in direct accordance with Property 4.4. Finally, we can see from (6.20) that the forward direction friction force component produced by $\dot{n}_i$ is positive when $\text{sgn}(\phi_{i-1} + \phi_i) = \text{sgn}(\dot{n}_i)$ and negative otherwise, which is in direct accordance with Property 4.3. In summary, since the ground friction model in (6.20) directly captures the qualitative properties derived in Sect. 4.5, we argue that the friction model in (6.20) is *qualitatively* similar to the friction model in (2.24) also for large link angles.

Note that the *quantitative* discrepancy between the models in (6.20) and (2.24) will also increase when the ratio between the friction coefficients c_n and c_t is increased. In particular, the ground friction model in (6.20) is based on the approximation $c_t \cos^2\theta_i + c_n \sin^2\theta_i \approx c_t$, which is valid for small link angles. However, this approximation becomes less valid when the ratio c_n/c_t is increased.

6.7.4 Accuracy Issues of the Rotational Dynamics

As explained in Sect. 6.5.2, the rotational dynamics of the snake robot in (6.33) was not derived from first principles. Instead, the model was developed to directly capture the *qualitative* behaviour of a snake robot during rotational motion. The rotation model in (6.33) is very general and does not specify how to set the parameters λ_1 and λ_2 as functions of the remaining parameters of the snake robot. However, we conjecture that the behaviour of this model will be *qualitatively* similar to the behaviour of a snake robot with revolute joints, and also *quantitatively* similar when the parameters λ_1 and λ_2 are properly chosen.

6.8 Stabilisability Analysis of the Simplified Model

In this section, we show that the simplified model maintains the stabilisability properties of snake robot locomotion that were derived in Sect. 4.3. In particular, we investigate the properties of an asymptotically stabilising control law for the simplified model to any equilibrium point $\mathbf{x}^e = [(\boldsymbol{\phi}^e)^T, \theta^e, p_x^e, p_y^e, \mathbf{v}_\phi^T = \mathbf{0}, v_\theta = 0, v_t = 0, v_n = 0]^T$. As explained in Sect. 4.3, a well-known result presented in Brockett (1983) states that a necessary condition for the existence of a *time-invariant* (i.e. not explicitly dependent on time) *continuous* state feedback law, $\mathbf{u} = \mathbf{u}(\mathbf{x})$, that makes $\mathbf{x}^e$ asymptotically stable, is that the image of the mapping $(\mathbf{x}, \mathbf{u}) \mapsto \dot{\mathbf{x}}$ contains some neighbourhood of $\dot{\mathbf{x}} = \mathbf{0}$. A result presented in Coron and Rosier (1994) states that a control system that can be asymptotically stabilised (in the Filippov sense) by a *time-invariant discontinuous* state feedback law can be asymptotically stabilised by a *time-varying continuous* state feedback law. If, moreover, the control system is *affine* (i.e. linear with respect to the control input), then it can be asymptotically stabilised by a *time-invariant continuous* state feedback law. We now employ these results to prove the following fundamental result:

Theorem 6.1 *An asymptotically stabilising feedback control law for a planar snake robot described by* (6.35a)–(6.35h) *to any equilibrium point must be time-varying, i.e. of the form* $\mathbf{u} = \mathbf{u}(\mathbf{x}, t)$.

Proof The result in Brockett (1983) states that the mapping $(\mathbf{x}, \mathbf{u}) \mapsto \dot{\mathbf{x}}$ must map an arbitrary neighbourhood of $\mathbf{x}^e$ onto a neighbourhood of $\dot{\mathbf{x}} = \mathbf{0}$. For this to be true, points of the form

$$\dot{\mathbf{x}} = \left[\dot{\boldsymbol{\phi}}^T = \mathbf{0}, \dot{\theta} = 0, \dot{p}_x = 0, \dot{p}_y = 0, \dot{\mathbf{v}}_\phi^T = \mathbf{0}, \dot{v}_\theta = 0, \dot{v}_t = \epsilon \neq 0, \dot{v}_n = 0\right]^T \tag{6.38}$$

must be contained in this mapping for some arbitrary $\epsilon \neq 0$ because points of this form are contained in every neighbourhood of $\dot{\mathbf{x}} = \mathbf{0}$. However, these points do not exist for the model (6.35a)–(6.35h) because $\dot{v}_t = 0 \neq \epsilon$ when all the other derivatives of the state vector are zero. Hence, the snake robot *cannot* be asymptotically stabilised to $\mathbf{x}^e$ by a *time-invariant continuous* state feedback law. Moreover, since the model is affine and *cannot* be asymptotically stabilised by a *time-invariant continuous* state feedback law, the result in Coron and Rosier (1994) proves that the system can neither be asymptotically stabilised by a *time-invariant discontinuous* state feedback law. We can therefore conclude that an asymptotically stabilising control law for the snake robot to any equilibrium point must be time-varying, i.e. of the form $\mathbf{u} = \mathbf{u}(\mathbf{x}, t)$. □

The stabilisability properties of the simplified model stated in Theorem 6.1 are similar to the stabilisability properties of the complex model stated in Theorem 4.3, which implies that there are fundamental properties of the simplified and the complex model which are similar.

6.9 Controllability Analysis of the Simplified Model

In this section, we show that the simplified model maintains the controllability properties of snake robot locomotion that were derived in Sect. 4.4.2. We assume that the joint dynamics has been linearised by the control law (6.36) so that $\dot{\mathbf{v}}_\phi = \overline{\mathbf{u}}$. This enables us to rewrite the model of the snake robot (6.35a)–(6.35h) in the standard form of a control affine system as

$$\dot{\mathbf{x}} = \mathbf{f}(\mathbf{x}) + \sum_{j=1}^{N-1} \mathbf{g}_j \overline{\mathbf{u}}_j, \tag{6.39}$$

where $\mathbf{f}(\mathbf{x})$ contains all the terms from (6.35a)–(6.35h) with $\overline{\mathbf{u}} = \mathbf{0}_{(N-1)\times 1}$, $\overline{\mathbf{u}}_j$ is the jth element of the control input vector $\overline{\mathbf{u}} \in \mathbb{R}^{N-1}$, and

$$\mathbf{g}_j = \begin{bmatrix} \mathbf{0}_{(N+2)\times 1} \\ \mathbf{e}_j \\ \mathbf{0}_{3\times 1} \end{bmatrix}, \tag{6.40}$$

where $\mathbf{e}_j$ denotes the jth standard basis vector in $\mathbb{R}^{N-1}$ (the jth column of $\mathbf{I}_{N-1}$). It can easily be shown that the linearisation of the model (6.39) about an equilibrium point $\mathbf{x}^e$ is not controllable since the *Kalman rank condition* is not satisfied (see Sect. 4.2). To study the controllability of the model in (6.39), we must therefore consider nonlinear controllability concepts.

In the following, we investigate the controllability of the simplified model (6.39) in terms of *strong accessibility* and *small-time local controllability* (STLC) (see Sect. 4.2) by following the same approach as for the complex model in Sect. 4.4.2. We assume that the snake robot consists of $N = 4$ links interconnected by $N - 1 = 3$ joints. The model of this robot has $2N + 4 = 12$ states. We argue that the following controllability results will also be valid for a snake robot with more links. In particular, a snake robot with $N > 4$ links can behave as a snake robot with

$N = 4$ links by fixing $(N-4)$ joint coordinates at zero and allowing the remaining joints to move. By calculating Lie brackets of the system vector fields in (6.39), we can construct the following *accessibility algebra* of the system evaluated at an equilibrium point $\mathbf{x}^e$:

$$\boldsymbol{\Delta}(\mathbf{x}^e) = [\boldsymbol{\Delta}_1, \ldots, \boldsymbol{\Delta}_{15}]_{\mathbf{x}^e} \in \mathbb{R}^{12\times 15}, \tag{6.41}$$

where

$$\begin{aligned}
&\boldsymbol{\Delta}_1 = \mathbf{g}_1, \qquad \boldsymbol{\Delta}_2 = \mathbf{g}_2, \qquad \boldsymbol{\Delta}_3 = \mathbf{g}_3, \\
&\boldsymbol{\Delta}_4 = [\mathbf{f}, \mathbf{g}_1], \qquad \boldsymbol{\Delta}_5 = [\mathbf{f}, \mathbf{g}_2], \qquad \boldsymbol{\Delta}_6 = [\mathbf{f}, \mathbf{g}_3], \\
&\boldsymbol{\Delta}_7 = \big[\mathbf{f}, [\mathbf{f}, \mathbf{g}_1]\big], \qquad \boldsymbol{\Delta}_8 = \big[\mathbf{f}, \big[\mathbf{f}, [\mathbf{f}, \mathbf{g}_1]\big]\big], \\
&\boldsymbol{\Delta}_9 = \big[\mathbf{f}, \big[\mathbf{f}, \big[\mathbf{f}, [\mathbf{f}, \mathbf{g}_1]\big]\big]\big], \\
&\boldsymbol{\Delta}_{10} = \big[\mathbf{g}_1, \big[\mathbf{f}, [\mathbf{f}, \mathbf{g}_2]\big]\big], \\
&\boldsymbol{\Delta}_{11} = \big[\mathbf{g}_1, \big[\mathbf{f}, \big[\mathbf{f}, [\mathbf{f}, \mathbf{g}_2]\big]\big]\big], \\
&\boldsymbol{\Delta}_{12} = \big[\mathbf{g}_1, \big[\mathbf{f}, \big[\mathbf{f}, \big[\mathbf{f}, [\mathbf{f}, \mathbf{g}_2]\big]\big]\big]\big], \\
&\boldsymbol{\Delta}_{13} = \big[\mathbf{g}_1, \big[\mathbf{f}, \big[\mathbf{f}, \big[\mathbf{f}, \big[\mathbf{f}, [\mathbf{f}, \mathbf{g}_2]\big]\big]\big]\big]\big], \\
&\boldsymbol{\Delta}_{14} = \big[\mathbf{g}_1, \big[\mathbf{f}, \big[\mathbf{f}, \big[\mathbf{f}, \big[\mathbf{f}, [\mathbf{f}, \mathbf{g}_3]\big]\big]\big]\big]\big], \\
&\boldsymbol{\Delta}_{15} = \big[\mathbf{g}_2, \big[\mathbf{f}, \big[\mathbf{f}, \big[\mathbf{f}, \big[\mathbf{f}, [\mathbf{f}, \mathbf{g}_3]\big]\big]\big]\big]\big].
\end{aligned}$$

The accessibility algebra satisfies the following property:

Property 6.1 The accessibility algebra $\boldsymbol{\Delta}(\mathbf{x}^e)$ in (6.41) has full rank ($\text{rank}(\boldsymbol{\Delta}(\mathbf{x}^e)) = 12$) as long as the sum of the joint coordinates is non-zero, i.e. as long as $\overline{\mathbf{e}}^T\boldsymbol{\phi} \neq 0$.

We do not present the expressions contained in each column of $\boldsymbol{\Delta}(\mathbf{x}^e)$ since the expressions are rather excessive. However, Property 6.1 can be shown to hold by employing a computer software for symbolic mathematics, such as *Matlab Symbolic Toolbox*. Note that we have included three more columns than rows in $\boldsymbol{\Delta}(\mathbf{x}^e)$ because different pairs of columns become linearly independent at certain configurations. Including three redundant columns ensures that $\boldsymbol{\Delta}(\mathbf{x}^e)$ does not drop rank at these configurations. We are now ready to state the following result:

Theorem 6.2 *A planar snake robot described by* (6.35a)–(6.35h) *with* $N = 4$ *links is* locally strongly accessible *from any equilibrium point* $\mathbf{x}^e$ *satisfying* $\overline{\mathbf{e}}^T\boldsymbol{\phi} \neq 0$.

Proof By Theorem 4.1, the system is locally strongly accessible from $\mathbf{x}^e$ if $\boldsymbol{\Delta}(\mathbf{x}^e)$ in (6.41) has full rank (i.e. spans a 12-dimensional space) and does not contain the drift vector field $\mathbf{f}$ by itself (i.e. unbracketed). By Property 6.1, the snake robot satisfies these conditions as long as $\overline{\mathbf{e}}^T\boldsymbol{\phi} \neq 0$. This completes the proof. □

We now show that the snake robot does *not* satisfy sufficient conditions for small-time local controllability (STLC). As described in Sect. 4.2, STLC requires that we classify the Lie brackets of the system vector fields in terms of *good* and *bad* brackets. A Lie bracket is said to be *bad* if it contains the drift vector field $\mathbf{f}$ an odd number of times and each control vector field $\mathbf{g}_j$ an even number of times (0 is even). This classification is motivated by the fact that a bad bracket *may* have directional constraints. For example, the drift vector $\mathbf{f}$ is *bad* because it only allows motion in its positive direction. According to Theorem 4.2, the snake robot is STLC from an equilibrium point $\mathbf{x}^e$ if it is accessible from $\mathbf{x}^e$ and all *bad* brackets of the system can be neutralised, i.e. written as linear combinations of *good* brackets of lower θ-*degree* or lower l-*degree*. The model of the snake robot satisfies the following property:

Property 6.2 The brackets $\mathbf{g}_j, [\mathbf{f}, \mathbf{g}_j], [\mathbf{g}_j, [\mathbf{f}, \mathbf{g}_k]], [\mathbf{g}_j, [\mathbf{f}, [\mathbf{f}, \mathbf{g}_k]]], [[\mathbf{f}, \mathbf{g}_j], [\mathbf{f}, \mathbf{g}_k]]$, $[\mathbf{f}, [\mathbf{f}, \mathbf{g}_j]], [\mathbf{f}, [\mathbf{f}, [\mathbf{f}, \mathbf{g}_j]]], \ldots, [\mathbf{f}, [\cdots [\mathbf{f}, \mathbf{g}_j]] \cdots]$, where $j, k \in \{1, 2, 3\}$ and $j \neq k$, are all *good* brackets, but do *not* span the entire 12-dimensional state space.

We do not present the expressions contained in the brackets in Property 6.2 due to their excessive nature. However, the property can be shown to hold by employing a computer software for symbolic mathematics, such as *Matlab Symbolic Toolbox*. Property 6.2 enables us to state the following result:

Theorem 6.3 *At any equilibrium point* $\mathbf{x}^e$, *a planar snake robot described by* (6.35a)–(6.35h) *with* $N = 4$ *links does* not *satisfy the sufficient conditions for* small-time local controllability (*STLC*) *stated in Theorem* 4.2.

Proof The bracket $[\mathbf{g}_j, [\mathbf{f}, [\mathbf{f}, [\mathbf{f}, \mathbf{g}_j]]]]$ of the system, where $j \in \{1, 2, 3\}$, is a *bad* bracket. The only *good* brackets of lower θ-degree or lower l-degree that can neutralise this bad bracket are of the form $\mathbf{g}_j, [\mathbf{f}, \mathbf{g}_j], [\mathbf{g}_j, [\mathbf{f}, \mathbf{g}_k]], [\mathbf{g}_j, [\mathbf{f}, [\mathbf{f}, \mathbf{g}_k]]]$, $[[\mathbf{f}, \mathbf{g}_j], [\mathbf{f}, \mathbf{g}_k]], [\mathbf{f}, [\mathbf{f}, \mathbf{g}_j]], [\mathbf{f}, [\mathbf{f}, [\mathbf{f}, \mathbf{g}_j]]], \ldots, [\mathbf{f}, [\cdots [\mathbf{f}, \mathbf{g}_j]] \cdots]$, where $j, k \in \{1, 2, 3\}$ and $j \neq k$. By Property 6.2, these brackets do *not* span the entire 12-dimensional state space. We therefore cannot express the bad bracket as a linear combination of good brackets of lower θ-degree or lower l-degree. Since there are bad brackets of the system that cannot be neutralised, the system does not satisfy the conditions for STLC stated in Theorem 4.2. □

The result in Theorem 6.3 is similar to the result concerning STLC of the complex model in Theorem 4.6. Regarding local accessibility of the complex model, Theorem 4.5 and Property 4.1 in Sect. 4.4.2 state that the accessibility algebra of the complex model has full rank except for configurations where all joint coordinates are equal ($\phi_1 = \phi_2 = \cdots = \phi_{N-1}$), which will be the case when the snake robot is lying straight or forming an arc. The condition $\bar{\mathbf{e}}^T \boldsymbol{\phi} \neq 0$ stated in Theorem 6.2 states that a configuration is singular when the sum of the relative linear link displacements is zero. Since the sum of the relative linear link displacements is zero for both straight and arc shaped snake robots with revolute joints, the singular configurations of the complex model revealed by Property 4.1 are actually contained in the singular configurations stated in Theorem 6.2. This similarity supports the conjecture that the simplified model captures the essential part of the dynamics of planar snake locomotion. Note that there are singular configurations of the simplified model that do not easily translate to the complex model. A wave shape where the sum of the relative linear link displacements is zero is a singular configuration in the simplified model but is not singular to a snake robot with revolute joints. These additional singular configurations of the simplified model arise since the sum of the joint coordinates is employed to model the rotation of the robot in accordance with Property 4.7 in Sect. 4.9. Nonetheless, the most important conclusion to be drawn from Theorem 6.2 is that the snake robot is locally strongly accessible from *almost* any equilibrium point, except for certain singular configurations. This conclusion is in accordance with Theorem 4.5.

6.10 Simulation Study: Comparison Between the Complex and the Simplified Model

This section presents simulation results in order to compare the complex snake robot model given by (2.47) with the simplified model given by (6.35a)–(6.35h).

Remark 6.6 As noted in Remark 6.5 in Sect. 6.5.2, we have not specified how to set the rotation parameters λ_1 and λ_2 as functions of the remaining parameters of the snake robot. However, knowing this mapping is not critical to the intended use of the simplified model for controller design purposes as long as we know that the *qualitative* and the approximate *quantitative* behaviour of the complex model is *contained* within the simplified model for some choice of numerical values of the parameters λ_1 and λ_2. The purpose of this section is to illustrate this qualitative and quantitative similarity between the two models.

6.10.1 Simulation Parameters

Both models were implemented and simulated in *Matlab R2008b* on a laptop running *Windows XP*. The dynamics was calculated using the *ode45* solver in Matlab with a relative and absolute error tolerance of 10^{-6}.

We considered a snake robot with $N = 10$ links of length $l = 0.14$ m and mass $m = 1$ kg. The links of the snake robot with revolute joints had moment of inertia $J = 0.0016$ kgm^2. The ground friction coefficients were $c_t = 1$ and $c_n = 3$, and the rotation parameters of the simplified model were $\lambda_1 = 0.5$ and $\lambda_2 = 20$. Both models were simulated with the joint controller given by (4.34) with controller gains $k_p = 20$ and $k_d = 5$. In order to compare the two models, we controlled the snake robot in both models according to the gait pattern lateral undulation defined in (4.32). The angular frequency and phase shift of the gait pattern were $\omega = 120°$/s and $\delta = 40°$, respectively, while the amplitude α and the joint offset ϕ_o are presented with each simulation result below.

6.10.2 Relationship Between the Joint Coordinates in the Complex and Simplified Models

In order to illustrate the range of link angles where the simplified model agrees with the complex model, we carried out simulations of lateral undulation where the amplitude of the link angles was $\theta_i = 10°$, $15°$, $20°$, and $30°$, respectively, for $i \in \{1, \ldots, N\}$. For the kinematic parameters and gait parameters listed in Sect. 6.10.1, these link angle amplitudes are achieved in the complex model by choosing the amplitude α in (4.32) as $\alpha = 7.1°$, $10.5°$, $13.9°$, and $21°$, respectively. Furthermore, these link angle amplitudes will cause the amplitudes of the normal direction distance between the links (i.e. the amplitude of the joint coordinates in the simplified model) to be $y_{i+1} - y_i = 2.3$ cm, 3.4 cm, 4.5 cm, and 6.6 cm, respectively. These relationships are shown in Fig. 6.6 and were found through trial and error by simulating the complex model of the snake robot with the parameters listed in Sect. 6.10.1. In particular, the figure shows a simulation of the complex model where the snake robot first conducts lateral undulation with $\alpha = 7.1°$ for 30 s, then with $\alpha = 10.5°$ for 30 s, then with $\alpha = 13.9°$ for 30 s, and finally with $\alpha = 21°$ for 30 s. Figure 6.6(a) shows the maximum value of all the N link angles at each timestep, while Fig. 6.6(b) shows the maximum value of the normal direction distance between the links at each timestep. The link angle amplitude and the normal direction link distance corresponding to each value of α were estimated as the average of the plotted values inside each 30-s time interval.

In summary, Fig. 6.6 shows that lateral undulation with the kinematic parameters and gait parameters listed in Sect. 6.10.1, where the amplitudes of the link angles are $\theta_i = 10°$, $15°$, $20°$, and $30°$, respectively, is achieved in the complex model by choosing the amplitude α in (4.32) as $\alpha = 7.1°$, $10.5°$, $13.9°$, and $21°$, respectively, and is achieved in the simplified model by choosing the amplitude α in (4.32) as $\alpha = 2.3$ cm, 3.4 cm, 4.5 cm, and 6.6 cm, respectively.

6.10.3 Comparison of Straight Motion

We begin by comparing the complex and simplified models during lateral undulation along a straight line. The amplitude α in (4.32) was set in intervals of 45 s according to the values derived for both models in the previous subsection, i.e. $\alpha = 7.1°$, $10.5°$, $13.9°$, and $21°$, respectively, for the complex model and $\alpha = 2.3$ cm, 3.4 cm, 4.5 cm, and 6.6 cm, respectively, for the simplified model. The joint offset ϕ_o in (4.32) was set to zero in both models. The snake robot in the complex model started moving from $(p_x = 0, p_y = -0.5)$, while the snake robot in the simplified model started moving from $(p_x = 0, p_y = 0.5)$. Both snake robots were initially headed along the global x axis with zero joint coordinates.

The simulation result is shown in Fig. 6.7. In particular, the motion of the CM of the two robots is plotted in Fig. 6.7(a), where the configuration of the snake robots is visualised in the middle of each 45-s time interval. Furthermore, Figs. 6.7(b)

Fig. 6.6 A simulation of the complex model where the snake robot conducts lateral undulation at four different joint angle amplitudes. The plots show the corresponding amplitudes of the link angles (*top*) and the amplitudes of the normal direction distances between the links (*bottom*)

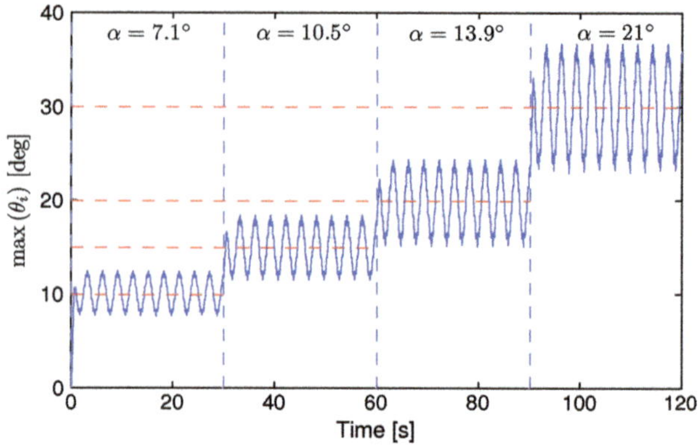

(a) The link angle amplitudes $\theta_i = 10°$, $15°$, $20°$, and $30°$ are achieved by choosing the joint angle amplitudes as $\alpha = 7.1°$, $10.5°$, $13.9°$, and $21°$, respectively.

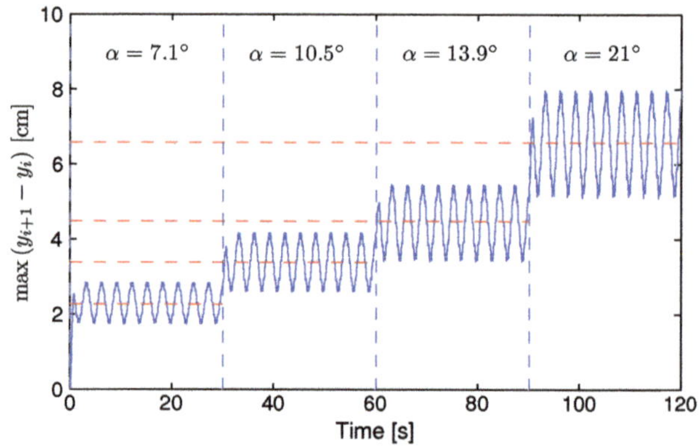

(b) The estimated amplitudes of the normal direction distance between the links at each of the four joint angle amplitudes are, respectively, $y_{i+1} - y_i = 2.3$ cm, 3.4 cm, 4.5 cm, and 6.6 cm.

and (c) show the CM velocities of the snake robots in the global x and y directions, respectively.

The simulation results indicate that the forward motion dynamics of the simplified model agrees well with the complex model as long as the link angles are limited. In particular, there is a good agreement between the two models in the two first time intervals, where the link angle amplitude is 10° and 15°, respectively. In the third time interval, where the link angle amplitude is increased to 20°, the forward velocity in the two models begin to deviate. This deviation is increased in the fourth and last time intervals, where the link angle amplitude is increased to 30°. These results agree very well with the predictions that were made in conjunction with the derivation of the simplified model. In particular, Fig. 6.4 suggests that the ground friction forces in the simplified model rely on approximations which are valid for

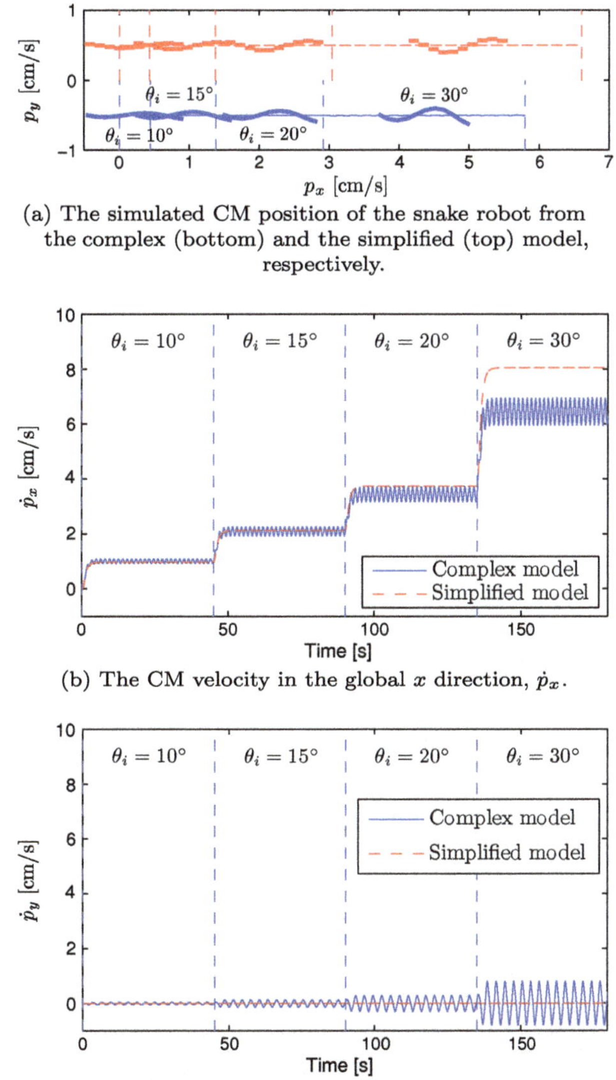

Fig. 6.7 Simulation result which compares the complex and simplified models during lateral undulation along a straight line at four different link angle amplitudes

(a) The simulated CM position of the snake robot from the complex (bottom) and the simplified (top) model, respectively.

(b) The CM velocity in the global x direction, $\dot{p}_x$.

(c) The CM velocity in the global y direction, $\dot{p}_y$.

link angles limited by approximately $|\theta_i| < 20°$. The simulation result in Fig. 6.7 supports this conjecture.

Notice that the velocity plots from the complex model contain high-frequency components which are not visible in the plots from the simplified model. This indicates that the complex model contains nonlinear components which are not included in the simplified model. However, the similar behaviour of the two models at limited link angles indicates that the simplified model contains the parts of the complex model that determine the overall motion of the snake robot. This suggests that we

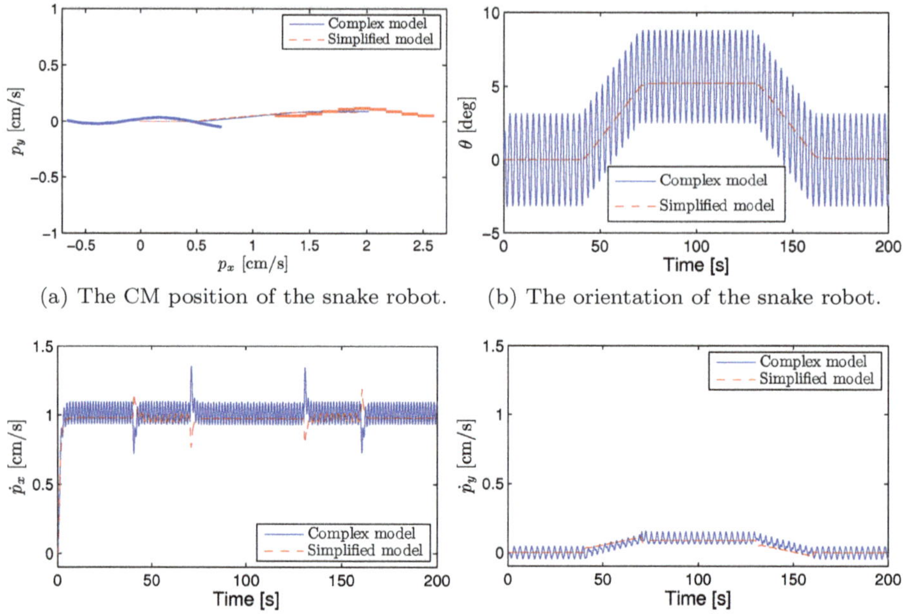

(a) The CM position of the snake robot.

(b) The orientation of the snake robot.

(c) The CM velocity in the global x direction of the snake robot, $\dot{p}_x$.

(d) The CM velocity in the global y direction of the snake robot, $\dot{p}_y$.

Fig. 6.8 Simulation result which compares the complex and simplified models during turning motion. The amplitude of the link angles is $\theta_i = 10°$

may use the simplified model to develop general analysis and control design results that will also apply to the complex model.

6.10.4 Comparison of Turning Motion

In the next simulation, we compare the complex and the simplified model during lateral undulation involving turning motion. In order to induce turning motion, the joint offset ϕ_o in (4.32) was set to $\phi_o = \frac{1}{3}\alpha$ in the time interval $t \in [40, 70]$ and $\phi_o = -\frac{1}{3}\alpha$ in the time interval $t \in [130, 160]$. The offset was zero outside these two time intervals. To compare the models at different link angle amplitudes, the simulation was carried out three times with link angle amplitude $\theta_i = 10°$, $20°$, and $30°$, respectively, for $i \in \{1, \dots, N\}$. The corresponding amplitude α in (4.32) was set according to the values derived in Sect. 6.10.2, i.e. $\alpha = 7.1°$, $13.9°$, and $21°$, respectively, for the complex model and $\alpha = 2.3$ cm, 4.5 cm, and 6.6 cm, respectively, for the simplified model. All initial state values of the snake robot were zero.

The simulation results from the three trials are shown in Figs. 6.8, 6.9, and 6.10, respectively. In all three figures, the motion of the CM of the two robots is plotted in subfigure (a), where the snake robot from the complex model is visualised at

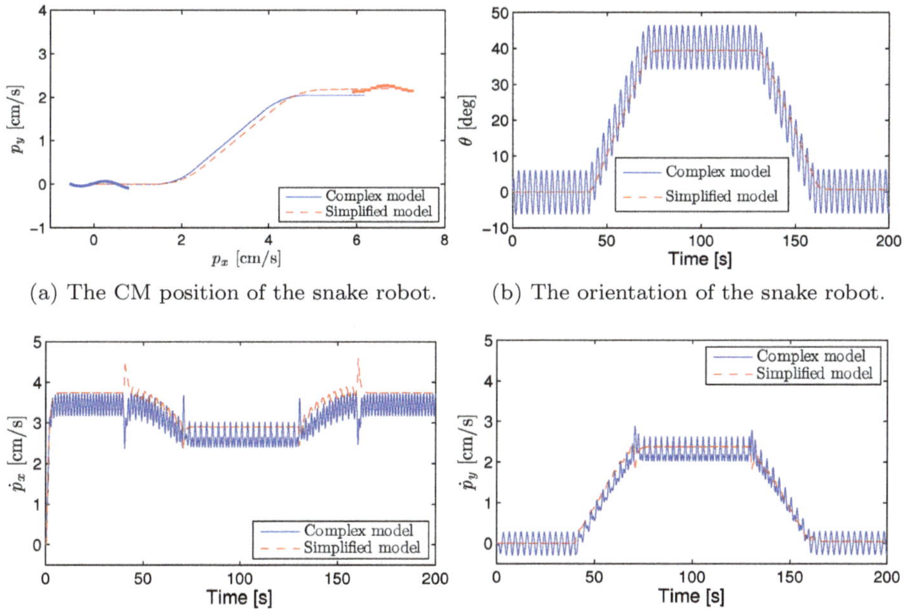

(a) The CM position of the snake robot.

(b) The orientation of the snake robot.

(c) The CM velocity in the global x direction of the snake robot, $\dot{p}_x$.

(d) The CM velocity in the global y direction of the snake robot, $\dot{p}_y$.

Fig. 6.9 Simulation result which compares the complex and simplified models during turning motion. The amplitude of the link angles is $\theta_i = 20°$

$t = 5$ s, and the snake robot from the simplified model is visualised at $t = 195$ s. Furthermore, subfigure (b) shows the orientation of the snake robots, which was given by θ in the simplified model, and which was estimated as the average of the link angles in the complex model, i.e. as $\overline{\theta} = \frac{1}{N}\sum_{i=1}^{N}\theta_i$. Subfigures (c) and (d) show the CM velocities of the snake robots in the global x and y directions, respectively.

In all three simulation trials, the *qualitative* behaviour of the snake robot from the simplified model is similar to the behaviour from the complex model. With the chosen numerical values of the rotation parameters λ_1 and λ_2 in the simplified model, we also achieved a good *quantitative* similarity between the two models. We can see that the quantitative similarity is best in the first trial, where the link angle amplitude is 10°, and that there is some deviation between the two models in the second trial, where the link angle amplitude is 20°. The deviation is even larger in the third trial, where the link angle amplitude is 30°. Again, these results agree with the predictions that were made in conjunction with the derivation of the simplified model, namely that the *quantitative* agreement between the complex and simplified modes is best for link angles limited by approximately $|\theta_i| < 20°$. Note that we also predicted that the two models will be *qualitatively* similar also for link angles above this approximate bound. The simulation result in Fig. 6.10 supports this conjecture since the *qualitative* behaviours of the two snake robots are similar even though there is a *quantitative* discrepancy between the plots.

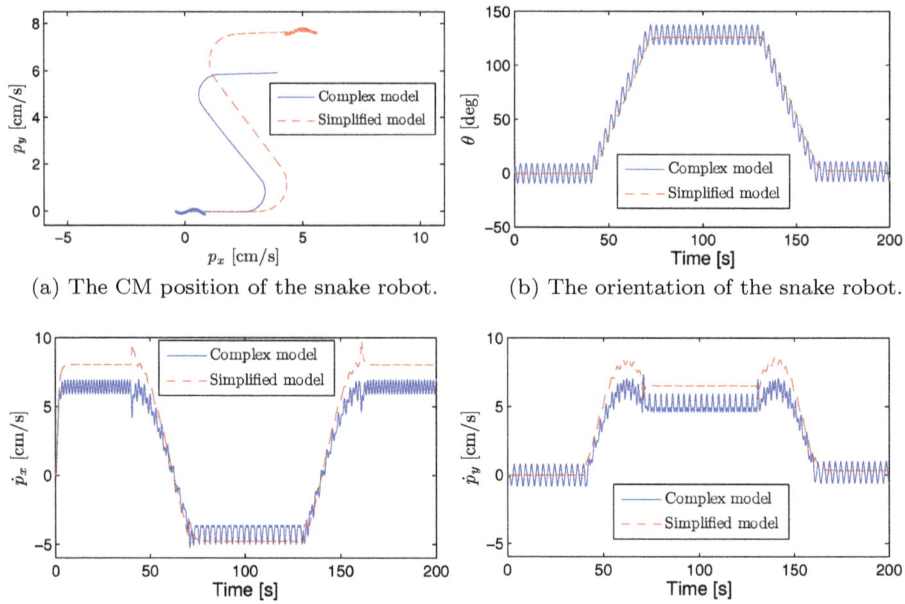

(a) The CM position of the snake robot.

(b) The orientation of the snake robot.

(c) The CM velocity in the global x direction of the snake robot, $\dot{p}_x$.

(d) The CM velocity in the global y direction of the snake robot, $\dot{p}_y$.

Fig. 6.10 Simulation result which compares the complex and simplified models during turning motion. The amplitude of the link angles is $\theta_i = 30°$

6.11 Chapter Summary

This chapter is summarised as follows:

- We have presented a simplified model of a planar snake robot in (6.35a)–(6.35h), which is intended for control design and stability analysis purposes.
- While the complex snake robot model given by (2.47) describes the body shape dynamics in terms of the *rotational* motion of the links, which results in complex equations of motion, the simplified model in (6.35a)–(6.35h) describes the body shape dynamics of the snake robot in terms of the *translational* motion of the links, which significantly simplifies the equations of motion.
- We have provided support of the claim that the simplified model captures the essential part of the dynamics of planar snake robot locomotion. In particular, we have shown that:
  - The stabilisability properties of the simplified model (stated in Theorem 6.1) are similar to the stabilisability properties of the complex model (stated in Theorem 4.3).
  - The simplified model is locally strongly accessible from *almost* any equilibrium point (see Theorem 6.2), which is also the case for the complex model (see Theorem 4.5).
  - The simplified model does not satisfy sufficient conditions for STLC (see Theorem 6.3), which is also the case for the complex model (see Theorem 4.6).

– The simulated behaviour of the snake robot from the simplified model is *qualitatively* similar to the behaviour from the complex model, and also *quantitatively* similar as long as the link angles are limited and the rotation parameters λ_1 and λ_2 are properly set.

Chapter 7
Analysis of Snake Robot Locomotion Based on Averaging Theory

In the previous chapter, we extended our toolbox with a new and more manageable model of snake robot locomotion. With this new model at our disposal, an intriguing question is whether we can use the model to derive new properties of snake robot dynamics. In this chapter, we will show that this indeed is the case.

The simplified model maps the periodic motion of the joints into the resulting propulsion of the snake robot. Since the joint motion is periodic, there must be some *average* effect of the joint motion that propels the robot. In this chapter, we use *averaging theory* (see Sanders et al. 2007) to study this average effect of the joint motion during the gait pattern lateral undulation. The analysis reveals new properties of snake robot locomotion that are both fundamental and useful from a motion planning perspective. In particular, we show that the average velocity of a snake robot during lateral undulation converges exponentially fast to a steady-state velocity, and an analytical expression for calculating the steady-state velocity is presented as a function of the gait pattern parameters. We also derive a set of relationships between the gait pattern parameters of lateral undulation and the resulting forward velocity of a planar snake robot. The material in this chapter is accompanied by simulation results and experimental results that support the theoretical findings.

The chapter is organised as follows. The relation between this chapter and previous literature is briefly discussed in Sect. 7.1. Section 7.2 gives a brief introduction to averaging theory. A model of the velocity dynamics of a snake robot is presented in Sect. 7.3, and the averaged form of this model is developed in Sect. 7.4. Section 7.5 studies the steady-state behaviour of the averaged velocity dynamics, while the relationships between the gait pattern parameters of lateral undulation and the resulting forward velocity of a snake robot are derived in Sect. 7.6. Simulation results are provided in Sects. 7.7 and 7.8 in order to support the validity of the theoretical findings, while Sect. 7.9 presents experimental results in order to investigate the validity of the derived properties of the velocity dynamics. Finally, the chapter is summarised in Sect. 7.10.

P. Liljebäck et al., *Snake Robots*, Advances in Industrial Control,
DOI 10.1007/978-1-4471-2996-7_7, © Springer-Verlag London 2013

7.1 The Relation Between This Chapter and Previous Literature

Averaging theory (see Sanders et al. 2007) is commonly used for studying dynamic systems with periodic and oscillatory behaviour, and is employed in several previous works to study snake-like locomotion. Averaging theory is, for instance, employed to study the motion of robotic fish and eel-like mechanisms in McIsaac and Ostrowski (2003a), Morgansen et al. (2007), Vela et al. (2002a, 2002b), and is employed to study elementary motions of planar snake robots in Chernousko (2003, 2005). The novel features of this chapter in relation to previous literature is therefore not the *use* of averaging theory, but rather the results that we develop by using this mathematical tool.

The first novel result of this chapter is an averaged model of the velocity dynamics of a snake robot during lateral undulation. As a second novel result, we show that the average velocity of a snake robot during lateral undulation converges exponentially fast to a steady-state velocity, and an analytical expression for calculating the steady-state velocity is presented as a function of the gait pattern parameters. To our best knowledge, this is the first time the steady-state velocity of a wheelless snake robot with anisotropic ground friction properties is derived analytically in terms of the gait pattern parameters. The third result is a set of fundamental relationships between the gait pattern parameters of lateral undulation and the resulting forward velocity of a planar snake robot. In particular, the derived properties state that the average forward velocity of a snake robot (1) is proportional to the squared amplitude of the sinusoidal motion of each joint, (2) is proportional to the angular frequency of the sinusoidal motion of each joint, (3) is proportional to a particular function of the constant phase shift between the joints, and (4) is maximised by the phase shift between the joints that also maximises the particular phase shift function. To our best knowledge, these fundamental properties of snake locomotion have never before been derived analytically.

7.2 Introduction to Averaging Theory

Consider a system of the form

$$\dot{\mathbf{x}} = \varepsilon \mathbf{f}(t, \mathbf{x}), \tag{7.1}$$

where ε is a small positive parameter characterising the magnitude of the perturbations of the system, $\mathbf{x} \in \mathbb{R}^n$, and $\mathbf{f}(t, \mathbf{x})$ is T-periodic, i.e. $\mathbf{f}(t + T, \mathbf{x}) = \mathbf{f}(t, \mathbf{x})$. A system that, in 'average', behaves similarly to the system in (7.1) is given by

$$\dot{\mathbf{x}} = \varepsilon \mathbf{f}_{\text{av}}(\mathbf{x}), \tag{7.2}$$

where

$$\mathbf{f}_{\text{av}}(\mathbf{x}) = \frac{1}{T} \int_0^T \mathbf{f}(\tau, \mathbf{x})\, \mathrm{d}\tau. \tag{7.3}$$

The above integral is calculated by treating the elements of the state vector $\mathbf{x}$ as constants since the underlying assumption is that the overall change of $\mathbf{x}$ is slow compared to the T-periodic fluctuations of $\mathbf{x}$. More specifically, the smallness requirement on ε ensures that $\mathbf{x}$ varies slowly with t relative to the periodic excitation of the system. The system response will thereby be determined predominantly by the average of the excitation. The following theorem follows directly from a more general theorem stated in Khalil (2002, Theorem 10.4):

Theorem 7.1 *Let* $\mathbf{f}(t,\mathbf{x})$ *and its partial derivatives with respect to* $\mathbf{x}$ *be continuous and bounded for* $(t,\mathbf{x}) \in [0,\infty) \times \mathbb{R}^n$. *Suppose that* $\mathbf{f}$ *is* T*-periodic in* t *for some* $T > 0$ *and* ε *is a positive parameter. Let* $\mathbf{x}(t,\varepsilon)$ *and* $\mathbf{x}_{\mathrm{av}}(t,\varepsilon)$ *denote the solutions of* (7.1) *and* (7.2), *respectively. If the average system* (7.2) *has a* globally exponentially stable *equilibrium point and* $\|\mathbf{x}(0,\varepsilon) - \mathbf{x}_{\mathrm{av}}(0,\varepsilon)\| \le k_0\varepsilon$ *for some* $k_0 > 0$, *then there exist* $k > 0$ *and* $\varepsilon^* > 0$ *such that for all* $0 < \varepsilon < \varepsilon^*$,

$$\left\|\mathbf{x}(t,\varepsilon) - \mathbf{x}_{\mathrm{av}}(t,\varepsilon)\right\| \le k\varepsilon \quad \text{for all } t \in [0,\infty). \tag{7.4}$$

This theorem basically says that, for sufficiently small ε, the solutions of the original system (7.1) and the average system (7.2) remain close (of order ε) for all time if the initial conditions of the systems are close and the average system is *globally exponentially stable*. This implies that the original system will remain close to a trajectory which converges exponentially to the equilibrium point.

7.3 The Velocity Dynamics During Lateral Undulation

We will now study the velocity dynamics of the snake robot during the gait pattern lateral undulation. As defined in (4.32), lateral undulation is achieved by controlling joint $i \in \{1,\dots,N-1\}$ according to

$$\phi_{i,\mathrm{ref}} = \alpha \sin\big(\omega t + (i-1)\delta\big) + \phi_o, \tag{7.5}$$

where α and ω are the amplitude and frequency, respectively, of the sinusoidal joint motion, δ determines the phase shift between the joints, and ϕ_o is a joint offset. In this chapter, we assume that ϕ_o is constant so that

$$\dot{\phi}_{i,\mathrm{ref}} = \alpha\omega \cos\big(\omega t + (i-1)\delta\big). \tag{7.6}$$

It was shown in Sect. 4.8.2 that we can achieve exponentially stable tracking of the joint reference coordinates (7.5) with the control law (4.34). In the following, we will therefore assume that the joint coordinates $\boldsymbol{\phi}$ and the joint velocities $\mathbf{v}_\phi = \dot{\boldsymbol{\phi}}$ are given by (7.5) and (7.6), respectively.

We define the velocity dynamics of the simplified model in terms of the dynamics of the forward direction velocity v_t, the normal direction velocity v_n, and the angular velocity v_θ of the snake robot. From (6.35f), (6.35g), and (6.35h), the velocity dynamics is thereby given as

$$\dot{v}_t = -\frac{c_t}{m} v_t + \frac{2c_p}{Nm} v_n \overline{\mathbf{e}}^T \boldsymbol{\phi} - \frac{c_p}{Nm} \boldsymbol{\phi}^T \mathbf{A}\overline{\mathbf{D}} \mathbf{v}_\phi, \tag{7.7a}$$

$$\dot{v}_n = -\frac{c_n}{m} v_n + \frac{2c_p}{Nm} v_t \overline{\mathbf{e}}^T \boldsymbol{\phi}, \tag{7.7b}$$

$$\dot{v}_\theta = -\lambda_1 v_\theta + \frac{\lambda_2}{N-1} v_t \overline{\mathbf{e}}^T \boldsymbol{\phi}. \tag{7.7c}$$

In order to arrive at a model of the velocity dynamics during lateral undulation which is in the standard averaging form (7.1), we assume that the amplitude α and frequency ω of the joint motion are always set according to the rule

$$\omega = \frac{k_{\alpha\omega}}{\alpha^2}, \tag{7.8}$$

where $k_{\alpha\omega} > 0$ is a controller parameter. Note that α and ω are still independent parameters since any choice of α and ω can be obtained by choosing $k_{\alpha\omega} = \alpha^2\omega$. Using (7.5), (7.6), and (7.8) and introducing the velocity state vector $\mathbf{v} = [v_t, v_n, v_\theta]^T \in \mathbb{R}^3$, the velocity dynamics of the snake robot during lateral undulation can be written as

$$\dot{\mathbf{v}} = \begin{bmatrix} \dot{v}_t \\ \dot{v}_n \\ \dot{v}_\theta \end{bmatrix} = \mathbf{f}(t, \mathbf{v}), \tag{7.9}$$

where

$$\mathbf{f}(t, \mathbf{v}) = \begin{bmatrix} -\frac{c_t}{m} v_t + \frac{2c_p}{Nm} v_n f_1(\omega t) - \frac{c_p}{Nm} f_2(\omega t) \\ -\frac{c_n}{m} v_n + \frac{2c_p}{Nm} v_t f_1(\omega t) \\ -\lambda_1 v_\theta + \frac{\lambda_2}{N-1} v_t f_1(\omega t) \end{bmatrix}, \tag{7.10}$$

$$f_1(\omega t) = (N-1)\phi_o + \sum_{i=1}^{N-1} \alpha \sin\bigl(\omega t + (i-1)\delta\bigr), \tag{7.11}$$

$$\begin{aligned} f_2(\omega t) = \sum_{i=1}^{N-1}\sum_{j=1}^{N-1} \biggl[&\frac{k_{\alpha\omega}}{\alpha} \phi_o a_{ij} \cos\bigl(\omega t + (j-1)\delta\bigr) \\ &+ k_{\alpha\omega} a_{ij} \sin\bigl(\omega t + (i-1)\delta\bigr) \cos\bigl(\omega t + (j-1)\delta\bigr) \biggr], \end{aligned} \tag{7.12}$$

and a_{ij} denotes element ij of the matrix $\mathbf{A}\overline{\mathbf{D}} \in \mathbb{R}^{(N-1)\times(N-1)}$ (the matrices $\mathbf{A}$ and $\overline{\mathbf{D}}$ were defined in Sect. 6.2). To transform the model (7.9) into the standard form of averaging (7.1), we change the time scale from t to $\tau = \omega t$ and define $\varepsilon = 1/\omega$.

Since $\frac{\mathrm{d}}{\mathrm{d}t} = \frac{1}{\varepsilon}\frac{\mathrm{d}}{\mathrm{d}\tau}$, the model (7.9) can now be written as

$$\frac{\mathrm{d}\mathbf{v}}{\mathrm{d}\tau} = \varepsilon \mathbf{f}(\tau, \mathbf{v}), \tag{7.13}$$

where

$$\mathbf{f}(\tau, \mathbf{v}) = \begin{bmatrix} -\frac{c_t}{m} v_t + \frac{2c_p}{Nm} v_n f_1(\tau) - \frac{c_p}{Nm} f_2(\tau) \\ -\frac{c_n}{m} v_n + \frac{2c_p}{Nm} v_t f_1(\tau) \\ -\lambda_1 v_\theta + \frac{\lambda_2}{N-1} v_t f_1(\tau) \end{bmatrix}. \tag{7.14}$$

This model is in the standard form defined in (7.1). Note that when we require ε to be small, we equivalently require that the frequency of the joint motion $\omega = 1/\varepsilon$ is large.

7.4 The Averaged Velocity Dynamics During Lateral Undulation

The averaged model of (7.13) is calculated in accordance with (7.2) as

$$\frac{\mathrm{d}\mathbf{v}}{\mathrm{d}\tau} = \varepsilon \frac{1}{2\pi} \int_0^{2\pi} \mathbf{f}(\tau, \mathbf{v})\, \mathrm{d}\tau. \tag{7.15}$$

It can be verified that

$$\frac{1}{2\pi} \int_0^{2\pi} f_1(\tau)\, \mathrm{d}\tau = (N-1)\phi_o, \tag{7.16}$$

$$\frac{1}{2\pi} \int_0^{2\pi} f_2(\tau)\, \mathrm{d}\tau = -\frac{1}{2} k_{\alpha\omega} k_\delta, \tag{7.17}$$

where the constant $k_\delta \in \mathbb{R}$ is defined as

$$k_\delta = \sum_{i=1}^{N-1} \sum_{j=1}^{N-1} a_{ij} \sin\big((j-i)\delta\big). \tag{7.18}$$

The averaged model can therefore be written as

$$\frac{\mathrm{d}\mathbf{v}}{\mathrm{d}\tau} = \varepsilon(\mathcal{A}\mathbf{v} + \mathbf{b}), \tag{7.19}$$

where

$$\mathcal{A} = \mathcal{A}(\phi_o) = \begin{bmatrix} -\frac{c_t}{m} & \frac{2(N-1)}{Nm} c_p \phi_o & 0 \\ \frac{2(N-1)}{Nm} c_p \phi_o & -\frac{c_n}{m} & 0 \\ \lambda_2 \phi_o & 0 & -\lambda_1 \end{bmatrix}, \tag{7.20}$$

$$\mathbf{b} = \mathbf{b}(\alpha, \omega, \delta) = \begin{bmatrix} \frac{c_p}{2Nm} k_{\alpha\omega} k_\delta \\ 0 \\ 0 \end{bmatrix}. \tag{7.21}$$

By changing time scale back to t using that $\frac{\mathrm{d}}{\mathrm{d}\tau} = \varepsilon \frac{\mathrm{d}}{\mathrm{d}t}$, the averaged model is given by

$$\dot{\mathbf{v}} = \mathcal{A}\mathbf{v} + \mathbf{b}. \tag{7.22}$$

We see that the averaged model of the velocity dynamics is a linear system characterised by the parameters of the joint reference coordinates, i.e. by α, ω, δ, and ϕ_o.

Remark 7.1 The term *average velocity* will hereafter be used to denote the velocity described by the averaged model (7.22). This average velocity evolves according to the average changes of the original model (7.9) and will typically correspond to the average of the velocities of the original model over the last cycle of the periodic gait pattern (i.e. the last T seconds). The average velocity can, in other words, be regarded as a low-pass filtered version of the original velocity.

7.5 The Steady-State Behaviour of the Velocity Dynamics During Lateral Undulation

In order to determine the stability properties of the averaged model (7.22), we remove the constant offset term $\mathbf{b}$ with the coordinate transformation $\mathbf{z} = \mathbf{v} + \mathcal{A}^{-1}\mathbf{b}$. This gives

$$\dot{\mathbf{z}} = \dot{\mathbf{v}} = \mathcal{A}\big(\mathbf{z} - \mathcal{A}^{-1}\mathbf{b}\big) + \mathbf{b} = \mathcal{A}\mathbf{z}. \tag{7.23}$$

By employing a computer software for symbolic mathematics, such as *Matlab Symbolic Toolbox*, the eigenvalues of $\mathcal{A}$ are easily calculated as

$$\operatorname{eig}(\mathcal{A}) = \begin{bmatrix} -\frac{c_t + c_n}{2m} - \frac{\sqrt{(c_n N - c_t N)^2 + (4(N-1)c_p \phi_o)^2}}{2Nm} \\ -\frac{c_t + c_n}{2m} + \frac{\sqrt{(c_n N - c_t N)^2 + (4(N-1)c_p \phi_o)^2}}{2Nm} \\ -\lambda_1 \end{bmatrix}. \tag{7.24}$$

The equilibrium point $\mathbf{z} = \mathbf{0}$ is *globally exponentially stable* if all eigenvalues of $\mathcal{A}$ are negative (see Khalil 2002), which is easily seen to be the case if

$$|\phi_o| < \frac{N}{2(N-1)} \sqrt{c_t c_n}. \tag{7.25}$$

The limit in (7.25) concerns the amplitude of the joint coordinate offset ϕ_o and is a function of the friction coefficients c_t and c_n. This limit is not relevant to a snake robot with revolute joints since the normal direction distance between the links of such mechanisms is physically constrained by the revolute joints. The instability issue in (7.25) therefore suggests that the approach of modelling the link motion as translational displacements breaks down when the displacements become large.

Assuming that we choose ϕ_o to satisfy the limit (7.25), then $\mathbf{z}$ will converge exponentially to zero, which means that $\mathbf{v}$ will converge exponentially to $-\mathcal{A}^{-1}\mathbf{b}$, which in turn means that the average velocity will converge exponentially to the steady-state velocity

$$\mathbf{v}^* = \begin{bmatrix} v_t^* \\ v_n^* \\ v_\theta^* \end{bmatrix} = -\mathcal{A}^{-1}\mathbf{b}, \tag{7.26}$$

given analytically by

$$v_t^* = k_{\alpha\omega} k_\delta \frac{N c_n c_p}{2(N^2 c_t c_n - (4N^2 - 8N + 4)c_p^2 \phi_o^2)}, \tag{7.27a}$$

$$v_n^* = k_{\alpha\omega} k_\delta \frac{\phi_o (N-1) c_p^2}{N^2 c_t c_n - (4N^2 - 8N + 4)c_p^2 \phi_o^2}, \tag{7.27b}$$

$$v_\theta^* = k_{\alpha\omega} k_\delta \frac{\phi_o N c_n c_p \lambda_2}{2\lambda_1 (N^2 c_t c_n - (4N^2 - 8N + 4)c_p^2 \phi_o^2)}. \tag{7.27c}$$

We can see that the resulting steady-state velocity of the snake robot is proportional to the controller parameters $k_{\alpha\omega} = \alpha^2\omega$ and k_δ, and that the velocity also depends on nonlinear terms involving the joint coordinate offset ϕ_o.

Since the averaged model of the velocity dynamics given by (7.22) is *globally exponentially stable* (assuming that (7.25) is satisfied), it follows from Theorem 7.1 that, for sufficiently small ε (i.e. for sufficiently large ω), the average velocity given by (7.22) will approximate the exact velocity (7.9) for all time and that the error of this approximation is of order ε, i.e. bounded in accordance with (7.4). In this book, we will not investigate the lower limit of ω corresponding to some maximum error bound. However, the simulation results presented in Sect. 7.7 show that the exact and average velocities agree well when ω is set to values that are commonly used for snake robot locomotion.

We now summarise the above conclusions.

Theorem 7.2 *Consider a planar snake robot described by* (6.35a)–(6.35h). *Suppose that the joint coordinates* $\boldsymbol{\phi}$ *are controlled in exact accordance with* (7.5) *and* (7.6), *and that the joint coordinate offset* ϕ_o *satisfies* (7.25). *Then there exist* $k > 0$ *and* $\omega^* > 0$ *such that for all* $\omega > \omega^*$,

$$\left\| \mathbf{v}(t) - \mathbf{v}_{\mathrm{av}}(t) \right\| \le \frac{k}{\omega} \quad \text{for all } t \in [0, \infty), \tag{7.28}$$

where $\mathbf{v}(t)$ *denotes the exact velocity of the snake robot given by* (7.9), *and* $\mathbf{v}_{\mathrm{av}}(t)$ *denotes the average velocity given by* (7.22). *Furthermore, the average velocity* $\mathbf{v}_{\mathrm{av}}(t)$ *of the snake robot will converge exponentially fast to the steady-state velocity* $\mathbf{v}^*$ *given by* (7.26).

7.6 Relationships Between the Gait Parameters and the Forward Velocity During Lateral Undulation

Theorem 7.2 is a powerful result. First of all, it proves mathematically that lateral undulation enables a wheel-less snake robot with anisotropic ground friction properties to achieve forward propulsion (under the assumption that the body shape motion is modelled as translational link displacements). Second, the result gives an analytical expression for the steady-state velocity as a function of the controller parameters α, ω, δ, and ϕ_o, i.e. the amplitude, frequency, phase shift, and offset of the joint motion during lateral undulation. This information is relevant for motion planning purposes. We can for example immediately see from (7.26) that the steady-state velocity of the snake robot when it conducts lateral undulation with zero joint offset ($\phi_o = 0$) is given by $v_t^* = \frac{c_p}{2Nc_t} k_{\alpha\omega} k_\delta$, $v_n^* = 0$, and $v_\theta^* = 0$. A final powerful feature of Theorem 7.2 is that it applies to snake robots with an arbitrary number of links N.

In the following, we will use Theorem 7.2 to deduce some fundamental relationships between the gait pattern parameters and the forward velocity of the snake robot. The forward velocity is seen from (7.27a) to be proportional to the controller parameter $k_{\alpha\omega} = \alpha^2\omega$, i.e. the forward velocity is proportional to the square of the amplitude of the joint motion, α^2, and also proportional to the angular frequency, ω, of the joint motion. This information is useful from a motion planning perspective since it tells us that an increase/decrease of the forward velocity by a certain factor can be achieved by increasing/decreasing ω by the same factor or by increasing/decreasing α by the square root of this factor.

It is also seen from (7.27a) that the forward velocity of the snake robot is proportional to the function k_δ defined in (7.18). Since k_δ is a function of the phase shift δ between the joints, this means that the phase shift δ that will maximise the forward velocity can be determined as the δ that maximises k_δ. This is particularly interesting since we are now able to use the analytical expression of k_δ to determine the optimal phase shift δ that maximises the forward velocity of a planar snake robot with an arbitrary number of links N. Figure 7.1 presents a plot of the maximum value of k_δ as a function of the number of links N. For each N, the maximum value of k_δ was found using the mathematical computer software *Matlab*. The optimal phase shift is e.g. $\delta = 90°$ for $N = 3$ links, $\delta = 50.4°$ for $N = 5$ links, $\delta = 24.1°$ for $N = 10$ links, and $\delta = 11.5°$ for $N = 20$ links.

The above results can be summarised as follows:

Theorem 7.3 *Consider a planar snake robot with N links modelled by* (6.35a)–(6.35h) *and controlled in exact accordance with* (7.5) *and* (7.6). *The average forward velocity of the snake robot given by* (7.22) *will converge exponentially to a value which is proportional to*:

- *The squared amplitude of the sinusoidal joint motion, α^2.*
- *The angular frequency of the sinusoidal joint motion, ω.*

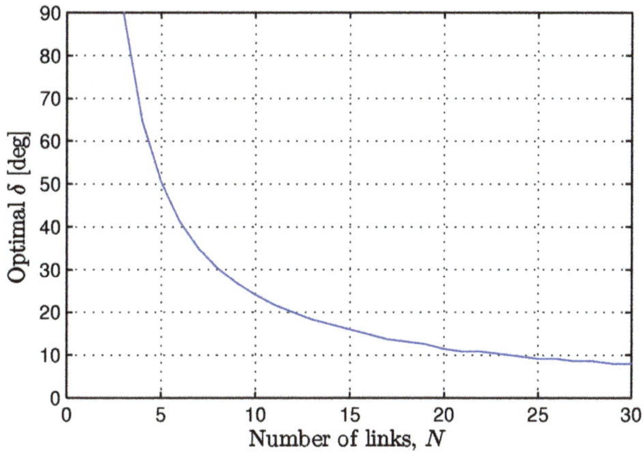

Fig. 7.1 The phase shift δ that maximises the forward velocity of a planar snake robot as a function of the number of links N

– *The function of the constant phase shift, δ, between the joints given by*

$$k_\delta = \sum_{i=1}^{N-1}\sum_{j=1}^{N-1} a_{ij} \sin\big((j-i)\delta\big), \tag{7.29}$$

where a_{ij} denotes the element ij of the matrix $\mathbf{A}\overline{\mathbf{D}}$. Moreover, for a given α and ω, the phase shift, δ, that maximises the average forward velocity is given by the δ that maximises k_δ.

7.7 Simulation Study: Comparison Between the Original and the Averaged Velocity Dynamics

This section presents simulation results in order to investigate the validity of Theorem 7.2, i.e. to validate the agreement between the original simplified model of the velocity dynamics (7.9) and the averaged model (7.22).

7.7.1 Simulation Parameters

The original model of the snake robot was given by (6.35a)–(6.35h) under the assumption that $\boldsymbol{\phi}$ was controlled in exact accordance with (7.5). The averaged model of the snake robot was given by (7.22). Both models were implemented and simulated in *Matlab R2008b* on a laptop running *Windows XP*. The dynamics was calculated using the *ode45* solver in Matlab with a relative and absolute error tolerance of 10^{-6}.

We considered a snake robot with $N = 10$ links of length $l = 0.14$ m and mass $m = 1$ kg. The ground friction coefficients were $c_t = 0.5$ and $c_n = 3$, and the rotation

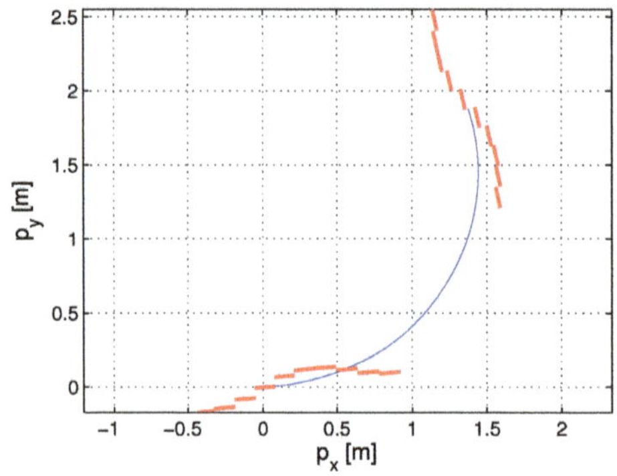

(a) The CM position of the snake robot.

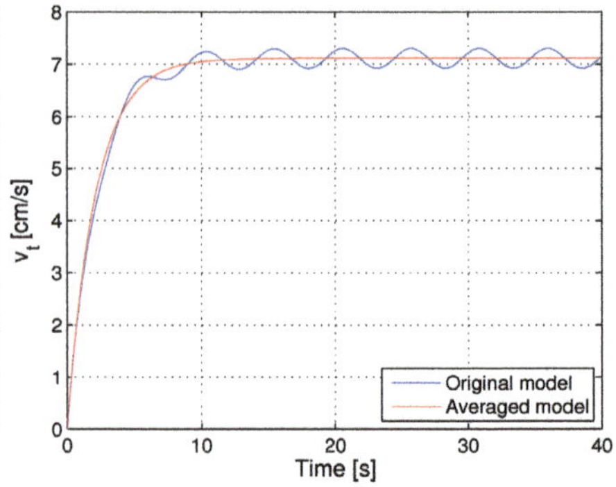

(b) The forward velocity v_t.

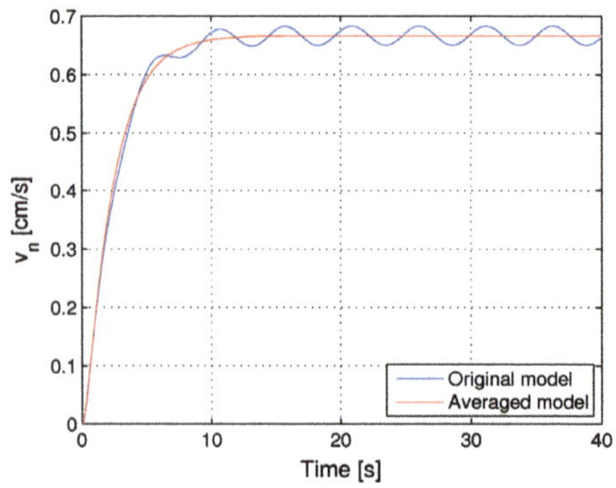

(c) The sideways velocity v_n.

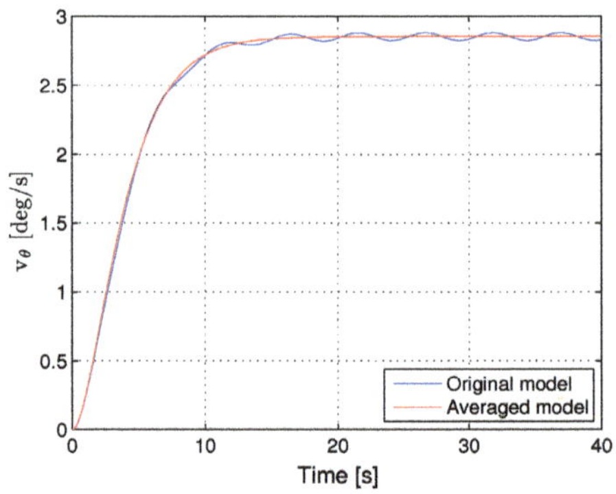

(d) The rotation velocity v_θ.

Fig. 7.2 Comparison between the original simplified model and the averaged model for counterclockwise turning motion with controller parameters $\alpha = 0.05$ m, $\omega = 70°$/s, $\delta = 40°$, and $\phi_o = l/8$ m

parameters were $\lambda_1 = 0.5$ and $\lambda_2 = 20$. The initial states of both models were set to the origin. The values of the gait pattern parameters α, ω, δ, and ϕ_o are presented with each simulation result below.

7.7.2 Simulation Results

The motion of the snake robot during lateral undulation was first simulated with the gait parameters $\alpha = 0.05$ m, $\omega = 70°$/s, $\delta = 40°$, and $\phi_o = l/8$ m. The joint coordinates were, in other words, offseted by $1/8$ of the link length l. In accordance with Theorem 7.2, the average velocity of the snake robot should then converge exponentially fast to the steady-state velocity $\overline{v}_t = 7.12$ cm/s, $\overline{v}_n = 0.67$ cm/s, and $\overline{v}_\theta = 2.86°$/s. This prediction agrees very well with the simulation result shown in Fig. 7.2. The top left plot shows the global CM position of the snake robot and also

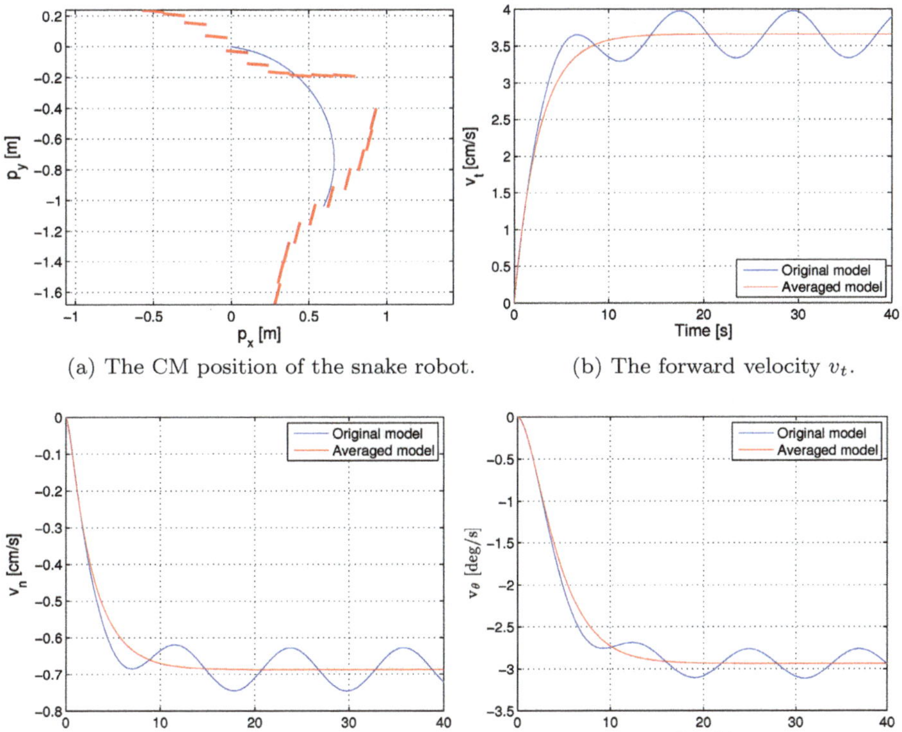

(a) The CM position of the snake robot. (b) The forward velocity v_t.

(c) The sideways velocity v_n. (d) The rotation velocity v_θ.

Fig. 7.3 Comparison between the original simplified model and the averaged model for clockwise turning motion with controller parameters $\alpha = 0.05$ m, $\omega = 30°/\text{s}$, $\delta = 40°$, and $\phi_o = -l/4$ m

the body shape at $t = 5$ s and $t = 40$ s, while the other three plots show the velocities from the original and the averaged model. The close overlap between the velocity plots from the two models suggests that $\omega = 70°/\text{s}$ is well above the (unknown) value of ω^* described in Theorem 7.2.

In the second simulation, the controller parameters were set to $\alpha = 0.05$ m, $\omega = 30°/\text{s}$, $\delta = 40°$, and $\phi_o = -l/4$ m. The joint coordinates were, in other words, offseted by $1/4$ of the link length l. In addition, we reduced the frequency of the sinusoidal motion from $\omega = 70°/\text{s}$ to $\omega = 30°/\text{s}$ to see how this affected the estimate of the average velocity. From Theorem 7.2, the average velocity should converge to $\overline{v}_t = 3.66$ cm/s, $\overline{v}_n = -0.69$ cm/s, and $\overline{v}_\theta = -2.94°/\text{s}$. This agrees very well with the simulation result shown in Fig. 7.3. The figure shows that there is still a good agreement between the velocities from the original and the averaged model even though we reduced ω considerably.

7.8 Simulation Study: Investigation of the Relationships Between Gait Parameters and Forward Velocity

This section presents simulation results in order to investigate the validity of Theorem 7.3, i.e. the validity of the relationships between the gait pattern parameters of lateral undulation and the resulting forward velocity of the snake robot. In addition to presenting simulation results from the simplified model (6.35a)–(6.35h), this section also shows that the relationships in Theorem 7.3 apply to the velocity from the complex model given by (2.47).

Remark 7.2 The joint coordinates of the simplified model (linear link displacements) are different from the joint coordinates of the complex model (joint angles). However, it still makes sense to investigate the validity of Theorem 7.3 for a snake robot with revolute joints since, as implied by Property 4.8, the rotational link motion is what produces the linear displacements captured by the simplified model. Note also that for limited link angles, there is approximately a linear relationship between the amplitude of the angular joint motion and the corresponding amplitude of the transversal link displacements. The relationship stated in Theorem 7.3 between the forward velocity and the squared amplitude of the sinusoidal joint motion, α^2, can therefore be expected to hold also when α denotes the amplitude of the angular joint motion of a snake robot with revolute joints.

7.8.1 Simulation Parameters

Both the complex model (2.47) and the simplified model (6.35a)–(6.35h) were implemented and simulated in *Matlab R2008b* on a laptop running *Windows XP*. The dynamics was calculated using the *ode45* solver in Matlab with a relative and absolute error tolerance of 10^{-3} and 10^{-6} in the complex and the simplified model, respectively.

We considered snake robots with $N = 3$, $N = 5$, $N = 10$, and $N = 20$ links of length $l = 0.14$ m, mass $m = 1$ kg, and moment of inertia $J = 0.0016$ kgm^2. The ground friction coefficients were $c_t = 1$ and $c_n = 3$, and the rotation parameters of the simplified model were $\lambda_1 = 0.5$ and $\lambda_2 = 20$. The linearised control input $\overline{\mathbf{u}}$ of both models was set according to the exponentially stable joint controller defined in (4.34) with controller gains $k_p = 20$ and $k_d = 5$. The joint reference coordinates were calculated according to the motion pattern lateral undulation defined in (7.5) with zero joint angle offset ($\phi_o = 0$). The values of the gait pattern parameters α, ω, and δ are presented with each simulation result below. The initial states of both models were set to the origin.

The simulation results below present the forward velocity of the snake robot, denoted by $\overline{v}$, for different sets of gait pattern parameters. The velocity was calculated at the end of each simulation trial as the linear distance travelled by the CM of the snake robot divided by the simulation time, which was chosen to be $t_{\text{sim}} = 10$ s. In other words, the velocity was calculated as

$$\overline{v} = \frac{\sqrt{(p_x(10) - p_x(0))^2 + (p_y(10) - p_y(0))^2}}{10}. \tag{7.30}$$

Remark 7.3 The essence of the simulation results presented in this section is contained in the particular shape of each graph, which means that a quantitative comparison between the graphs from the complex and the simplified model is not relevant. The values of the amplitude α of the joint motion in the complex model were therefore set independently from the values of α in the simplified model.

7.8.2 Simulation Results

Relationship Between the Forward Velocity and α

Theorem 7.3 states that the average forward velocity of a planar snake robot is proportional to the squared amplitude of the sinusoidal joint motion, α^2. We investigated the validity of this result by simulating the snake robot with different values of α and calculating the resulting average forward velocity according to (7.30). The simulation results from the simplified model are shown in Fig. 7.4(a), while the simulation results from the complex model with viscous and Coulomb ground friction, respectively, are shown in Fig. 7.4(b). The number of links N and the corresponding values of ω and δ are shown at the top of each plot. In the complex model, the range of α values is shorter for large N compared to for small N since a large angle amplitude would cause a 'collision' between the head and the tail of the snake when N is large. The plots clearly show an exponential increase in the forward speed $\overline{v}$ as the amplitude α increases. This is in accordance with Theorem 7.3.

Note that the amplitude of the joint motion cannot be increased indefinitely in the complex model. For sufficiently large α, the relative link velocity components that are *tangential* to the forward direction will no longer be negligible, which is assumed in the simplified model of the snake robot. Moreover, the effective length (from head to tail) of a snake robot with revolute joints is decreased significantly for large α, while the simplified model assumes that this length is constant. It is therefore reasonable to expect that the increase in the forward velocity from the complex model will decay for large α. This decay can be seen in the plots with viscous friction in Fig. 7.4(b), which shows that the increase of the velocity has a more linear character when α becomes large.

Relationship Between the Forward Velocity and ω

Theorem 7.3 states that the average forward velocity of a planar snake robot is proportional to the angular frequency, ω, of the joint motion. This result was investigated by simulating the snake robot with different values of ω and calculating the resulting average forward velocity. The simulation results from the simplified model

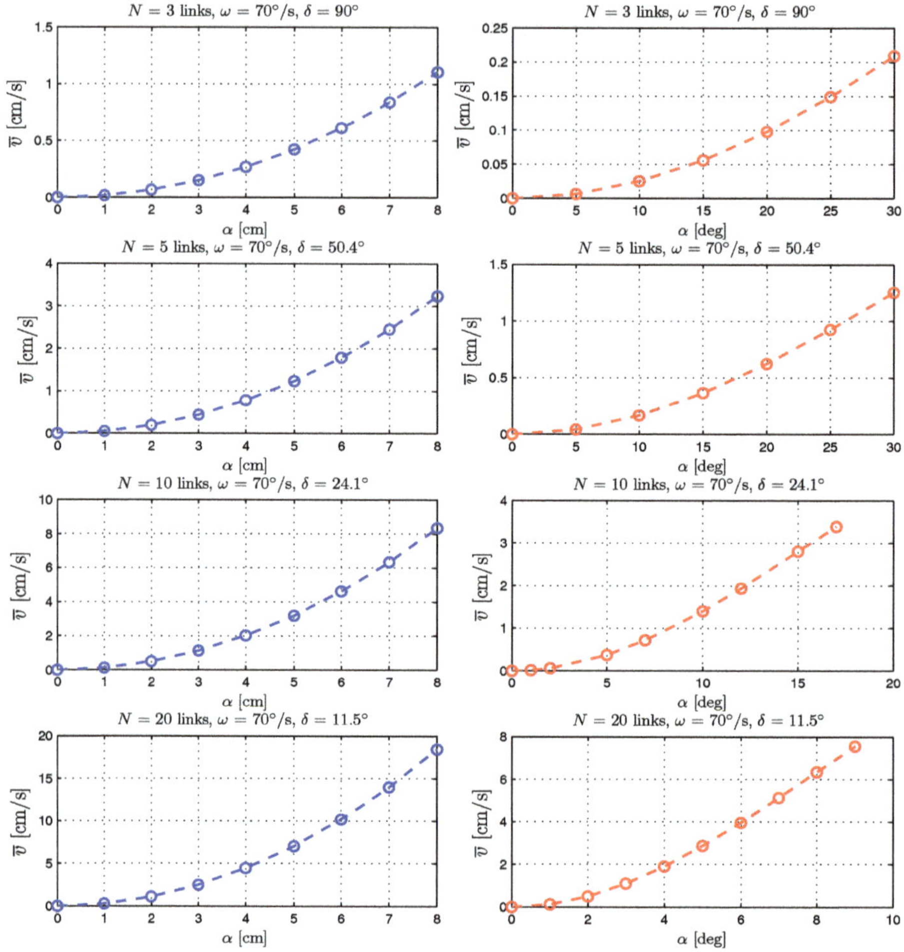

(a) Simulation with the simplified model. (b) Simulation with the complex model.

Fig. 7.4 The forward velocity of the snake robot for different values of α. The number of links N and the corresponding values of ω and δ are shown at the *top* of each plot

are shown in Fig. 7.5(a), while the simulation results from the complex model are shown in Fig. 7.5(b). The number of links N and the corresponding values of α and δ are shown at the top of each plot. The linear increase in the forward speed $\overline{v}$ as the frequency ω increases is clearly present in the plots from both models, which is in accordance with Theorem 7.3.

Relationship Between the Forward Velocity and δ

The final result stated in Theorem 7.3 is that the average forward velocity is maximised by the phase shift δ that maximises the function k_δ. To investigate the validity

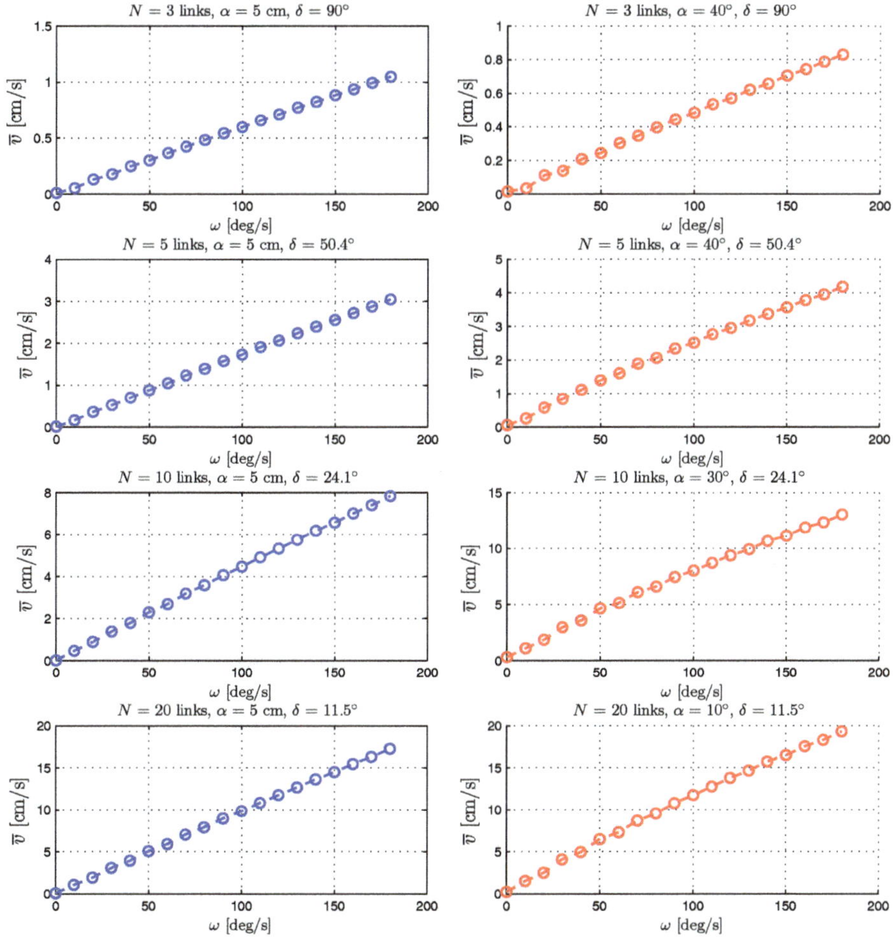

(a) Simulation with the simplified model. (b) Simulation with the complex model.

Fig. 7.5 The forward velocity of the snake robot for different values of ω. The number of links N and the corresponding values of α and δ are shown at the *top* of each plot

of this result, we simulated the snake robot with different values of δ to identify the phase shift that produced the highest forward velocity. The simulation results from the simplified model are shown in Fig. 7.6(a), while the simulation results from the complex model with viscous and Coulomb ground friction, respectively, are shown in Fig. 7.6(b). The number of links N and the corresponding values of α and ω are shown at the top of each plot. The δ value that maximises k_δ is indicated with a vertical dashed line in each plot. The maximum velocity of each plot in Fig. 7.6 seems to agree well with the δ value that maximises k_δ, which supports the prediction in Theorem 7.3 concerning the phase shift δ that maximises the average forward velocity of a snake robot during lateral undulation.

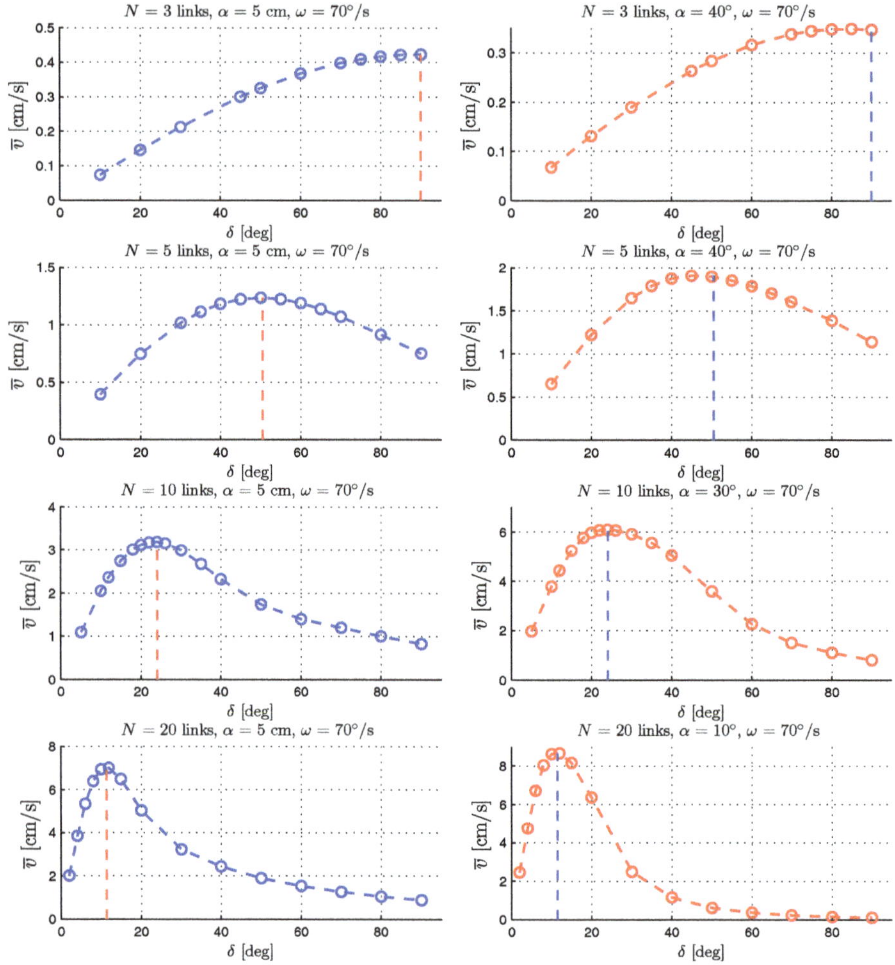

(a) Simulation with the simplified model. (b) Simulation with the complex model.

Fig. 7.6 The forward velocity of the snake robot for different values of δ. The number of links N and the corresponding values of α and ω are shown at the *top* of each plot. The *vertical dashed line* indicates the estimated δ value that, by Theorem 7.3, maximises the forward velocity

7.9 Experimental Study: Investigation of the Relationships Between Gait Parameters and Forward Velocity

In order to provide further support of the validity of Theorem 7.3, we present in this section results from an experimental investigation with the snake robot Wheeko. The design of the robot and the experimental setup (including the camera-based position measurement system) were presented in Chap. 3.

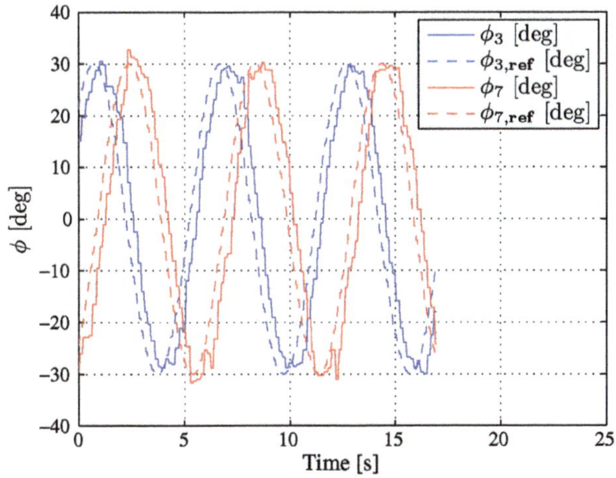

Fig. 7.7 A plot of the measured (*solid*) and the corresponding reference angles (*dashed*) of joint 3 and joint 7 during lateral undulation

7.9.1 Layout of the Experiment

Controlling the Joints According to Lateral Undulation

During the experiment, the joint reference angles were calculated on an external computer and sent to the snake robot through a wireless Bluetooth connection. The reference angles corresponding to the horizontal joint motion of the robot were calculated according to (7.5) with $N = 10$ links. The reference angles corresponding to the vertical joint motion were set to zero. The joint torque controller given by (4.34) was not employed since accurate torque control is not supported by the servo motors installed in the snake robot. Instead, the joint angles in the robot were controlled according to a proportional controller implemented in the microcontroller of each joint module. In order to show that the joint modules were able to track their joint reference angles, we provide in Fig. 7.7 a plot of the measured and the corresponding reference angles of two arbitrarily chosen joints (joint 3 and joint 7) during a run of lateral undulation with the snake robot. The plot indicates that the tracking of the joint reference angles was satisfactory.

Calculating the Forward Velocity of the Robot

To visualise how lateral undulation was carried out by the snake robot, we provide in Fig. 7.8 a few screen shots from a video recording of the robot. The position of the robot was recorded by the camera system described in Sect. 3.5, and the average forward velocity was calculated after each run as the travelled distance divided by the travel time. A typical plot of the measured position of the snake robot from a single run is shown in Fig. 7.9, which shows that the foremost joint module moves from side to side along the x direction but has a steady increase in the position along the y direction. The markers $\mathbf{p}_{\text{start}}$ and $\mathbf{p}_{\text{stop}}$ in the plot have been placed near

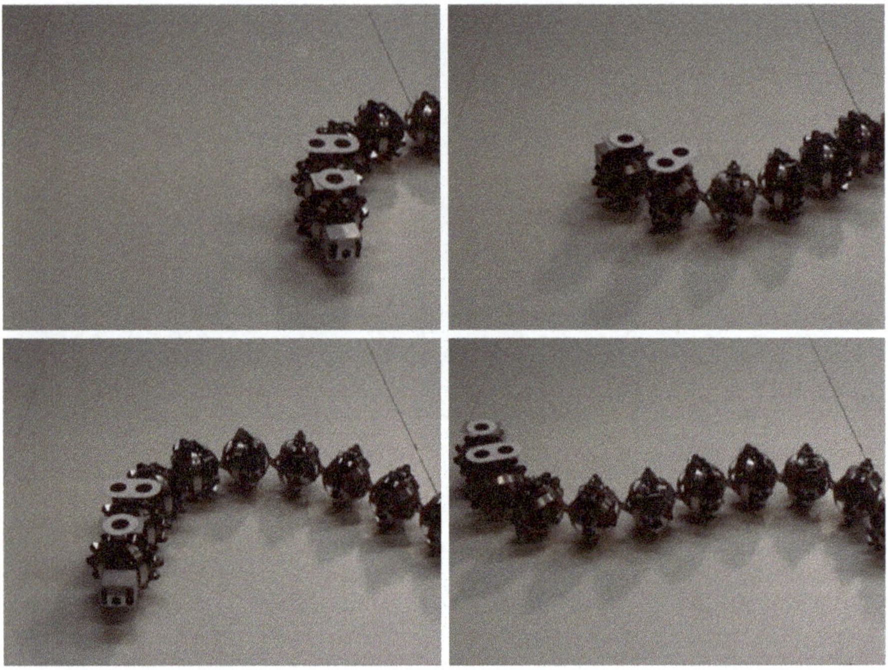

Fig. 7.8 The motion of the snake robot during a single run of lateral undulation

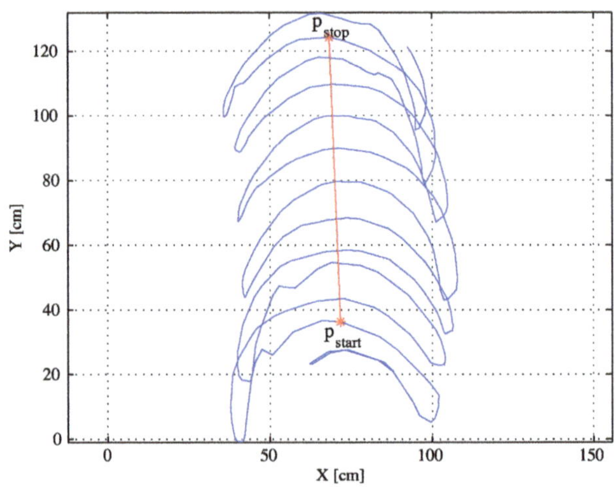

Fig. 7.9 A typical plot of the measured position of the snake robot during lateral undulation. The distance between the markers p_{start} and p_{stop} represents the distance travelled by the snake robot

the beginning and near the end of the dataset, respectively, at the approximate centre point of the cyclic sideways motion of the snake. We used the distance between these two markers to represent the distance travelled by the snake robot and calculated the travel time as the difference in sample time between the position measurements corresponding to the two markers. The average forward velocity of the snake robot

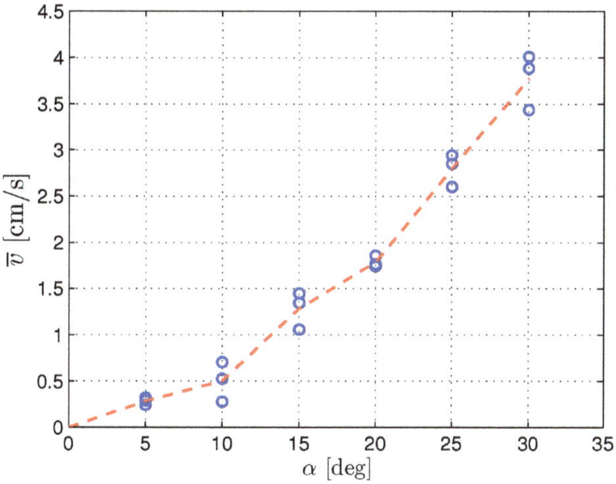

Fig. 7.10 The average forward velocity of the snake robot from three trials at different values of α. The remaining controller parameters were $\omega = 80°/\text{s}$, $\delta = 25°$, and $\phi_o = 0°$

was then calculated as

$$\overline{v} = \frac{\sqrt{(p_{\text{stop},x} - p_{\text{start},x})^2 + (p_{\text{stop},y} - p_{\text{start},y})^2}}{t_{\text{stop}} - t_{\text{start}}}. \tag{7.31}$$

When the duration, $t_{\text{stop}} - t_{\text{start}}$, of a single run of the robot is long, we conjecture that the accuracy of this velocity estimate will be sufficient for investigating the validity of Theorem 7.3. We developed a special software based on *Matlab* in order to easily identify the markers $\mathbf{p}_{\text{start}}$ and $\mathbf{p}_{\text{stop}}$ in the position plot from each run of the robot.

7.9.2 Experimental Results

Relationship Between the Forward Velocity and α

Theorem 7.3 states that the average forward velocity of a planar snake robot is proportional to the squared amplitude of the sinusoidal joint motion, α^2. We investigated the validity of this result by running the snake robot with different values of α and calculating the resulting average forward velocity according to (7.31). For each value of α, we ran the snake robot three times in order to get multiple velocity measurements. The remaining controller parameters were $\omega = 80°/\text{s}$, $\delta = 25°$, and $\phi_o = 0°$. Figure 7.10 presents the experimental results together with a dashed line between the average of the three velocities measured for each value of α. The plot clearly shows an exponential increase in the forward speed $\overline{v}$ as the amplitude α increases. This is in accordance with Theorem 7.3.

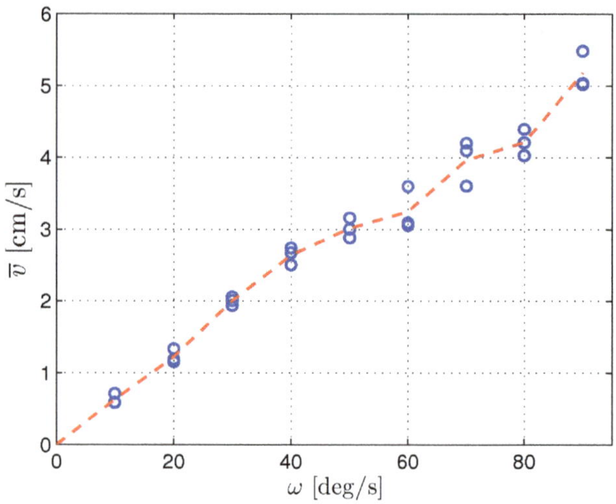

Fig. 7.11 The average forward velocity of the snake robot from three trials at different values of ω. The remaining controller parameters were $\alpha = 30°$, $\delta = 25°$, and $\phi_o = 0°$

Relationship Between the Forward Velocity and ω

Theorem 7.3 states that the average forward velocity of a planar snake robot is proportional to the angular frequency, ω, of the joint motion. This result was investigated by running the snake robot with different values of ω and calculating the resulting average forward velocity according to (7.31). For each value of ω, we ran the snake robot three times in order to get multiple velocity measurements. The remaining controller parameters were $\alpha = 30°$, $\delta = 25°$, and $\phi_o = 0°$. Figure 7.11 presents the experimental results together with a dashed line between the average of the three velocities measured for each value of ω. The plot clearly shows a linear increase in the forward speed $\overline{v}$ as the frequency ω increases. This is in accordance with Theorem 7.3.

Relationship Between the Forward Velocity and δ

The final result stated in Theorem 7.3 is that the average forward velocity is maximised by the phase shift δ that maximises the function k_δ. To investigate the validity of this result, we ran the snake robot with different values of δ to identify the phase shift that produced the highest forward velocity. For each value of δ, we ran the snake robot eight times in order to get multiple velocity measurements. The remaining controller parameters were $\alpha = 30°$, $\omega = 80°/\text{s}$, and $\phi_o = 0°$. Figure 7.12 presents the experimental results together with a dashed line between the average of the eight velocities measured for each value of δ. The δ value that maximises k_δ for $N = 10$ links is $\delta = 24.1°$ and is indicated with a vertical dashed line in Fig. 7.12. The plot indicates that the phase shift $\delta = 25°$ produced the highest forward velocity. This agrees well with the phase shift $\delta = 24.1°$ that maximises k_δ. The average velocity of the eight trials at $\delta = 25°$ was slightly below 4 cm/s. In summary, the

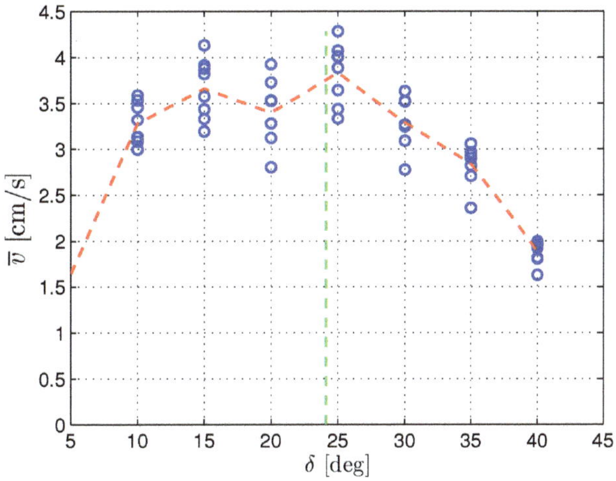

Fig. 7.12 The average forward velocity of the snake robot from eight trials at different values of δ. The remaining controller parameters were $\alpha = 30°$, $\omega = 80°/\text{s}$, and $\phi_o = 0°$

experimental results indicate that Theorem 7.3 provides a reasonable prediction of the phase shift, δ, that maximises the average forward velocity of a planar snake robot during lateral undulation.

7.10 Chapter Summary

This chapter is summarised as follows:

- We have in (7.22) presented an averaged model of the velocity dynamics of a planar snake robot during lateral undulation.
- We have shown that the upper bound of the error between the exact and the averaged velocity during lateral undulation is proportional to the inverse of the frequency of the sinusoidal joint motion, ω (see Theorem 7.2).
- We have shown that the average velocity of a snake robot during lateral undulation converges exponentially fast to the steady-state velocity given analytically in (7.26) as a function of the gait pattern parameters.
- We have derived fundamental relationships between the gait pattern parameters of lateral undulation and the resulting forward velocity of a snake robot (see Theorem 7.3). In particular, the derived properties state that the average forward velocity of a snake robot (1) is proportional to the squared amplitude of the sinusoidal motion of each joint, (2) is proportional to the angular frequency of the sinusoidal motion of each joint, (3) is proportional to the phase shift function given by (7.18), and (4) is maximised by the phase shift between the joints that also maximises the phase shift function in (7.18).
- We have presented simulation results that support the theoretical findings.
- We have presented experimental results with the snake robot Wheeko that support the derived relationships between the gait pattern parameters of lateral undulation and the resulting forward velocity of the robot.

Chapter 8
Path Following Control of Snake Robots Through a Cascaded Approach

In this chapter, we return to the problem of enabling a snake robot to track a planar path. In Chap. 5, we proposed a straight line path following controller and employed a Poincaré map to prove that the state variables of the snake robot, except for the position along the path, trace out a *locally exponentially stable* periodic orbit during motion along the desired path. The drawback of the analysis based on the Poincaré map, however, is that we are only able to infer conclusions regarding the *local* stability properties in the vicinity of the desired path. Moreover, since the analysis is based on simulating the model of the snake robot, the stability proof is only valid for the given choice of numerical controller parameters.

In order to elude the shortcomings of the previous analysis, this chapter extends the path following controller from Chap. 5 based on the simplified model of the snake robot. Using cascaded systems theory, we prove that the modified path following controller $\mathcal{K}$-exponentially stabilises the snake robot to any desired straight path. In particular, under the assumption that the forward velocity of the snake robot is non-zero and positive, we show that the model of the snake robot and the controller can be written as a cascaded system where the body shape changes affect the global orientation of the robot, which subsequently affects the cross-track error between the robot and the desired path. The $\mathcal{K}$-exponential stability of the cascaded system guarantees that the cross-track error and the heading of the snake robot with respect to the direction of the path converge to zero. The performance of the path following controller is investigated through simulations and through experiments with the snake robot Wheeko. The simulations and experimental results show that the proposed controller successfully steers the snake robot towards and along the desired straight path.

This chapter also considers path following of more general paths. In particular, we propose a waypoint guidance strategy which provably steers a snake robot along a path defined by waypoints interconnected by straight lines. In addition, we describe how the straight line path following controller can be extended to path following of curved paths by employing the approach previously presented in Børhaug (2008) in the context of path following control of marine vessels.

P. Liljebäck et al., *Snake Robots*, Advances in Industrial Control,
DOI 10.1007/978-1-4471-2996-7_8,

Note that this chapter considers path following, in contrast to trajectory tracking, where the goal is additionally to control the position of the system *along* the path. During path following, we steer the system towards and along the path but do not consider the temporal position of the system along the path. We will not consider trajectory tracking in this book.

The chapter is organised as follows. The relation between this chapter and previous literature is considered in Sect. 8.1. Section 8.2 presents some mathematical preliminaries. The straight line path following controller is presented and analysed in Sect. 8.3, while Sect. 8.4 describes how the controller can be extended to path following of general curved paths. The waypoint guidance strategy is proposed in Sect. 8.5. Simulation results and experimental results concerning the straight line path following controller are presented in Sect. 8.7, while Sect. 8.8 presents simulation results concerning the waypoint guidance strategy. Finally, the chapter is summarised in Sect. 8.9.

8.1 The Relation Between This Chapter and Previous Literature

Since this chapter extends the path following controller from Chap. 5, the discussion in Sect. 5.1 regarding previous literature on position and path following control of snake robots applies also to the current chapter. The extensions in this chapter with respect to Chap. 5 are all novel in relation to previous literature.

In particular, the first and main novel result of this chapter is the straight line path following controller, which, using cascaded systems theory, is proved to $\mathcal{K}$-exponentially stabilise a snake robot to any desired straight path. To our best knowledge, this is the first time the stability properties of a path following controller for a snake robot without sideslip (nonholonomic) constraints are formally proved. The experimental investigation of the path following controller using the snake robot Wheeko is also considered a novel result of this chapter. The second result is the description of how the straight line path following controller can be extended to path following of general curved paths. Finally, the third result is the waypoint guidance strategy for steering a snake robot along a path defined by waypoints interconnected by straight lines. Waypoint guidance has, to our best knowledge, not previously been considered for motion control of snake robots. The waypoint guidance strategy builds on the straight line path following controller and represents an operator-friendly framework for steering a snake robot between arbitrary locations on a flat surface.

8.2 Mathematical Preliminaries

We begin by presenting some stability concepts that will be employed to analyse the straight line path following controller of the snake robot. The stability concepts make use of class $\mathcal{K}$ and class $\mathcal{KL}$ functions. A function being of class $\mathcal{K}$ basically means that the function is strictly increasing with respect to its argument. A function of class $\mathcal{KL}$ has two arguments, is strictly increasing with respect to the first argument when the second argument is fixed, and is decreasing with respect to the second argument when the first argument is fixed. Class $\mathcal{K}$ and class $\mathcal{KL}$ functions are formally defined as follows.

Definition 8.1 (Class $\mathcal{K}$ functions, see Definition 4.2 in Khalil 2002) A continuous function $\alpha : \mathbb{R}^+ \to \mathbb{R}^+$ is said to belong to class $\mathcal{K}$ if it is strictly increasing and $\alpha(0) = 0$.

Definition 8.2 (Class $\mathcal{KL}$ functions, see Definition 4.3 in Khalil 2002) A continuous function $\beta : \mathbb{R}^+ \times \mathbb{R}^+ \to \mathbb{R}^+$ is said to belong to class $\mathcal{KL}$ if, for each fixed s, the mapping $\beta(r, s)$ belongs to class $\mathcal{K}$ with respect to r and, for each fixed r, the mapping $\beta(r, s)$ is decreasing with respect to s and $\beta(r, s) \to 0$ as $s \to \infty$.

We now present some stability concepts for systems of the form

$$\dot{\mathbf{x}} = \mathbf{f}(t, \mathbf{x}), \tag{8.1}$$

where $\mathbf{f} : \mathbb{R}_{\geq 0} \times \mathbb{R}^n \to \mathbb{R}^n$ is piecewise continuous in t and locally Lipschitz in $\mathbf{x}$.

Definition 8.3 (UGAS, see Lemma 4.5 in Khalil 2002) The equilibrium point $\mathbf{x} = \mathbf{0}$ of system (8.1) is uniformly globally asymptotically stable (UGAS) if there exists a class $\mathcal{KL}$ function β such that for any initial state $\mathbf{x}(t_0)$,

$$\|\mathbf{x}(t)\| \leq \beta\big(\|\mathbf{x}(t_0)\|, t - t_0\big) \quad \forall t \geq t_0 \geq 0. \tag{8.2}$$

A system being UGAS basically means that the state $\mathbf{x}$ converges to zero as $t \to \infty$. A special case of UGAS arises when the class $\mathcal{KL}$ function β takes the form of an exponential function as follows.

Definition 8.4 (UGES, see Definition 4.5 in Khalil 2002) The equilibrium point $\mathbf{x} = \mathbf{0}$ of system (8.1) is uniformly globally exponentially stable (UGES) if there exist positive constants k and λ such that for any initial state $\mathbf{x}(t_0)$,

$$\|\mathbf{x}(t)\| \leq k\|\mathbf{x}(t_0)\|e^{-\lambda(t-t_0)} \quad \forall t \geq t_0 \geq 0. \tag{8.3}$$

A slightly weaker form of stability than exponential stability is $\mathcal{K}$-exponential stability, which is defined as follows.

Definition 8.5 (Global $\mathcal{K}$-exponential stability, see Definition 2 in Sørdalen and Egeland 1995) The equilibrium point $\mathbf{x} = \mathbf{0}$ of system (8.1) is globally $\mathcal{K}$-exponentially stable if there exist a positive constant λ and a class $\mathcal{K}$ function α such that for any initial state $\mathbf{x}(t_0)$,

$$\|\mathbf{x}(t)\| \leq \alpha\big(\|\mathbf{x}(t_0)\|\big)e^{-\lambda(t-t_0)} \quad \forall t \geq t_0 \geq 0. \tag{8.4}$$

As first noted in Lefeber (2000), the following corollary holds.

Corollary 8.1 *Global $\mathcal{K}$-exponential stability is equivalent to the system being both UGAS and ULES (uniformly locally exponentially stable).*

Remark 8.1 For simplicity, if the equilibrium point $\mathbf{x} = \mathbf{0}$ of a system is UGAS/UGES/globally $\mathcal{K}$-exponentially stable, we often say that the system itself is UGAS/UGES/globally $\mathcal{K}$-exponentially stable.

Next, we present some stability concepts for cascaded systems of the form

$$\dot{\mathbf{x}} = \mathbf{f}_1(t, \mathbf{x}) + \mathbf{g}(t, \mathbf{x}, \mathbf{y})\mathbf{y}, \tag{8.5}$$

$$\dot{\mathbf{y}} = \mathbf{f}_2(t, \mathbf{y}), \tag{8.6}$$

where $\mathbf{x} \in \mathbb{R}^n$, $\mathbf{y} \in \mathbb{R}^m$, $\mathbf{f}_1(t, \mathbf{x})$ is continuously differentiable in $(t, \mathbf{x})$, and $\mathbf{f}_2(t, \mathbf{y})$ and $\mathbf{g}(t, \mathbf{x}, \mathbf{y})$ are continuous in their arguments and locally Lipschitz in $\mathbf{y}$ and $(\mathbf{x}, \mathbf{y})$, respectively. Many dynamical systems can be written in this cascaded form, where we see that the $\mathbf{y}$-dynamics in (8.6) perturbs the $\mathbf{x}$-dynamics in (8.5) through the interconnection term $\mathbf{g}(t, \mathbf{x}, \mathbf{y})\mathbf{y}$.

Theorem 8.1 (See Theorem 2 in Panteley and Loria 1998) *The cascaded system* (8.5), (8.6) *is UGAS if the following three assumptions are satisfied*:

(A1) *The system $\dot{\mathbf{x}} = \mathbf{f}_1(t, \mathbf{x})$ is UGAS with a radially unbounded Lyapunov function satisfying*

$$\left\| \frac{\partial V}{\partial \mathbf{x}} \right\| \|\mathbf{x}\| \leq cV(t, \mathbf{x}) \quad \forall \|\mathbf{x}\| \geq \eta, \tag{8.7}$$

where $c > 0$ and $\eta > 0$ are constants.

(A2) *The function $\mathbf{g}(t, \mathbf{x}, \mathbf{y})$ satisfies*

$$\left\|\mathbf{g}(t, \mathbf{x}, \mathbf{y})\right\| \leq \theta_1\left(\|\mathbf{y}\|\right) + \theta_2\left(\|\mathbf{y}\|\right)\|\mathbf{x}\|, \tag{8.8}$$

where $\theta_1, \theta_2 : \mathbb{R}_{\geq 0} \to \mathbb{R}_{\geq 0}$ are continuous.

(A3) *The system $\dot{\mathbf{y}} = \mathbf{f}_2(t, \mathbf{y})$ is UGAS, and for all $t_0 \geq 0$,*

$$\int_{t_0}^{\infty} \left\|\mathbf{y}(t)\right\| \mathrm{d}t \leq \kappa\left(\left\|\mathbf{y}(t_0)\right\|\right), \tag{8.9}$$

where the function $\kappa(\cdot)$ is a class $\mathcal{K}$ function.

Lemma 8.1 (See Lemma 8 in Panteley et al. 1998) *If in addition to the assumptions in Theorem* 8.1, *both $\dot{\mathbf{x}} = \mathbf{f}_1(t, \mathbf{x})$ and $\dot{\mathbf{y}} = \mathbf{f}_2(t, \mathbf{y})$ are globally $\mathcal{K}$-exponentially stable, then the cascaded system* (8.5), (8.6) *is globally $\mathcal{K}$-exponentially stable.*

8.3 Straight Line Path Following Control of Snake Robots

In this section, we design and analyse a straight line path following controller for a snake robot described by the simplified model in (6.35a)–(6.35h).

8.3.1 Control Objective

The control objective is to steer the snake robot so that it converges to and subsequently tracks a straight path while maintaining a heading which is parallel to the path. To this end, we define the global coordinate system so that the global x axis is aligned with the desired straight path. The position of the snake robot along the global y axis, p_y, is thereby the shortest distance from the robot to the desired path (i.e. the cross-track error), and the orientation of the snake robot, θ, is the angle that the robot forms with the desired path. The control objective is thus to regulate p_y and θ to zero.

Since snake robot locomotion is a slow form of robotic mobility, which is generally employed for traversability purposes, we consider it less important to accurately control the forward velocity of the robot. During path following with a snake robot, it therefore makes sense to focus all the control efforts on converging to the path and subsequently progressing along the path at some non-zero forward velocity $v_t \in [V_{\min}, V_{\max}]$, where $V_{\min}$ and $V_{\max}$ represent the boundaries of some positive interval in which we would like the forward velocity to be contained.

From the above discussion, the control problem is to design a feedback control law

$$\mathbf{u} = \mathbf{u}(t, \boldsymbol{\phi}, \theta, p_y, \mathbf{v}_\phi, v_\theta, v_t, v_n) \in \mathbb{R}^{N-1} \tag{8.10}$$

such that the following control objectives are reached:

$$\lim_{t\to\infty} p_y(t) = 0, \tag{8.11}$$

$$\lim_{t\to\infty} \theta(t) = 0. \tag{8.12}$$

Remark 8.2 The path following controller proposed in Chap. 5 did not attempt to suppress the oscillatory behaviour of the heading and position of the snake robot during motion along the desired path. However, since the path following controller proposed in the following is based on the simplified model of the snake robot, circumventing this oscillating behaviour is a more manageable task. In this chapter, we therefore do not just attempt to regulate the cross-track error and the heading so that they oscillate about zero, but also so that they converge to zero.

8.3.2 Assumptions

Similar to the approach in Chap. 5, we will base the path following controller on the gait pattern lateral undulation, which was defined in (4.32). In Chap. 7, we investigated the velocity dynamics of a snake robot during lateral undulation. In particular, Theorem 7.3 implies that the forward velocity during lateral undulation oscillates around a positive non-zero average velocity that can be predetermined based on the gait pattern parameters. In other words, when the snake robot conducts lateral undulation, Theorem 7.3 suggests that the forward velocity is contained in some non-zero and positive interval $[V_{\min}, V_{\max}]$ that can be scaled based on the gait pattern parameters. We therefore choose to base the path following controller of the snake robot on the following assumption:

Assumption 8.1 *The snake robot moves by lateral undulation and has a forward velocity which is always non-zero and positive, i.e.* $v_t \in [V_{\min}, V_{\max}]$ $\forall t \geq 0$, *where* $V_{\max} \geq V_{\min} > 0$.

Remark 8.3 In addition to the support from Theorem 7.3, the validity of Assumption 8.1 can be seen by inspecting the equations of motion in (6.35a)–(6.35h). The dynamics of the forward velocity in (6.35g) contains three terms. As implied by the averaging analysis in Chap. 7, the term $-\frac{c_p}{Nm}\boldsymbol{\phi}^T\mathbf{A}\overline{\mathbf{D}}\mathbf{v}_\phi$ is positive and accelerates the robot forward during lateral undulation, while the term $-\frac{c_t}{m}v_t$ is the ground friction force. The combined effect of these two terms can never make the forward velocity v_t become zero during lateral undulation. This leaves $\frac{2c_p}{Nm}v_n\overline{\mathbf{e}}^T\boldsymbol{\phi}$ as the only term that can produce a negative forward acceleration that forces v_t to zero. This term is negative when the sideways velocity v_n and the sum of the joint coordinates $\overline{\mathbf{e}}^T\boldsymbol{\phi}$ have opposite signs. However, it can be seen from (6.35h) that v_n and $\overline{\mathbf{e}}^T\boldsymbol{\phi}$ will always tend in the same rather than the opposite direction when $v_t > 0$. It is therefore unlikely for v_n and $\overline{\mathbf{e}}^T\boldsymbol{\phi}$ to have opposite signs over the long period required to force v_t to zero.

8.3.3 Model Transformation

On the basis of the preceding discussion and Assumption 8.1, we will not control the dynamics of the forward velocity v_t given by (6.35g) and instead treat the forward velocity as a positive parameter satisfying $v_t \in [V_{\min}, V_{\max}]$.

As seen in (6.35f) and (6.35h), the joint coordinates $\boldsymbol{\phi}$ are present in the dynamics of both the angular velocity v_θ and the sideways velocity v_n of the snake robot. This complicates the controller design since the body shape changes will affect both the heading and the sideways motion of the robot. Motivated by Do and Pan (2003), we see that it is possible to remove the effect of $\boldsymbol{\phi}$ on the sideways velocity by a coordinate transformation. In particular, we move the point that determines the position of the snake robot a distance ϵ along the tangential direction of the robot from the CM to a new location, which is precisely where the body shape changes of the robot

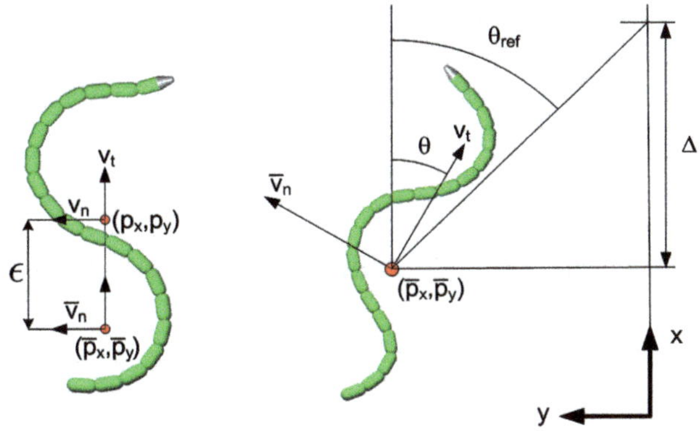

Fig. 8.1 *Left*: The coordinate transformation of the snake robot. *Right*: The Line-of-Sight (LOS) guidance law

(characterised by $\bar{\mathbf{e}}^T\boldsymbol{\phi}$) generate a pure rotational motion and no sideways force. This coordinate transformation is illustrated to the left in Fig. 8.1 and is defined as

$$\bar{p}_x = p_x + \epsilon\cos\theta, \tag{8.13a}$$

$$\bar{p}_y = p_y + \epsilon\sin\theta, \tag{8.13b}$$

$$\bar{v}_n = v_n + \epsilon v_\theta, \tag{8.13c}$$

where ϵ is a constant parameter defined as

$$\epsilon = -\frac{2(N-1)}{Nm}\frac{c_p}{\lambda_2}. \tag{8.14}$$

With the new coordinates in (8.13a)–(8.13c), the model(6.35a)–(6.35h) is transformed into

$$\dot{\boldsymbol{\phi}} = \mathbf{v}_\phi, \tag{8.15a}$$

$$\dot{\theta} = v_\theta, \tag{8.15b}$$

$$\dot{\bar{p}}_y = v_t\sin\theta + \bar{v}_n\cos\theta, \tag{8.15c}$$

$$\dot{\mathbf{v}}_\phi = -\frac{c_n}{m}\mathbf{v}_\phi + \frac{c_p}{m}v_t\mathbf{A}\mathbf{D}^T\boldsymbol{\phi} + \frac{1}{m}\mathbf{D}\mathbf{D}^T\mathbf{u}, \tag{8.15d}$$

$$\dot{v}_\theta = -\lambda_1 v_\theta + \frac{\lambda_2}{N-1}v_t\bar{\mathbf{e}}^T\boldsymbol{\phi}, \tag{8.15e}$$

$$\dot{\bar{v}}_n = Xv_\theta + Y\bar{v}_n, \tag{8.15f}$$

where, by Assumption 8.1, the parameter $v_t \in [V_{\min}, V_{\max}]$ with $V_{\max} \geq V_{\min} > 0$, and where

$$X = \epsilon\left(\frac{c_n}{m} - \lambda_1\right), \tag{8.16a}$$

$$Y = -\frac{c_n}{m}. \tag{8.16b}$$

The two scalar constants X and Y have been introduced in (8.15f) for simplicity of notation in the following sections. Note also that (6.35c) is not included in (8.15a)–(8.15f) since we do not consider the temporal position of the system along the path during path following.

8.3.4 The Straight Line Path Following Controller

The path following controller of the snake robot consists of two main components. The first component is the gait pattern controller, which propels the snake robot forward according to the gait pattern lateral undulation (as stated in Assumption 8.1). The second component is the heading controller, which steers the snake robot towards and subsequently along the desired path. The two components of the path following controller are now presented.

Gait Pattern Controller

As stated in Definition 4.1, a snake robot moves by lateral undulation by controlling joint $i \in \{1, \dots, N-1\}$ according to

$$\phi_{i,\mathrm{ref}} = \alpha \sin\bigl(\omega t + (i-1)\delta\bigr) + \phi_o, \tag{8.17}$$

where α and ω are the amplitude and frequency, respectively, of the sinusoidal joint motion, δ determines the phase shift between the joints, and ϕ_o is a joint offset, which the heading controller will use to control the direction of the locomotion. The average forward velocity v_t^* of the snake robot during straight path motion is given from Theorem 7.3 as

$$v_t^* = \frac{c_p}{2Nc_t}\alpha^2\omega k_\delta, \tag{8.18}$$

where k_δ is a constant parameter determined by the phase shift δ. This relation can be used to choose the gait parameters α, ω, and δ in order to achieve the desired average forward velocity.

As proposed in Sect. 6.6, we set the actuator forces according to the linearising control law

$$\mathbf{u} = m\bigl(\mathbf{D}\mathbf{D}^T\bigr)^{-1}\left(\overline{\mathbf{u}} + \frac{c_n}{m}\dot{\boldsymbol{\phi}} - \frac{c_p}{m}v_t\mathbf{A}\mathbf{D}^T\boldsymbol{\phi}\right), \tag{8.19}$$

where $\overline{\mathbf{u}} \in \mathbb{R}^{N-1}$ is a new set of control inputs. This control law transforms the joint dynamics (8.15d) into $\dot{\mathbf{v}}_\phi = \ddot{\boldsymbol{\phi}} = \overline{\mathbf{u}}$. In order to make the joints track the joint reference coordinates given by (8.17), we employ the joint controller from Sect. 4.8.2, i.e. we set the new control input $\overline{\mathbf{u}}$ according to

$$\overline{\mathbf{u}} = \ddot{\boldsymbol{\phi}}_{\mathrm{ref}} + k_{v_\phi}(\dot{\boldsymbol{\phi}}_{\mathrm{ref}} - \dot{\boldsymbol{\phi}}) + k_\phi(\boldsymbol{\phi}_{\mathrm{ref}} - \boldsymbol{\phi}), \tag{8.20}$$

where $k_\phi > 0$ and $k_{v_\phi} > 0$ are scalar controller gains, and $\boldsymbol{\phi}_{\text{ref}} \in \mathbb{R}^{N-1}$ are the joint reference coordinates given by (8.17). By introducing the error variable

$$\widetilde{\boldsymbol{\phi}} = \boldsymbol{\phi} - \boldsymbol{\phi}_{\text{ref}}, \tag{8.21}$$

the joint dynamics given by (8.15a) and (8.15d) can be written in terms of the error dynamics

$$\ddot{\widetilde{\boldsymbol{\phi}}} + k_{v_\phi} \dot{\widetilde{\boldsymbol{\phi}}} + k_\phi \widetilde{\boldsymbol{\phi}} = \mathbf{0}, \tag{8.22}$$

which is clearly *exponentially stable* (see Khalil 2002). This means that the joint coordinates exponentially track the reference coordinates given by (8.17).

Heading Controller

In order to steer the snake robot towards the desired straight path, we employ the Line-of-Sight (LOS) guidance law that was also employed in Chap. 5 and is defined as

$$\theta_{\text{ref}} = -\arctan\left(\frac{\overline{p}_y}{\Delta}\right), \tag{8.23}$$

where $\overline{p}_y$ is the cross-track error, and $\Delta > 0$ is a design parameter referred to as the *look-ahead distance*. This LOS guidance law is commonly used during e.g. path following control of marine surface vessels (see e.g. Fossen 2002; Fredriksen and Pettersen 2006). As illustrated to the right in Fig. 8.1, the LOS angle θ_{ref} corresponds to the orientation of the snake robot when it is headed towards the point located a distance Δ ahead of the snake robot along the desired path. The value of Δ is important since it determines the rate of convergence to the desired path.

Based on Property 4.7, we will use the joint offset coordinate ϕ_o in (8.17) to ensure that the heading of the snake robot θ tracks the LOS angle given by (8.23). Motivated by Fredriksen and Pettersen (2006) and Pettersen and Lefeber (2001), we conjecture that making θ track the LOS angle θ_{ref} will make the snake converge to the desired path and subsequently follow the path with its heading parallel to the path. In other words, we conjecture that a control law that makes θ track θ_{ref} will fulfill the control objectives (8.11) and (8.12). To derive the control law for ϕ_o, we first rewrite the dynamics of v_θ given by (8.15e) with the new coordinates $\widetilde{\boldsymbol{\phi}}$ in (8.21), which gives the dynamics of v_θ as a function of the joint reference coordinates given by (8.17). From (8.21) we have that $\boldsymbol{\phi} = \boldsymbol{\phi}_{\text{ref}} + \widetilde{\boldsymbol{\phi}}$. Using (8.17), we can therefore rewrite (8.15e) as

$$\dot{v}_\theta = -\lambda_1 v_\theta + \lambda_2 v_t \phi_o + \frac{\lambda_2}{N-1} v_t \left(\sum_{i=1}^{N-1} \alpha \sin\bigl(\omega t + (i-1)\delta\bigr) + \overline{\mathbf{e}}^T \widetilde{\boldsymbol{\phi}} \right). \tag{8.24}$$

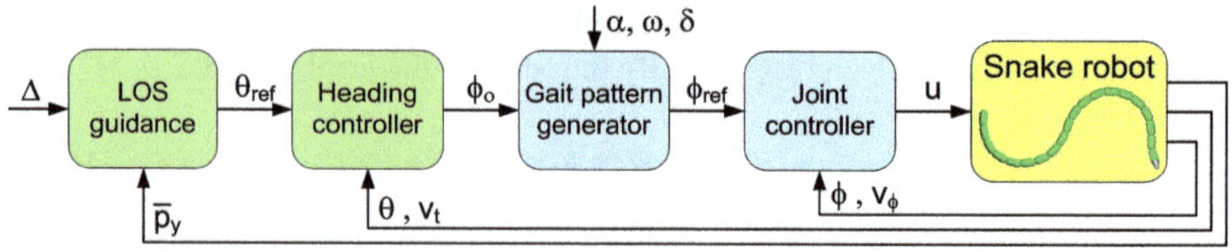

Fig. 8.2 The structure of the straight line path following controller

Consequently, choosing ϕ_o as

$$\phi_o = \frac{1}{\lambda_2 v_t}\left(\ddot{\theta}_{\mathrm{ref}} + \lambda_1 \dot{\theta}_{\mathrm{ref}} - k_\theta(\theta - \theta_{\mathrm{ref}}) - \frac{\lambda_2}{N-1} v_t \sum_{i=1}^{N-1} \alpha \sin\big(\omega t + (i-1)\delta\big)\right), \tag{8.25}$$

where $k_\theta > 0$ is a scalar controller gain, enables us to write the dynamics of the heading angle θ, which is given by (8.15b) and (8.15e), in terms of the error dynamics

$$\ddot{\widetilde{\theta}} + \lambda_1 \dot{\widetilde{\theta}} + k_\theta \widetilde{\theta} = \frac{\lambda_2}{N-1} v_t \bar{\mathbf{e}}^T \widetilde{\boldsymbol{\phi}}, \tag{8.26}$$

where we have introduced the error variable

$$\widetilde{\theta} = \theta - \theta_{\mathrm{ref}}. \tag{8.27}$$

Remark 8.4 The joint coordinate offset in (8.25) depends on the inverse of the forward velocity v_t. This does not represent a problem since, by Assumption 8.1, the forward velocity is always non-zero. When implementing the path following controller, this issue can be avoided by activating the controller *after* the snake robot has obtained a positive forward velocity.

Remark 8.5 The error dynamics of the joints in (8.22) and the error dynamics of the heading in (8.26) represent a cascaded system. In particular, system (8.22) perturbs system (8.26) through the interconnection term $\frac{\lambda_2}{N-1} v_t \bar{\mathbf{e}}^T \widetilde{\boldsymbol{\phi}}$. Using cascaded systems theory, it will be shown in Sect. 8.3.6 that the origin of this cascaded system is globally $\mathcal{K}$-exponentially stable.

We have now presented the complete path following controller of the snake robot. The structure of the complete controller is summarised in Fig. 8.2.

8.3.5 The Stability Properties of the Path Following Controller

Based on the guidance and control laws presented in the previous subsection, we now state the main result concerning the straight line path following controller. The result specifies a lower bound on the look-ahead distance Δ employed in (8.23). This bound is given a physical interpretation in Remark 8.7 below and is derived in the proof presented in Sect. 8.3.6.

Theorem 8.2 *Consider a planar snake robot described by the model* (8.15a)–(8.15f) *and suppose that Assumption* 8.1 *is satisfied. If the look-ahead distance Δ of the LOS guidance law* (8.23) *is chosen such that*

$$\Delta > \frac{|X|}{|Y|}\left(1 + \frac{V_{\max}}{V_{\min}}\right), \tag{8.28}$$

then the path following controller defined by (8.17), (8.19), (8.20), (8.23), *and* (8.25) *guarantees that the control objectives* (8.11) *and* (8.12) *are achieved for any set of initial conditions satisfying $v_t \in [V_{\min}, V_{\max}]$.*

Proof The proof of this theorem is given in Sect. 8.3.6. □

Remark 8.6 Theorem 8.2 does not specify the boundary values $V_{\min}$ and $V_{\max}$ of the interval in which the forward velocity v_t is contained. By Assumption 8.1, however, there exists a positive interval that contains v_t for all time $t \geq 0$. In order to be able to quantify the bound on Δ, it would be helpful if we could specify $V_{\min}$ and $V_{\max}$ as a function of the gait pattern parameters α, ω, δ, and ϕ_o. This is an interesting, but still unsolved, problem.

Remark 8.7 The lower bound on the look-ahead distance Δ in (8.28) ensures that the sideways velocity $\overline{v}_n$ of the snake robot in (8.15f) is well behaved under the perturbations from the angular velocity v_θ. In particular, the magnitude of v_θ during convergence to the desired path is determined by the look-ahead distance Δ, i.e. the robot rotates fast when Δ is small (and vice versa). We see from (8.15f) that v_θ only has a small influence on $\overline{v}_n$ when $|X| \ll |Y|$, which means that we then can allow the magnitude of v_θ to be large, i.e. Δ can be small. Similarly, v_θ has a great influence on $\overline{v}_n$ when $|X| \gg |Y|$, which means that the magnitude of v_θ must be restricted, i.e. Δ must be large. These conditions are directly reflected by the lower bound in (8.28).

Remark 8.8 As explained in Remark 6.1 in Sect. 6.2, the assumptions underlying the simplified model are only valid as long as the link angles with respect to the forward direction are limited. The stability result in Theorem 8.2 is therefore claimed only for snake robots conducting lateral undulation with limited link angles.

8.3.6 Proof of Theorem 8.2

We will prove Theorem 8.2 in three steps. In the first step, we show that the complete system, including the path following controller, can be written as a cascaded system. In the second step, we prove stability of the nominal systems in the cascade. Finally, we derive bounds on the interconnection terms between the nominal systems, which, by Theorem 8.1 and Lemma 8.1, allow us to conclude the stability of the complete cascaded system. We will follow the steps of a similar proof presented in Pavlov et al. (2007).

We begin by rewriting the dynamics of the cross-track error $\overline{p}_y$ and the sideways velocity $\overline{v}_n$ in terms of the heading error $\widetilde{\theta}$. From (8.27) and (8.23) we have that

$$\theta = -\arctan\left(\frac{\overline{p}_y}{\Delta}\right) + \widetilde{\theta}. \tag{8.29}$$

By using the relations

$$\begin{aligned} \sin\left(-\arctan\left(\frac{\overline{p}_y}{\Delta}\right)\right) &= -\frac{\overline{p}_y}{\sqrt{\overline{p}_y^2 + \Delta^2}}, \\ \cos\left(-\arctan\left(\frac{\overline{p}_y}{\Delta}\right)\right) &= \frac{\Delta}{\sqrt{\overline{p}_y^2 + \Delta^2}}, \end{aligned} \tag{8.30}$$

it can be verified that (8.15c) can be written in terms of the heading error $\widetilde{\theta}$ as

$$\dot{\overline{p}}_y = -\frac{v_t}{\sigma}\overline{p}_y + \frac{\Delta}{\sigma}\overline{v}_n + \gamma\widetilde{\theta}, \tag{8.31}$$

where

$$\sigma = \sqrt{\overline{p}_y^2 + \Delta^2}, \tag{8.32}$$

$$\gamma = \frac{\sin\widetilde{\theta}}{\widetilde{\theta}}\frac{(v_t\Delta + \overline{v}_n\overline{p}_y)}{\sigma} + \frac{1-\cos\widetilde{\theta}}{\widetilde{\theta}}\frac{(v_t\overline{p}_y - \overline{v}_n\Delta)}{\sigma}. \tag{8.33}$$

Through similar manipulations, we can rewrite (8.15f) in the new coordinates as

$$\dot{\overline{v}}_n = \frac{X\Delta v_t}{\sigma^3}\overline{p}_y + \left(Y - \frac{X\Delta^2}{\sigma^3}\right)\overline{v}_n - \frac{X\Delta}{\sigma^2}\gamma\widetilde{\theta} + X\dot{\widetilde{\theta}}. \tag{8.34}$$

Collecting the error variables as

$$\eta = \begin{bmatrix} \widetilde{\boldsymbol{\phi}} \\ \dot{\widetilde{\boldsymbol{\phi}}} \end{bmatrix} \in \mathbb{R}^{2N-2}, \qquad \xi = \begin{bmatrix} \widetilde{\theta} \\ \dot{\widetilde{\theta}} \end{bmatrix} \in \mathbb{R}^2, \tag{8.35}$$

and using (8.22), (8.26), (8.31), and (8.34), we can write the model of the snake robot (8.15a)–(8.15f) during path following as

$$\begin{bmatrix} \dot{\overline{p}}_y \\ \dot{\overline{v}}_n \end{bmatrix} = \mathbf{C}(\overline{p}_y) \begin{bmatrix} \overline{p}_y \\ \overline{v}_n \end{bmatrix} + \mathbf{H}_\xi(\overline{p}_y, \overline{v}_n, \boldsymbol{\xi})\boldsymbol{\xi}, \tag{8.36a}$$

$$\dot{\boldsymbol{\xi}} = \begin{bmatrix} 0 & 1 \\ -k_\theta & -\lambda_1 \end{bmatrix} \boldsymbol{\xi} + \mathbf{H}_\eta \boldsymbol{\eta}, \tag{8.36b}$$

$$\dot{\boldsymbol{\eta}} = \begin{bmatrix} \mathbf{0}_{(N-1)\times(N-1)} & \mathbf{I}_{N-1} \\ -k_\phi \mathbf{I}_{N-1} & -k_{v_\phi} \mathbf{I}_{N-1} \end{bmatrix} \boldsymbol{\eta}, \tag{8.36c}$$

where

$$\mathbf{H}_\eta = \begin{bmatrix} \mathbf{0}_{1\times(N-1)} & \mathbf{0}_{1\times(N-1)} \\ \frac{\lambda_2}{N-1} v_t \bar{\mathbf{e}}^T & \mathbf{0}_{1\times(N-1)} \end{bmatrix} \in \mathbb{R}^{2\times(2N-2)}, \tag{8.37}$$

$$\mathbf{H}_\xi(\overline{p}_y, \overline{v}_n, \boldsymbol{\xi}) = \begin{bmatrix} \gamma & 0 \\ -\frac{X\Delta}{\sigma^2}\gamma & X \end{bmatrix} \in \mathbb{R}^{2\times 2}, \tag{8.38}$$

$$\mathbf{C}(\overline{p}_y) = \begin{bmatrix} -\frac{v_t}{\sigma} & \frac{\Delta}{\sigma} \\ \frac{X\Delta v_t}{\sigma^3} & (Y - \frac{X\Delta^2}{\sigma^3}) \end{bmatrix} \in \mathbb{R}^{2\times 2}. \tag{8.39}$$

System (8.36a)–(8.36c) is a cascaded system. In particular, the $\boldsymbol{\eta}$-dynamics in (8.36c) perturbs the $\boldsymbol{\xi}$-dynamics in (8.36b) through the interconnection term $\mathbf{H}_\eta\boldsymbol{\eta}$, and the $\boldsymbol{\xi}$-dynamics perturbs the $(\overline{p}_y, \overline{v}_n)$-dynamics in (8.36a) through the interconnection term $\mathbf{H}_\xi(\overline{p}_y, \overline{v}_n, \boldsymbol{\xi})\boldsymbol{\xi}$.

We now investigate the stability of the nominal systems of the cascade, i.e. all parts of (8.36a)–(8.36c) except the interconnection terms. The origin $\boldsymbol{\eta} = \mathbf{0}$ of the linear system (8.36c) and the origin $\boldsymbol{\xi} = \mathbf{0}$ of the linear nominal system in (8.36b) are UGES (see Definition 8.4) since the system matrices clearly are Hurwitz (Khalil 2002) for $k_\theta, \lambda_1, k_\phi, k_{v_\phi} > 0$. The nominal system of (8.36a) is given by

$$\begin{bmatrix} \dot{\overline{p}}_y \\ \dot{\overline{v}}_n \end{bmatrix} = \mathbf{C}(\overline{p}_y) \begin{bmatrix} \overline{p}_y \\ \overline{v}_n \end{bmatrix} \tag{8.40}$$

and has the stability properties established by the following two lemmas.

Lemma 8.2 *Under the conditions of Theorem* 8.2, *the origin of the system* (8.40) *is UGAS with a quadratic Lyapunov function.*

Proof The proof of this lemma has previously been presented in Pavlov et al. (2007) and is included in Appendix A for completeness. □

Lemma 8.3 *Under the conditions of Theorem* 8.2, *the origin of the system* (8.40) *is globally $\mathcal{K}$-exponentially stable.*

Proof The proof of this lemma is presented in Appendix B. □

Since exponential stability implies $\mathcal{K}$-exponential stability, we can conclude that all nominal systems of the cascade (8.36a)–(8.36c) are globally $\mathcal{K}$-exponentially stable. Next, we derive bounds on the interconnection terms in the cascade. The induced 2-norm of the matrix $\mathbf{H}_\eta$ satisfies (see Appendix A in Khalil 2002)

$$\|\mathbf{H}_\eta\|_2 \le \sqrt{2N-2}\max_j \sum_{i=1}^{2}\{\mathbf{H}_\eta\}_{ij} \le \frac{\sqrt{2}\lambda_2 V_{\max}}{\sqrt{N-1}}, \tag{8.41}$$

while the induced 2-norm of the matrix $\mathbf{H}_\xi(\overline{p}_y, \overline{v}_n, \boldsymbol{\xi})$ satisfies

$$\begin{aligned} \|\mathbf{H}_\xi\|_2 &\le \sqrt{2}\max_j \sum_{i=1}^{2}\{\mathbf{H}_\xi\}_{ij} \le \sqrt{2}\max\left(|\gamma| + \frac{|X|\Delta}{\sigma^2}|\gamma|, |X|\right) \\ &\le \sqrt{2}\left(|\gamma| + \frac{|X|\Delta}{\sigma^2}|\gamma| + |X|\right). \end{aligned} \tag{8.42}$$

The function γ given by (8.33) is bounded according to

$$\begin{aligned} \gamma &\le \left|\frac{\sin\widetilde{\theta}}{\widetilde{\theta}}\right| \frac{V_{\max}\Delta + |\overline{v}_n||\overline{p}_y|}{\sigma} + \left|\frac{1-\cos\widetilde{\theta}}{\widetilde{\theta}}\right| \frac{V_{\max}|\overline{p}_y| + |\overline{v}_n|\Delta}{\sigma} \\ &\le \frac{V_{\max}\Delta}{\sigma} + \frac{|\overline{v}_n||\overline{p}_y|}{\sigma} + \frac{V_{\max}|\overline{p}_y|}{\sigma} + \frac{|\overline{v}_n|\Delta}{\sigma} \le 2V_{\max} + 2|\overline{v}_n|. \end{aligned} \tag{8.43}$$

By inserting (8.43) into (8.42), it is straightforward to verify that

$$\|\mathbf{H}_\xi\|_2 \le \mathcal{F}_1 + \mathcal{F}_2\left\|\begin{bmatrix}\overline{p}_y \\ \overline{v}_n\end{bmatrix}\right\|_2, \tag{8.44}$$

where

$$\mathcal{F}_1 = \sqrt{2}\left(2V_{\max}\left(1 + \frac{|X|}{\Delta}\right) + |X|\right), \tag{8.45}$$

$$\mathcal{F}_2 = 2\sqrt{2}\left(1 + \frac{|X|}{\Delta}\right). \tag{8.46}$$

We are now ready to apply Theorem 8.1 to the cascaded system (8.36a)–(8.36c). We first consider the cascade of (8.36b) and (8.36c), for which it is straightforward to verify that Assumptions A1 and A3 of Theorem 8.1 are satisfied since system (8.36c) and the nominal system of (8.36b) are both UGES (see Definition 8.4). Furthermore, Assumption A2 is trivially satisfied since $\|\mathbf{H}_\eta\|_2$ is bounded by the constant derived in (8.41). The cascaded system (8.36b), (8.36c) is therefore UGAS and, by Lemma 8.1, also globally $\mathcal{K}$-exponentially stable.

Next, we consider the cascade of (8.36a) and (8.36b), for which Assumption A1 of Theorem 8.1 is satisfied since, by Lemma 8.2, the nominal system of (8.36a) is UGAS with a quadratic Lyapunov function. Furthermore, it follows directly from (8.44) that Assumption A2 is satisfied. Finally, since the perturbing system (8.36b) is globally $\mathcal{K}$-exponentially stable, Assumption A3 is satisfied since the bound in Assumption A3 is easily shown to hold for any $\mathcal{K}$-exponentially stable system by integrating both sides of (8.4) from t_0 to ∞. The cascaded system (8.36a), (8.36b) is therefore UGAS and, by Lemma 8.1, also globally $\mathcal{K}$-exponentially stable since the nominal system of (8.36a) and the perturbing system (8.36b) are both globally $\mathcal{K}$-exponentially stable.

In summary, the complete cascaded system (8.36a)–(8.36c) is globally $\mathcal{K}$-exponentially stable. This means that $\overline{p}_y(t) \to 0$ and $\widetilde{\theta}(t) \to 0$, which, by (8.29), implies that $\theta(t) \to 0$, which means that control objective (8.12) is achieved. It subsequently follows from (8.13b) that $p_y(t) \to 0$, which means that control objective (8.11) is achieved. This completes the proof of Theorem 8.2.

Remark 8.9 Any gait pattern controller that exponentially stabilises the error variable (8.21), i.e. not just the joint controller proposed in (8.19) and (8.20), makes the complete cascaded system globally $\mathcal{K}$-exponentially stable. This is a nice feature of cascaded systems theory.

8.4 Path Following Control of Snake Robots Along Curved Paths

The straight line path following controller presented in Sect. 8.3.4 can be extended to path following of general curved paths by following an approach presented in Børhaug (2008) in the context of path following control of marine vessels. In this section, we describe how this extension can be achieved.

8.4.1 Comments on the Curved Path Following Controller

The stability proof regarding the convergence of the snake robot to the curved path will not be detailed in this book since the proof is developed by following similar steps as the approach presented in Børhaug (2008). Moreover, we do not consider curved path following control to be particularly relevant to our long-term goal of achieving snake robot locomotion in cluttered environments. In particular, we would like our control design efforts for snake robots on flat surfaces to be extendable to the case where the environment is uneven and no longer flat. The idea of requiring a snake robot to follow a nice and smooth curved path in an unknown and cluttered environment (i.e. where the geometry of the environment constrains the motion significantly) seems unrealistic. A more realistic and applicable approach for path following of general paths is, in our opinion, to characterise the general

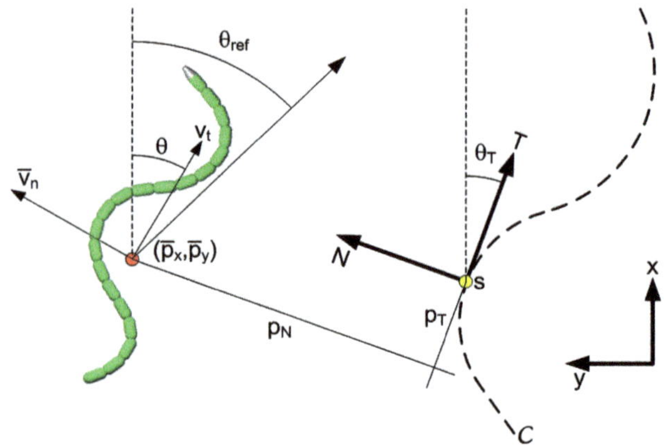

Fig. 8.3 The guidance strategy for path following of general curved paths

path in terms of waypoints interconnected by straight lines, where the straight line reference path between each waypoint basically tells the robot to take the shortest possible path to each waypoint. We will elaborate this approach in Sect. 8.5 and will further develop the approach in Chap. 13 in order to enable snake robot locomotion along general paths in a cluttered environment with obstacles.

8.4.2 The Curved Path Following Controller

The desired curved path that the snake robot should follow is a continuously differentiable curve denoted by $\mathcal{C}$ (see Fig. 8.3). The idea behind the controller is to steer the snake robot towards a *virtual* particle that moves along the path. The distance travelled by the particle along the curve is denoted by s, which means that $\dot{s}$ is the instantaneous speed of the particle along the curve. Furthermore, we define a moving coordinate frame with axes denoted by T and N such that the origin of the frame coincides with the particle and the T axis is always tangential to the curve. This is called a *Serret–Frenet* coordinate frame (see e.g. Egeland and Gravdahl 2002). As visualised in Fig. 8.3, the angle of the T axis with respect to the global x axis is denoted by θ_T, and the position of the snake robot in the T–N frame is denoted by (p_T, p_N).

Since the goal is to make the snake robot converge to and follow the desired path $\mathcal{C}$, we state the control objective as

$$\lim_{t\to\infty} p_T(t) = 0, \qquad \lim_{t\to\infty} p_N(t) = 0. \tag{8.47}$$

Based on the curved path following controller for marine vessels proposed in Børhaug (2008), we claim that control objective (8.47) is achieved by steering the heading θ of the snake robot according to the guidance law

$$\theta_{\text{ref}} = \theta_T - \arctan\left(\frac{\overline{v}_n}{v_t}\right) - \arctan\left(\frac{p_N}{\sqrt{\Delta^2 + p_T^2}}\right) \tag{8.48}$$

and updating the position of the virtual particle along the curve according to

$$\dot{s} = U \frac{\sqrt{\Delta^2 + p_T^2} + p_T}{\sqrt{\Delta^2 + p_T^2 + p_N^2}}, \qquad U = \sqrt{v_t^2 + \overline{v}_n^2}. \tag{8.49}$$

The curved path following controller is, in other words, defined by (8.17), (8.19), (8.20), and (8.25), where θ_{ref} is given by (8.48), and where s is updated according to (8.49). The proof of the achievement of control objective (8.47) can be developed based on the results in Børhaug (2008) by taking into account a few differences between the model of the snake robot and the model of the vessel considered in Børhaug (2008). The proof also requires us to make specific assumptions regarding the minimum forward velocity of the robot, the maximum forward acceleration of the robot, the maximum curvature of the path, and the look-ahead distance Δ.

8.5 Waypoint Guidance Control of Snake Robots

In this section, we employ the straight line path following controller presented in Sect. 8.3 in order to propose a guidance strategy for steering a snake robot between a set of reference locations, or *waypoints*, in the environment of the robot. Waypoint guidance is, in our opinion, an approach which is well suited for general motion control of snake robots. The waypoint guidance strategy proposed in the following represents an operator-friendly framework for steering a snake robot between arbitrary locations on a flat surface.

8.5.1 Description of the Approach

Future applications of snake robots will generally involve motion in challenging and cluttered environments where the aim is to bring sensors and/or tools to a single or several specified target location(s). In these situations, the exact path taken by the robot as it moves towards the target(s) is generally of less interest as long as the robot reaches the target(s) within a reasonable amount of time. Specifying the motion in terms of waypoints supports this target-oriented control approach. Waypoint guidance is a commonly used approach for control of e.g. marine surface vessels (see e.g. Fossen 2002).

In accordance with the target-oriented approach discussed above, we choose to interconnect the waypoints by straight lines and employ the path following controller presented in Sect. 8.3 in order to steer the snake robot towards the straight line leading to the next waypoint. This approach is illustrated in Fig. 8.4. As argued in Sect. 8.4.1, the reason for considering straight lines instead of curved paths is our long-term goal of also employing the guidance strategy in uneven and cluttered environments. The idea of requiring a snake robot to follow a nice and smooth curved

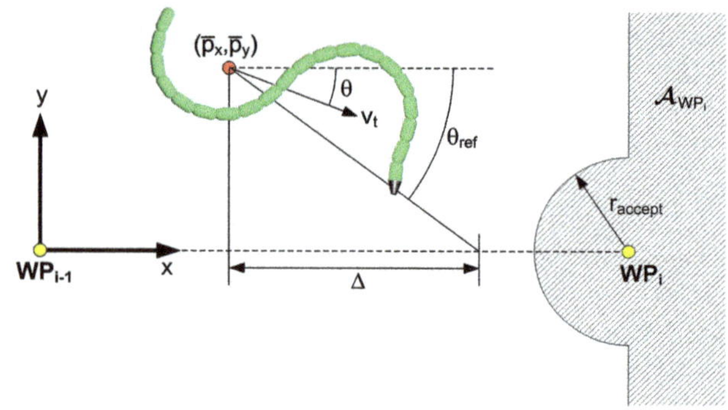

Fig. 8.4 The waypoint guidance strategy

path in an unknown and uneven environment seems unrealistic, while a straight line reference path between each waypoint basically tells the robot to take the shortest possible path to the next waypoint.

A common rule for switching between the waypoints is to proceed towards the next waypoint as soon as the position of the system enters inside an *acceptance circle* enclosing the current waypoint (see Fossen 2002). In the present work, we propose that the acceptance circle is replaced by an *acceptance region* composed of an acceptance circle and also the right half-plane of a coordinate system with origo in the current waypoint and x axis pointing away from the previous waypoint (see illustration in Fig. 8.4). With this definition, we are guaranteed that the robot will reach the acceptance region of the current waypoint no matter how the waypoints are defined. With only acceptance circles enclosing each waypoint, there would be the risk that the robot misses a waypoint which is placed too close to the previous waypoint, which would make the robot proceed indefinitely along the path away from the waypoint that was missed. Note that although the acceptance region is infinitely large, the path following controller presented in Sect. 8.3 guarantees rapid convergence to the straight path between two waypoints.

8.5.2 The Waypoint Guidance Strategy

In the following, we formalise the guidance strategy described in the previous subsection.

Definition 8.6 (Waypoint) A *waypoint* is a reference position along the path of the snake robot. There are k waypoints, and the ith waypoint is denoted by WP_i, where $i \in \{1, \dots, k\}$.

Definition 8.7 (Acceptance region) The *acceptance region* of WP_i, denoted $\mathcal{A}_{WP_i}$, is the union of all points inside a circle centred in WP_i with radius r_{accept} and the right half-plane of a coordinate system with origo in WP_i and x axis aligned with the vector from WP_{i-1} to WP_i.

Definition 8.8 (The waypoint guidance problem) Given a set of k waypoints $WP_1, \dots, WP_k$, the *waypoint guidance problem* is the task of steering the position of the snake robot into the acceptance region of each of the waypoints $WP_1, \dots, WP_k$ in consecutive order.

In accordance with the above definitions and the description in the previous subsection, we now present the waypoint guidance strategy for the snake robot.

Algorithm 8.1 (The waypoint guidance strategy)

1. Define the initial position of the snake robot as WP_0.
2. Repeat for all $i \in \{0, \dots, k-1\}$:
   a. Move the origin of the global frame to WP_i and orient the global x axis towards WP_{i+1}.
   b. Conduct straight line path following according to the controller from Sect. 8.3 until $(p_x, p_y) \in \mathcal{A}_{WP_{i+1}}$.

The guidance strategy proposed in Algorithm 8.1 satisfies the following result:

Theorem 8.3 *The waypoint guidance problem presented in Definition* 8.8 *is solved by Algorithm* 8.1 *for a planar snake robot described by the model* (8.15a)–(8.15f) *under the conditions of Theorem* 8.2.

Proof Given any waypoint WP_i that the snake robot is crawling towards, where $i \in \{1, \dots, k\}$, we are ensured by Definition 8.7 that the desired straight path of the robot points into the acceptance region of WP_i. By Theorem 8.2, the snake robot will eventually reach the desired straight path and progress along the path indefinitely, which means that the position of the snake robot will eventually reach the acceptance region of WP_i. This completes the proof. □

8.6 Simulation Study: The Performance of the Straight Line Path Following Controller

In this section, we present simulation results that illustrate the performance of the straight line path following controller proposed in Sect. 8.3.

8.6.1 Simulation Parameters

The model of the snake robot (6.35a)–(6.35h) and the path following controller defined by (8.17), (8.19), (8.20), (8.23), and (8.25) were implemented and simulated in *Matlab R2008b* on a laptop running *Windows XP*. The model dynamics was calculated using the *ode45* solver in Matlab with a relative and absolute error tolerance of 10^{-6}.

We considered a snake robot with $N = 10$ links of length $l = 0.14$ m and mass $m = 1$ kg. Furthermore, we chose the ground friction coefficients as $c_t = 0.5$ and $c_n = 3$, the rotation parameters as $\lambda_1 = 0.5$ and $\lambda_2 = 20$, the controller gains as $k_\phi = 20$, $k_{v_\phi} = 5$, and $k_\theta = 0.05$, and we calculated the coordinate transformation distance according to (8.14) as $\epsilon = -80.4$ cm. The gait parameters were $\alpha = 4.5$ cm, $\omega = 120°/\text{s}$, and $\delta = 40°$, which by (8.18) corresponds to the average forward velocity $v_t^* = \frac{c_p}{2Nc_t}\alpha^2\omega k_\delta = 9.4$ cm/s. By making the conjecture that the forward velocity will always be contained in the interval $v_t \in [V_{\min}, V_{\max}] = [0.5v_t^*, 2v_t^*] = [4.7\text{ cm/s}, 18.8\text{ cm/s}]$, the lower bound on the look-ahead distance Δ is given by (8.28) as $\Delta > 0.15$ m. During the simulations, we chose the look-ahead distance equal to twice the length of the snake robot, i.e. $\Delta = 2.8$ m, which is well above the estimated lower limit.

The derivatives $\dot{\phi}_o$, $\ddot{\phi}_o$, $\dot{\theta}_{\text{ref}}$, and $\ddot{\theta}_{\text{ref}}$, which are needed for the calculation of the control input in (8.20) and (8.25), were obtained by passing ϕ_o and θ_{ref} through a 3rd-order low-pass filtering reference model (see Appendix C.2). The parameters of the reference model were set to $\omega = \pi/2$ and $\zeta = 1$.

The initial state of the snake robot was chosen as $\boldsymbol{\phi} = \mathbf{0}°$, $\theta = 90°$, $p_x = 0$ m, $p_y = 1$ m, $\mathbf{v}_\phi = \mathbf{0}°/\text{s}$, $v_\theta = 0°/\text{s}$, $v_t = 2$ cm/s, and $v_n = 0$ cm/s, i.e. the snake robot was initially oriented along the global y axis and located 1 m away from the x axis with an initial forward velocity of 0.1 m/s, i.e. moving away from the desired path.

8.6.2 Simulation Results

The simulation results are shown in Fig. 8.5. From Figs. 8.5(a)–(b) we see that the position of the snake robot converges nicely to the desired path (i.e. the x axis). Figure 8.5(a) shows the configuration of the snake robot at $t = 10$ s, $t = 50$ s, $t = 100$ s, and $t = 150$ s. Note that Fig. 8.5(b) shows the cross-track error in terms of the y axis coordinate of the CM of the robot, not the transformed y axis coordinate given by (8.13b). The heading of the snake robot, shown in Fig. 8.5(c), also converges nicely to zero, i.e. to the direction of the desired path. As seen in Fig. 8.5(e), the forward velocity is always non-zero and positive, as required by Assumption 8.1, and converges to the velocity $v_t^* = \frac{c_p}{2Nc_t}\alpha^2\omega k_\delta = 9.4$ cm/s, which was estimated above. Figure 8.5(f) shows the joint coordinate of an arbitrarily chosen joint (joint 5) during the path following. The plot shows a very good tracking of the corresponding joint reference coordinates. In summary, the simulation results illustrate that the proposed path following controller successfully steers the snake robot towards and along the desired straight path.

8.7 Experimental Study: The Performance of the Straight Line Path Following Controller

We have experimentally investigated the performance of the straight line path following controller proposed in Sect. 8.3 by use of the snake robot Wheeko, which

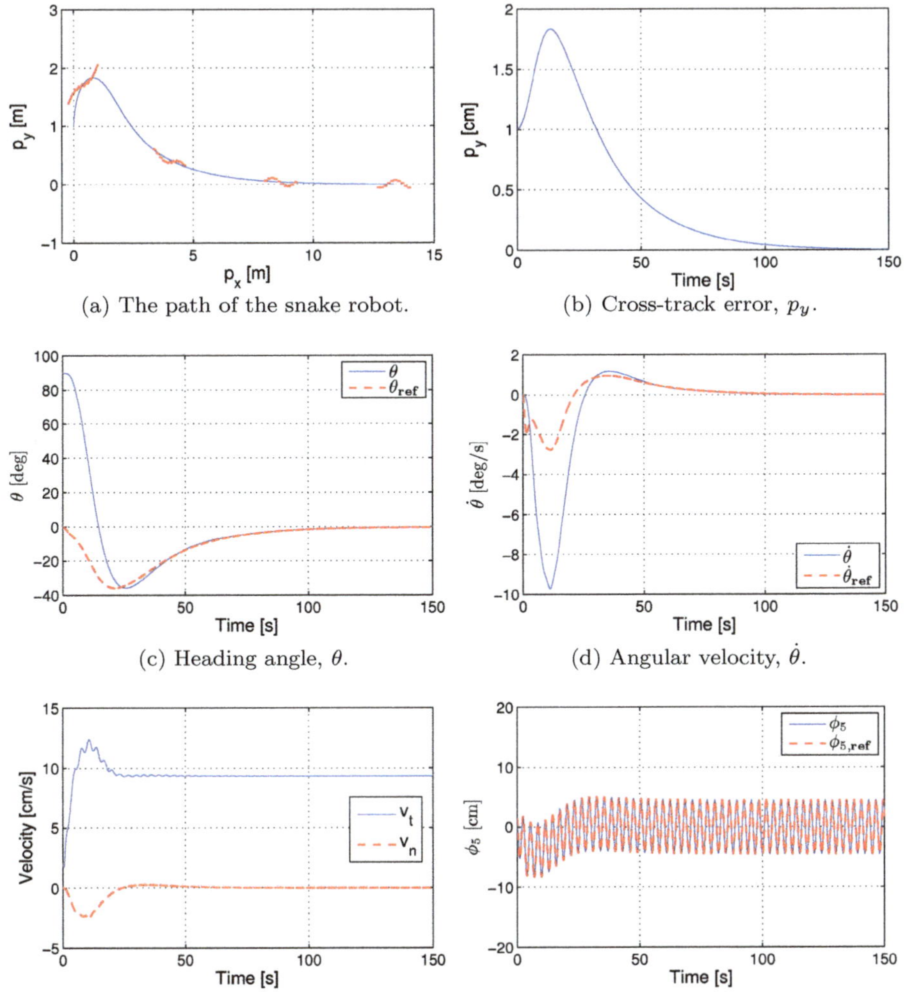

(a) The path of the snake robot.

(b) Cross-track error, p_y.

(c) Heading angle, θ.

(d) Angular velocity, $\dot{\theta}$.

(e) Velocity, v_t and v_n.

(f) Coordinate of joint 5, ϕ_5.

Fig. 8.5 Simulation of straight line path following with the snake robot initially headed away from the desired path

was presented in Chap. 3. The experimental results are presented in the following.

8.7.1 Implementation Issues

The joint coordinates of the simplified model (linear link displacements), on which the path following controller is based, are different from the joint coordinates of the physical snake robot (joint angles). However, it still makes sense to employ the path following controller to control the physical snake robot since, as implied by Property 4.8 in Sect. 4.10, the rotational link motion is what produces the linear displacements captured by the simplified model.

The main purpose of the simplified model is to describe the *qualitative* behaviour of a snake robot with revolute joints. The *quantitative* behaviour of the model is determined by the choice of numerical values of the ground friction coefficients c_t and c_n, and the rotation parameters λ_1 and λ_2. Unfortunately, the values of these parameters, which reflect the specific properties of the physical snake robot used in the experiments, are not known (see for instance Remark 6.5 in Sect. 6.5.2). Since λ_1 and λ_2 appear in the equations of the path following controller, we therefore chose to treat these coefficients as controller gains in order to implement the controller of the physical robot.

Furthermore, the unspecified values of the ground friction coefficients and the rotation parameters prevented us from determining the coordinate transformation distance ϵ in (8.14), which depends on c_p and λ_2. During the experiments, we therefore set this coordinate transformation distance to $\epsilon = 0$, i.e. we measured the cross-track error as $\overline{p}_y = p_y$. Note that the value of ϵ could anyhow be expected to be small based on the interpretation of this parameter given in Sect. 8.3.3, which means that measuring the cross-track error from the CM of the robot or from a point located a small distance ϵ away from the CM can be expected to produce similar controller performance. Note also that since the ϵ transformation is tangential to the robot, the value of ϵ has approximately no effect on the cross-track error when the heading of the snake robot with respect to the path is close to zero.

8.7.2 Implementation of the Path Following Controller of the Physical Snake Robot

The experimental setup, which consisted of the snake robot Wheeko and the camera-based position measurement system, is presented in Chap. 3. The path following controller of the snake robot was implemented on an external computer according to (8.17), (8.23), and (8.25). We did not implement the joint torque controller given by (8.19) and (8.20) since accurate torque control is not supported by the servo motors installed in the snake robot. The joint angles were instead controlled according to a proportional controller implemented in the microcontroller of each joint module. Note that we can experimentally validate Theorem 8.2 without implementing the joint controller in (8.19) and (8.20) since, as stated in Remark 8.9, the global $\mathcal{K}$-exponential stability of the complete system only requires that the error dynamics of the joints is exponentially stabilised.

The orientation $\overline{\theta}$ of the snake robot was estimated according to (2.2), i.e. as the average of the individual link angles. Furthermore, the forward velocity $\overline{v}_t$ of the robot, which is needed to calculate the joint angle offset in (8.25), was estimated at 0.5 Hz as the displacement of the CM of the robot divided by the sampling interval (i.e. 2 s). The sampling interval was chosen to be large to obtain a reasonably accurate velocity estimate but was sufficiently short for the experiments since the robot was moved at a slow pace. As explained in Sect. 8.7.1, the coordinate transformation distance in (8.14) was set to $\epsilon = 0$, i.e. we measured the cross-track error as $\overline{p}_y = p_y$.

The LOS angle θ_{ref} given by (8.23) was calculated with a look-ahead distance Δ equal to half the length of the snake robot, i.e. $\Delta = 0.7$ m. We conjecture that this value is well above the lower limit of Δ given by (8.28). The actual values of $V_{\min}$ and $V_{\max}$ are not known a priori, and as noted in Remark 8.6, specifying the bounds on Δ as a function of the gait pattern parameters α, ω, δ, and ϕ_o remains an unsolved problem. To ensure a smooth control input, the LOS angle θ_{ref} was passed through a 3rd-order low-pass filtering reference model (see Appendix C.2). The parameters of the reference model were set to $\omega = \pi/2$ and $\zeta = 1$. The output from this filter also provided the derivatives $\dot{\theta}_{\text{ref}}$ and $\ddot{\theta}_{\text{ref}}$, which are required in the calculation of ϕ_o in (8.25). The evolution of the reference values from the filter were calculated with a first-order numerical integration scheme.

The joint angle offset ϕ_o given by (8.25) was calculated with the gains set to $k_\theta = 1$, $\lambda_1 = 0.5$, and $\lambda_2 = 20$. The rotation parameters λ_1 and λ_2 were treated as controller gains, as explained in Sect. 8.7.1. We saturated the joint angle offset according to $\phi_o \in [-25°, 25°]$ in order to keep the joint reference angles within reasonable bounds with respect to the maximum allowable joint angles of the physical snake robot. This saturation also avoided the singularity in (8.25) at $v_t = 0$ (see Remark 8.4). Furthermore, to ensure that the joint angle offset was smooth despite any steps in the estimate of the forward velocity $\overline{v}_t$, we filtered ϕ_o with a first-order low-pass filter with cutoff frequency at 1.25 Hz.

The reference angles corresponding to the horizontal joint motion of the robot were calculated according to (8.17) with $N = 10$ links and with gait parameters $\alpha = 30°$, $\omega = 50°/\text{s}$, and $\delta = 36°$. The reference angles corresponding to the vertical joint motion were set to zero to achieve a purely planar locomotion.

8.7.3 Experimental Results

The straight line path following controller was experimentally investigated from three different sets of initial conditions. In the first trial of the experiment, the initial state of the snake robot was approximately $\boldsymbol{\phi} = \mathbf{0}°$, $\theta = 0°$, $p_x = 0$ m, $p_y = 1.3$ m, $\mathbf{v}_\phi = \mathbf{0}°/\text{s}$, $v_\theta = 0°/\text{s}$, $v_t = 0$ m/s, and $v_n = 0$ m/s, i.e. the snake robot was initially headed *along* the desired path (the x axis), and the initial distance from the CM to the desired path was 1.3 m. In the second trial, the initial state of the snake robot was approximately $\boldsymbol{\phi} = \mathbf{0}°$, $\theta = 90°$, $p_x = 0$ m, $p_y = 0.5$ m, $\mathbf{v}_\phi = \mathbf{0}°/\text{s}$, $v_\theta = 0°/\text{s}$, $v_t = 0$ m/s, and $v_n = 0$ m/s, i.e. the snake robot was initially headed *away* from the desired path (the x axis), and the initial distance from the CM to the desired path was 0.5 m. In the third and final trial, the initial state of the robot was approximately $\boldsymbol{\phi} = \mathbf{0}°$, $\theta = -90°$, $p_x = 0$ m, $p_y = 0.9$ m, $\mathbf{v}_\phi = \mathbf{0}°/\text{s}$, $v_\theta = 0°/\text{s}$, $v_t = 0$ m/s, and

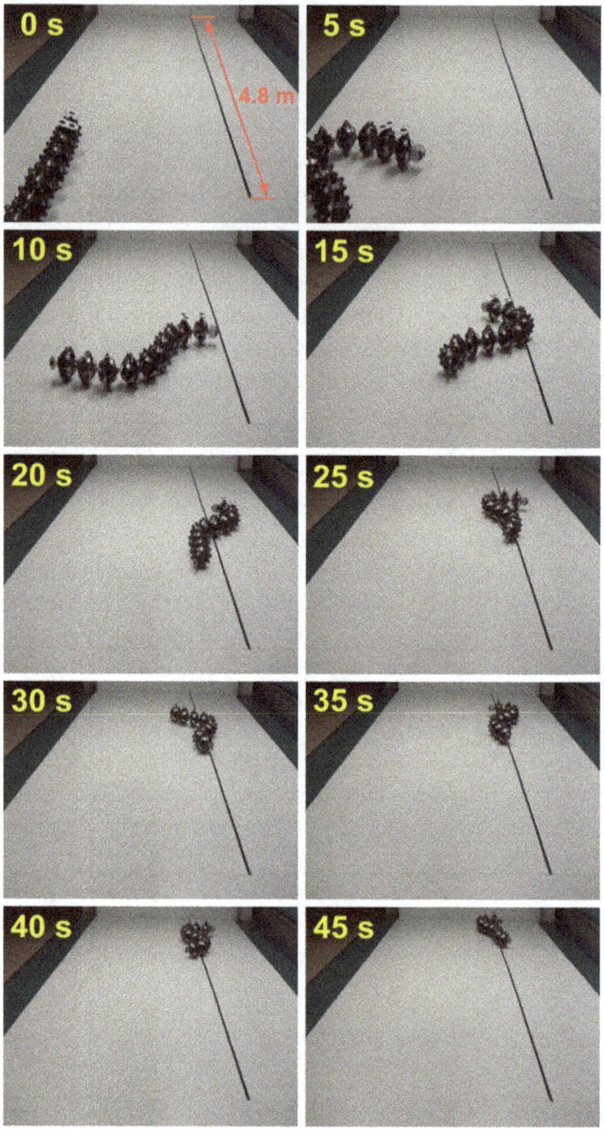

Fig. 8.6 The motion of the snake robot during path following with initial heading along the desired path. The *black line* on the floor indicates the desired path, i.e. the global x axis

$v_n = 0$ m/s, i.e. the snake robot was initially headed *towards* the desired path (the x axis), and the initial distance from the CM to the desired path was 0.9 m.

The experimentally measured motion of the snake robot from the first trial is presented in Figs. 8.6 and 8.7, from the second trial in Figs. 8.8 and 8.9, and from the third trial in Figs. 8.10 and 8.11. The desired path, i.e. the global x axis, is indicated with a black line on the floor in the pictures of the snake robot during the three trials.

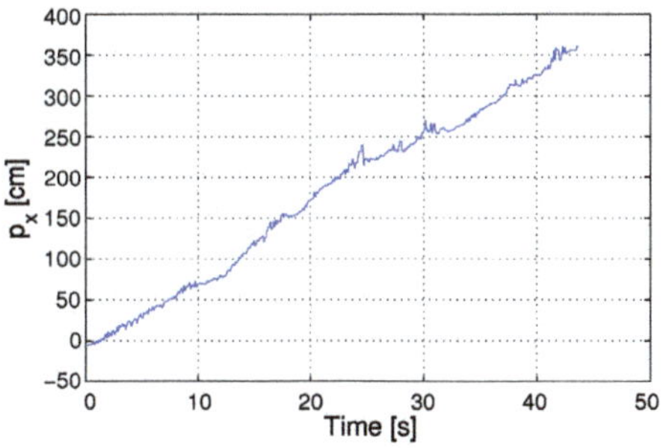

(a) Position along the path, p_x.

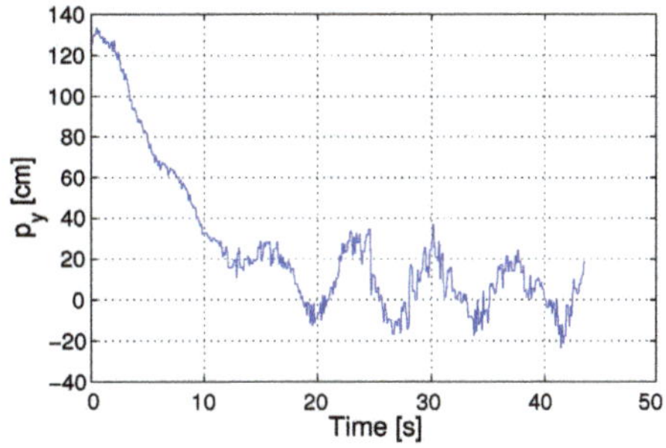

(b) Cross-track error, p_y.

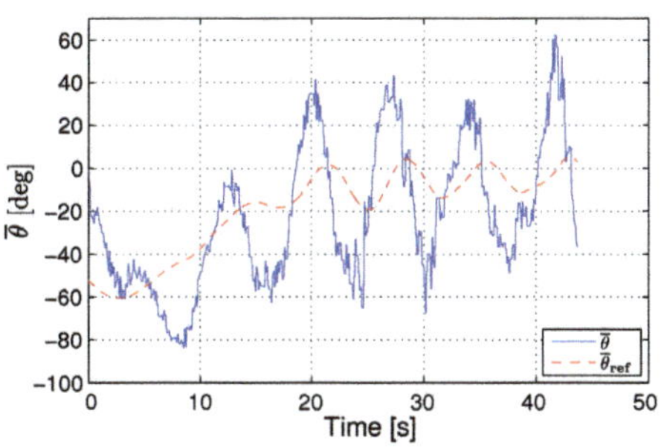

(c) Heading angle, $\overline{\theta}$.

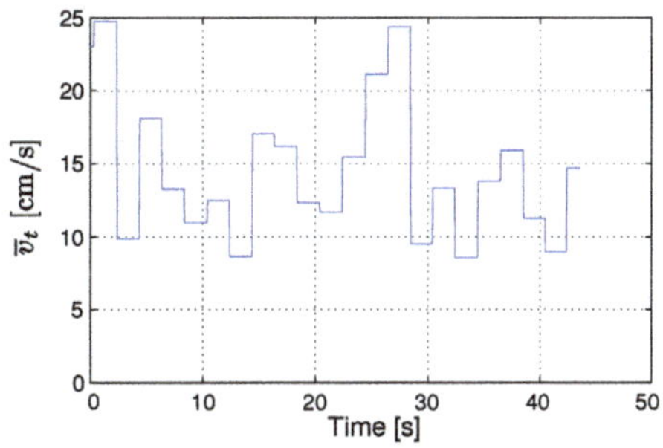

(d) Forward velocity, $\overline{v}_t$.

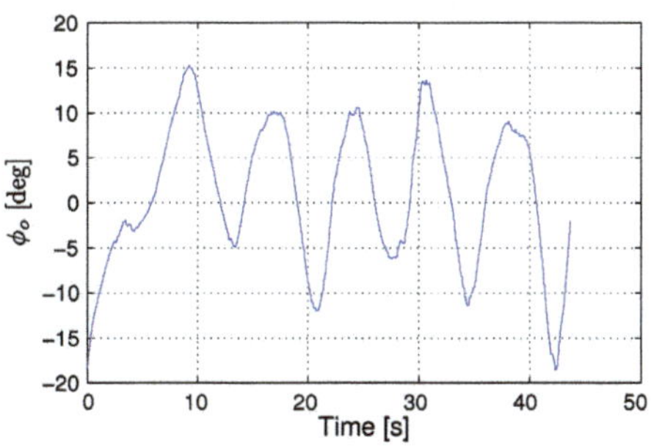

(e) Joint angle offset, ϕ_o.

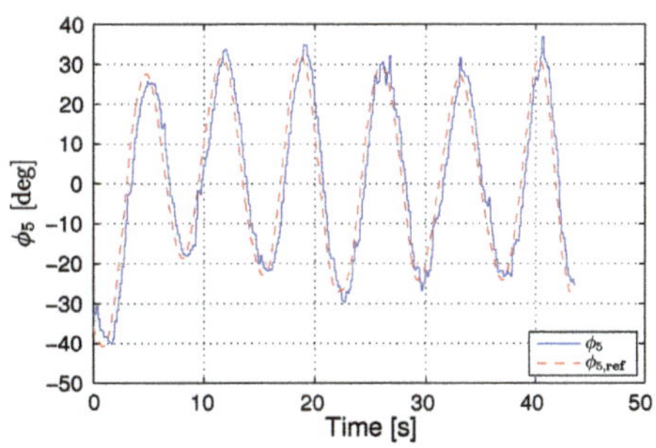

(f) Angle of joint 5, ϕ_5.

Fig. 8.7 Straight line path following with the physical snake robot initially headed along the desired path

The visualisations in Figs. 8.6, 8.8, and 8.10 indicate that the snake robot converged nicely towards and along the desired path during all three trials. This claim is supported by the plots of the cross-track error in Figs. 8.7(b), 8.9(b), and 8.11(b), respectively, which show that the cross-track error converges to and oscillates about zero. For a snake robot with revolute joints, it is difficult to achieve a purely non-oscillating motion of the CM, which was achieved in the simulation results based on the simplified model in Sect. 8.6. We therefore expected the cross-track error to oscillate about zero, as seen in the plots, rather than converge to zero.

Fig. 8.8 The motion of the snake robot during path following with initial heading away from the desired path. The *black line* on the floor indicates the desired path, i.e. the global x axis

Similar to the oscillatory behaviour of the CM, the heading $\overline{\theta}$ of the snake robot was also expected to oscillate. In particular, while θ provides an explicit representation of the heading in the simplified model, such a representation is not available for a snake robot with revolute joints, which forced us to estimate the heading according to $\overline{\theta}$, i.e. as the average of the link angles (see Remark 6.3). The oscillatory behaviour of $\overline{\theta}$ was thereby expected since the average of the link angles will not always be identically zero during forward locomotion. The heading during the trials is shown in Figs. 8.7(c), 8.9(c), and 8.11(c), respectively, which clearly

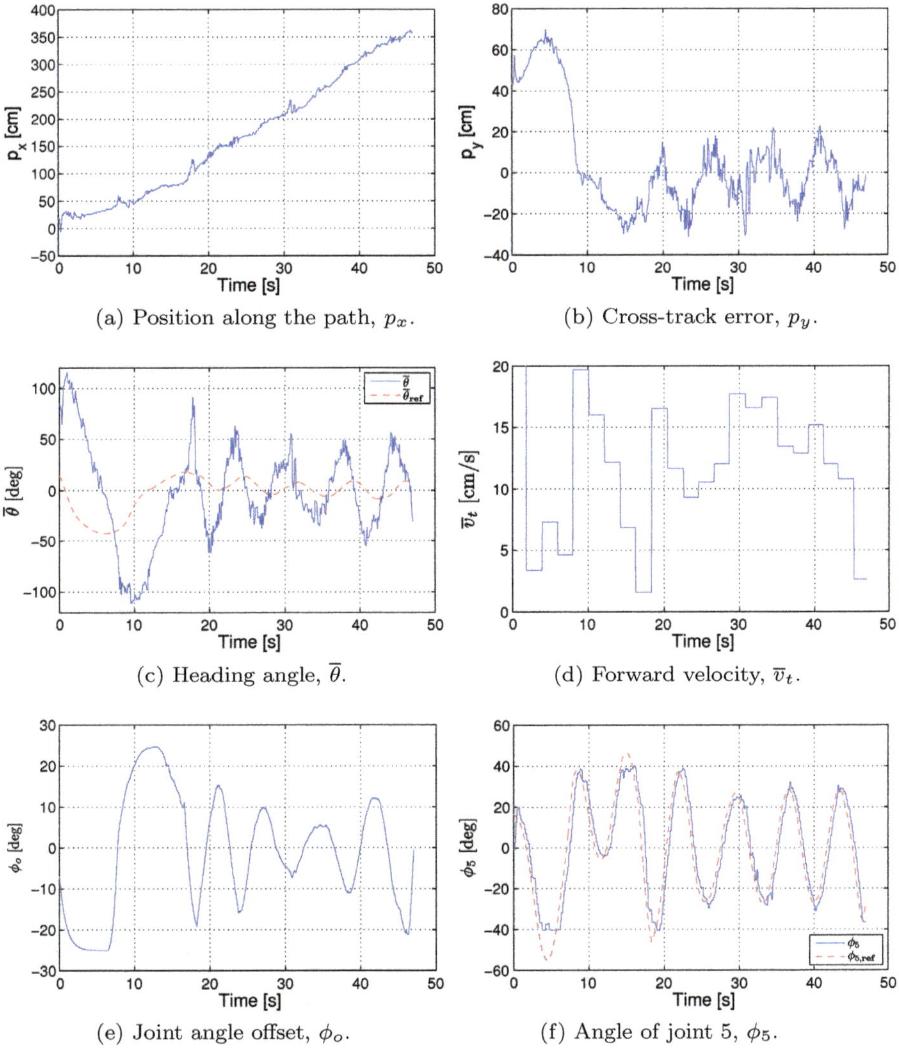

(a) Position along the path, p_x.

(b) Cross-track error, p_y.

(c) Heading angle, $\overline{\theta}$.

(d) Forward velocity, $\overline{v}_t$.

(e) Joint angle offset, ϕ_o.

(f) Angle of joint 5, ϕ_5.

Fig. 8.9 Straight line path following with the physical snake robot initially headed away from the desired path

show that $\overline{\theta}$ oscillates nicely about the reference heading $\overline{\theta}_{\text{ref}}$. In all three trials, the heading converges to and oscillates about zero, i.e. the direction of the desired path.

The forward velocity of the robot during each trial is shown in Figs. 8.7(d), 8.9(d), and 8.11(d), respectively. The variations in the velocity were primarily caused by the joint angle offset ϕ_o during turning motion, which sometimes interfered with the oscillatory body wave motion and caused the robot to lose momentum.

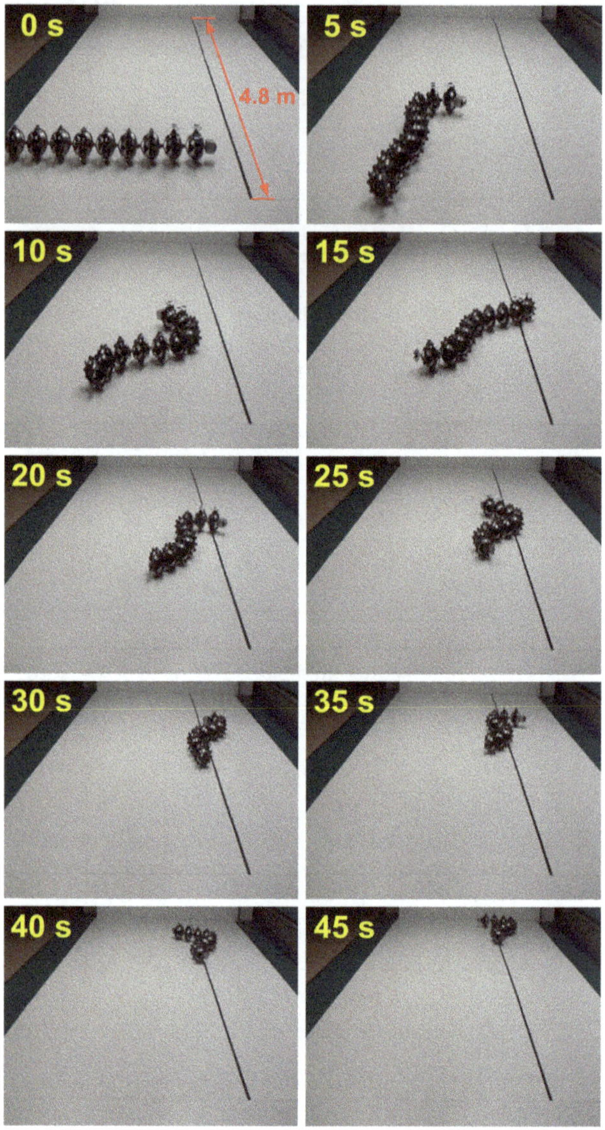

Fig. 8.10 The motion of the snake robot during path following with initial heading towards the desired path. The *black line* on the floor indicates the desired path, i.e. the global x axis

The joint angle of an arbitrarily chosen joint (joint 5) during each trial is shown in Figs. 8.7(f), 8.9(f), and 8.11(f), respectively, which indicate that the snake robot tracked its joint reference coordinates very well.

In summary, the proposed path following controller successfully steered the snake robot towards and along the desired straight path during all three trials of the experiment.

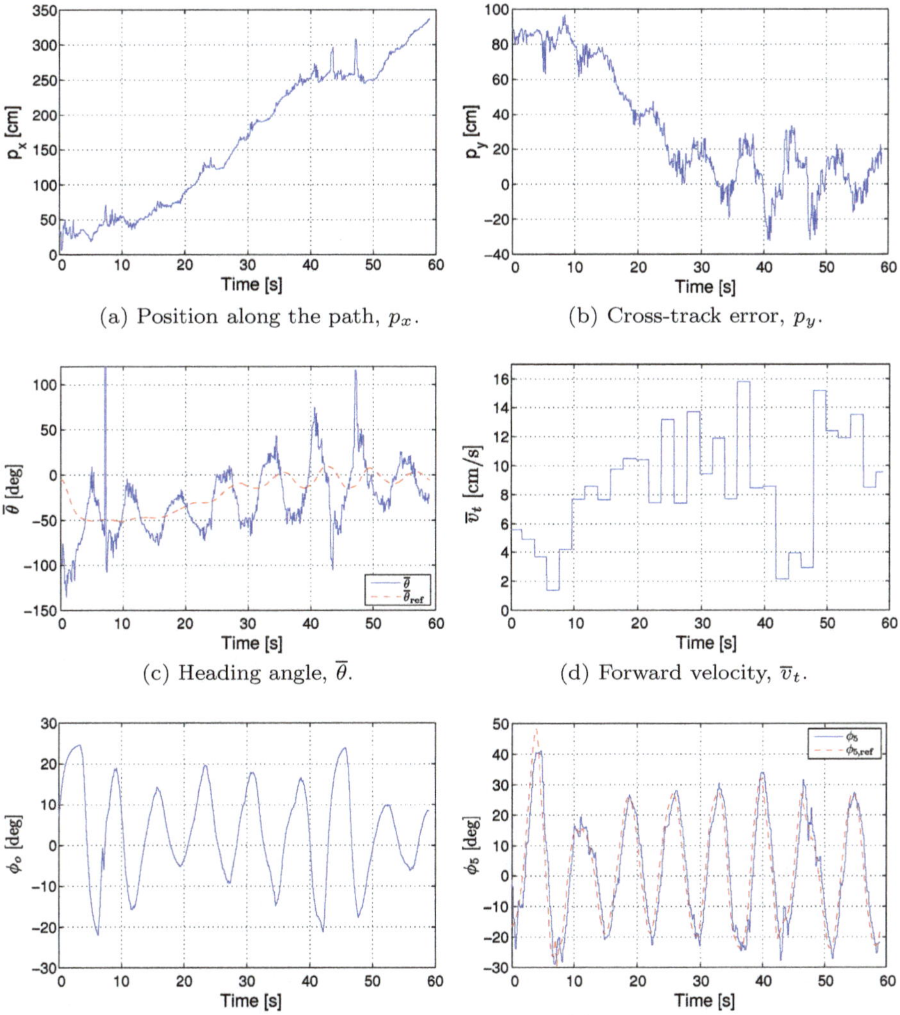

(a) Position along the path, p_x.

(b) Cross-track error, p_y.

(c) Heading angle, $\overline{\theta}$.

(d) Forward velocity, $\overline{v}_t$.

(e) Joint angle offset, ϕ_o.

(f) Angle of joint 5, ϕ_5.

Fig. 8.11 Straight line path following with the physical snake robot initially headed towards the desired path

8.8 Simulation Study: The Performance of the Waypoint Guidance Strategy

This section presents simulation results in order to investigate the performance of the waypoint guidance strategy proposed in Algorithm 8.1. In addition to simulation results from the simplified model of the snake robot given by (6.35a)–(6.35h), we also include simulation results from the complex model (2.47) to show that the applicability of the guidance strategy does not rely on the simplifications of the simplified model. Both models were implemented in *Matlab R2008b*, and the dynamics was calculated using the *ode45* solver in Matlab.

8.8.1 Implementation of the Guidance Strategy with the Simplified Model

We considered a snake robot with $N = 10$ links of length $l = 0.14$ m, mass $m = 1$ kg, and moment of inertia $J = 0.0016$ kgm^2. All initial state values were set to zero. Furthermore, we chose the ground friction coefficients as $c_t = 1$ and $c_n = 3$, and the rotation parameters as $\lambda_1 = 0.5$ and $\lambda_2 = 20$.

The radius of the acceptance circle enclosing each waypoint was $r_{\text{accept}} = 0.5$ m. The path following controller was implemented according to (8.17), (8.19), (8.20), (8.23), and (8.25), and with the coordinate transformation distance in (8.14) set to $\epsilon = -64.3$ cm. The controller gains were $k_\phi = 20$, $k_{v_\phi} = 5$, and $k_\theta = 0.05$, and the gait parameters were $\alpha = 3.4$ cm, $\omega = 150°/\text{s}$, and $\delta = 40°$. We chose the look-ahead distance as $\Delta = 1.4$ m, which corresponds to the length of the snake robot, and we conjecture that this value is well above the lower bound given by (8.28).

The derivatives of θ_{ref} and ϕ_o with respect to time, which are required in (8.20) and (8.25), were calculated by using a 3rd-order low-pass filtering reference model (see Appendix C). The parameters of the reference model were set to $\omega = \pi/2$ and $\zeta = 1$. We chose to saturate the joint angle offset according to $\phi_o \in [-\alpha, \alpha] = [-3.4\text{ cm}, 3.4\text{ cm}]$ in order to avoid the singularity in (8.25) at $v_t = 0$.

8.8.2 Implementation of the Guidance Strategy with the Complex Model

The ground friction in the complex model was defined by the viscous friction forces in (2.25) with friction coefficients $c_t = 1$ and $c_n = 3$. The orientation $\overline{\theta}$ and the forward velocity $\overline{v}_t$ were calculated according to (2.2) and (2.5), respectively. Based on the argumentation given for the physical snake robot in Sect. 8.7.2, the coordinate transformation distance ϵ in (8.14) was set to zero in the complex model (i.e. we measured the cross-track error as $\overline{p}_y = p_y$).

The shape variables of the simplified model (i.e. the transversal distance between the links) and the complex model (i.e. the joint angles) are different. The controller parameters of the complex model were therefore scaled to account for this difference. In particular, it was shown in Sect. 6.10.2 that lateral undulation in the simplified model with $\alpha = 0.034$ m and $\delta = 40°$ (which are the simulation parameters of the simplified model presented in the previous subsection) corresponds to lateral undulation in the complex model with $\alpha = 10.5°$ and $\delta = 40°$. For the complex model, we therefore set $\alpha = 10.5°$, and we scaled the controller gain k_θ with respect to the controller gain of the simplified model according to the scaling factor

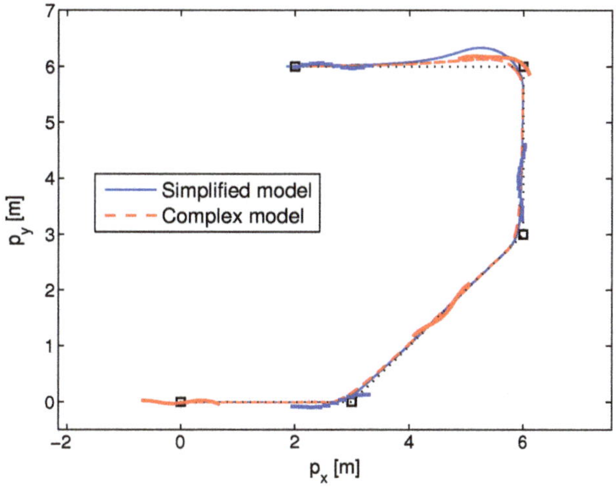

Fig. 8.12 The simulated path of the snake robot from the simplified model (*solid line*) and the complex model (*dashed line*) during waypoint guidance

10.5°/0.034 m, i.e. we set $k_\theta = 0.05\frac{10.5\pi/180}{0.034} = 0.27$. Furthermore, we chose to saturate the joint offset according to $\phi_o \in [-\alpha, \alpha] = [-10.5°, 10.5°]$ in order to avoid the singularity in (8.25) at $v_t = 0$. The remaining controller parameters of the complex model were set equal to the controller parameters of the simplified model.

8.8.3 Simulation Results

We defined $k = 4$ waypoints with global frame coordinates $(3, 0)$, $(6, 3)$, $(6, 6)$, and $(2, 6)$, respectively. Figure 8.12 shows the motion of the CM of the snake robot from the simplified model (solid line) and the complex model (dashed line), where each waypoint is indicated with a black square. The figure also shows the shape and position of the snake robot at $t = 100$ s, $t = 300$ s, and $t = 500$ s for the simplified model and at $t = 1$ s, $t = 200$ s, and $t = 400$ s for the complex model. Furthermore, Fig. 8.13 shows the cross-track error (in terms of the y axis coordinate of the CM of the robot), the heading angle, and the forward velocity from the two models. The heading angle and forward velocity from the complex model were calculated according to (2.2) and (2.5), respectively. The vertical lines in the plots indicate time instants where the guidance strategy switches to the next waypoint. We see that the state of the robot experiences a jump at each waypoint switch since, by Algorithm 8.1, the global frame is redefined at a waypoint switch.

There is a good agreement between the simulation results from the complex and the simplified model, and the waypoint guidance strategy performed well in both models. As seen in Fig. 8.12, the snake robot has a nice and smooth motion towards each waypoint in both models. Moreover, Figs. 8.13(a)–(b) and Figs. 8.13(c)–(d) show that the cross-track error and the heading, respectively, converge nicely to zero after each waypoint switch. In summary, the simulation results illustrate that the proposed waypoint guidance strategy successfully steers the snake robot towards each of the specified waypoints.

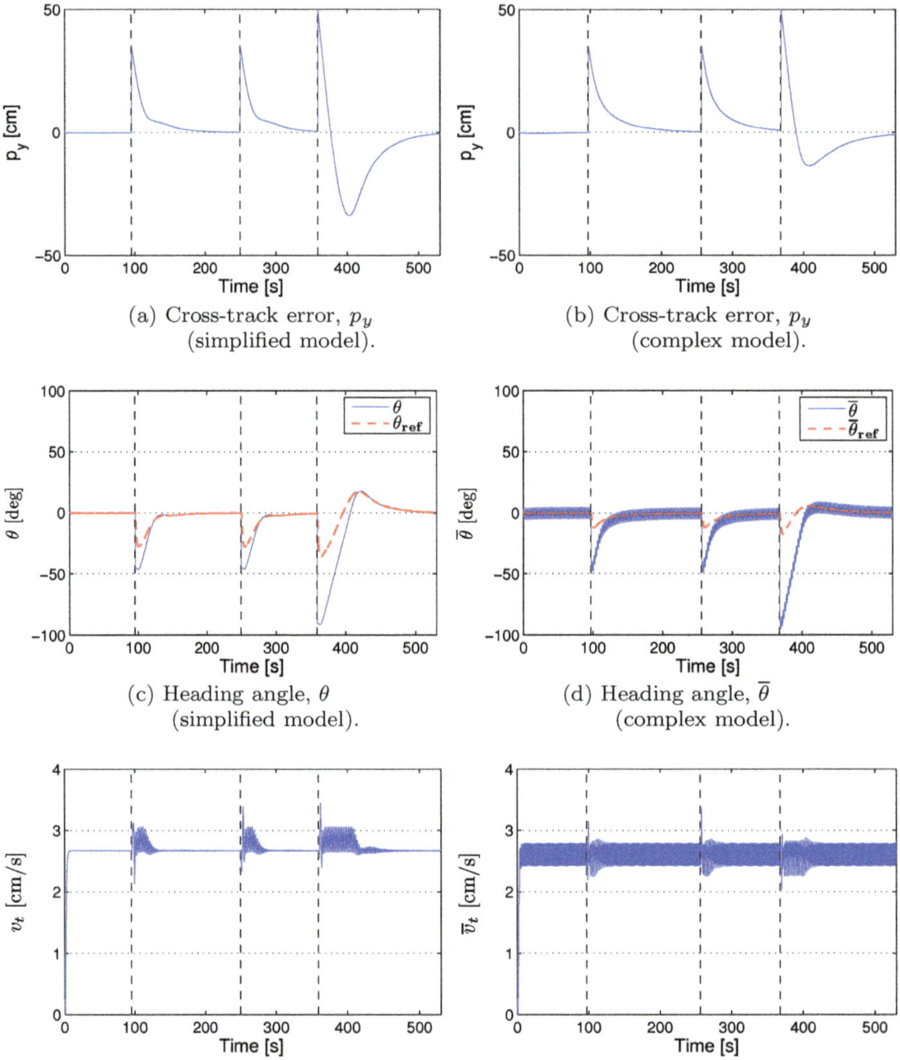

(a) Cross-track error, p_y (simplified model).

(b) Cross-track error, p_y (complex model).

(c) Heading angle, θ (simplified model).

(d) Heading angle, $\overline{\theta}$ (complex model).

(e) Forward velocity, v_t (simplified model).

(f) Forward velocity, $\overline{v}_t$ (complex model).

Fig. 8.13 Simulation of the waypoint guidance strategy with the simplified (*left*) and complex (*right*) models of the snake robot

8.9 Chapter Summary

This chapter is summarised as follows:

- We have proposed a path following controller that enables snake robots to track straight paths.

- Using cascaded systems theory, we have proven that the proposed path following controller $\mathcal{K}$-exponentially stabilises the snake robot to any desired straight path (see Theorem 8.2).
- The proof relies on the assumption that the forward velocity of the robot is contained in some non-zero and positive interval. Specifying the bounds of this interval as a function of the gait pattern parameters is an interesting, but still unsolved, problem.
- We have investigated the performance of the path following controller through simulations and through experiments with the snake robot Wheeko, where the proposed controller was shown to successfully steer the snake robot towards and along the desired straight path.
- We have described how the straight line path following controller can be extended to path following of general curved paths by employing an approach previously proposed in the marine control literature for path following control of marine vessels.
- We have proposed a waypoint guidance strategy for steering a snake robot along a path defined by waypoints interconnected by straight lines (see Algorithm 8.1).
- We have proven that the waypoint guidance strategy is guaranteed to steer the position of the snake robot into the acceptance region of each waypoint (see Theorem 8.3).
- We have presented simulation results that illustrated the successful performance of the waypoint guidance strategy.

Part II
Snake Robot Locomotion in Cluttered Environments

Chapter 9
Introduction to Part II

The long-term goal of our research activities on snake robot locomotion is to enable these mechanisms to move intelligently and efficiently in unknown and cluttered environments (see Fig. 9.1). The second part of this book is therefore devoted to snake robot locomotion in environments that are no longer assumed to be flat, which is more in line with practical applications of snake robots. The second part of the book is also in line with current trends in robotic research, which aim at making robots more 'aware' of their environment (e.g. for grasping and object manipulation purposes) and enabling them to work in unknown and unstructured environments (see EUROP/CARE 2009). Although snake robots have the potential of making a huge impact in applications which require robots with robust mobility, practical applications of mobile snake robots have not yet been demonstrated. An important reason for this limitation is that the majority of current snake robots are unable to 'feel', which prevents them from operating in unknown and challenging environments. In the second part of this book, we emphasise the ability to 'feel' as the key to realising the potential of these mechanisms. In particular, the results presented in Part II of this book have been developed on the basis of the following fundamental hypothesis:

Hypothesis 9.1 *Intelligent and efficient snake robot locomotion in unknown and cluttered environments requires that the snake robot can* sense *its environment and* adapt *its body shape and movements accordingly.*

We can provide both *theoretical* and *empirical* justifications for Hypothesis 9.1. A theoretical justification for the hypothesis follows from the analysis in Chap. 4, which enables us to conclude that the fundamental control principle of snake robot locomotion is to produce body shape changes that induce external contact forces whose sum points in the desired direction of motion. Since controlled body shape changes for inducing desired contact forces are equivalent to environment *adaptation*, and since environment adaptation is not possible without *sensing* the environment in some way, we have established a theoretical justification for Hypothesis 9.1.

An empirical justification for Hypothesis 9.1 follows from observations of biological snakes in nature. As described in Bauchot (1994), the sensory information

P. Liljebäck et al., *Snake Robots*, Advances in Industrial Control,
DOI 10.1007/978-1-4471-2996-7_9, © Springer-Verlag London 2013

Fig. 9.1 Snake robot locomotion in a cluttered environment

transmitted by the skin of a snake influences the shape to which the body adapts in a given situation. The body adaptation of snakes with respect to external objects is studied analytically and empirically in Gray (1946), Hirose (1993), Moon and Gans (1998). Environment sensing and adaptation is, in other words, inherently present in the motion of any snake, which serves as an empirical justification of Hypothesis 9.1.

Note that environment sensing and adaptation was not necessary in the control design efforts of Part I of this book since the surface beneath the snake robot was assumed to be *flat*, which allowed us to employ predetermined gait patterns in open loop without sensing the external forces.

It is implied by Hypothesis 9.1 that we recognise the interaction between the snake robot and its environment as the propelling force. We also made this recognition in Part I of this book. However, while ground friction represents the propelling forces during flat surface locomotion, the propelling forces in Part II will be contact forces from external objects (or obstacles) in the environment of the snake robot. At this point in the discussion, we are actually at the very core of the principle underlying snake robot locomotion. In particular, by considering Hypothesis 9.1 at a glance, one realises that an interesting and unique feature of snake robot locomotion compared to other forms of robotic mobility is that irregularities on the ground are actually beneficial for the propulsion since they provide push points for the robot. The word *obstacle* will be used in the following chapters to denote an object or an irregular surface in the path of the snake robot that can be utilised for propulsion. This may seem like a contradiction since it is, in fact, not an obstacle from the point of view of the snake robot. However, the characterisation is valid in the sense of mobile robotics in general. The use of this denotation therefore helps emphasise one of the fundamental differences between snake locomotion and other traditional means of mobility, such as wheeled, tracked, and legged mobility. While in traditional mobile robotics, the aim is typically to avoid obstacles, a snake robot should rather seek out and make contact with obstacles since they represent push points that can be utilised for more efficient propulsion. Hence, for snake robots, the aim is not *obstacle avoidance*, but rather *obstacle utilisation*. To fully embrace this concept, the term *obstacle-aided locomotion* was previously introduced in Transeth et al. (2008b).

We can now summarise the material in Part II of this book as modelling and control strategies for obstacle-aided locomotion. In accordance with the scope of the book described in Sect. 1.4, we will maintain a planar perspective throughout the remaining chapters, i.e. we consider the motion of the snake robot to be purely horizontal on a surface with vertical obstacles that induce horizontal contact forces on the robot. The reason for only considering planar motion is, as explained in Sect. 1.4, that we believe the essential control principles of snake robot locomotion are contained in a planar perspective. In particular, since the fully three-dimensional motion of a snake robot consists of motion components in a horizontal and vertical plane, respectively, we conjecture that control laws that fulfil some control objective in a planar perspective can be extended to fulfil a similar control objective in a fully three-dimensional perspective. Moreover, it makes no sense to attack the problem of three-dimensional motion in a cluttered environment before the simpler case of planar obstacle-aided locomotion is well understood.

We end this introductory text with a note on the literature review presented in Sect. 1.3. An interesting observation that can be made from this literature review (one might even call this a paradox) is that the majority of previous research on snake robots has focused on flat surface locomotion even though the main advantage of snake robots are their potential ability to move in uneven and cluttered environments. As described in Sect. 1.3, only a few published works consider snake robot locomotion in situations where the surface is no longer assumed to be flat. We therefore hope that the material presented in the second part of this book addresses a small part of the, in our opinion, large research gap that must be closed before we will ever see useful snake robots outside the laboratory.

Chapter 10
A Hybrid Model of Snake Robot Locomotion in Cluttered Environments

We begin the second part of this book by extending the model of the snake robot presented in (2.35) to include contact forces from external obstacles in the environment around the robot. Since the interaction with an obstacle represents a discrete event that only occurs when a link of the robot comes into contact with an obstacle, the snake robot will be subjected to both *continuous* and *discontinuous* dynamics in this environment. We will therefore describe the dynamics of the snake robot in terms of a *hybrid model* by employing the hybrid modelling framework described in Goebel et al. (2009).

An important difference between models of continuous dynamical systems, such as the model in (2.35), and models of hybrid dynamical systems, is that while most continuous models always exhibit a unique solution to the evolution of the state vector, a hybrid model may have a single solution, several solutions, or no solution at all. For the hybrid model of the snake robot, we will handle this existence and uniqueness issue by formulating the equations governing the obstacle contact forces as a *linear complementarity problem* (LCP). This formulation enables us to apply existing general results concerning existence and uniqueness of solutions to LCPs (see Cottle et al. 1992) to the model of the snake robot.

A long-term goal of the model proposed in this chapter is to facilitate development of model-based control laws for obstacle-aided locomotion with provable stability properties. We will therefore make several simplifying assumptions during the modelling process so that the environment interaction model maintains a simple and analytical form. In particular, we will model the interaction with obstacles by introducing a *unilateral velocity constraint* on each contacted link of the snake robot. This approach simplifies the equations of motion since the shape of the obstacles does not have to be considered explicitly.

In order to illustrate the validity of the proposed modelling approach, this chapter includes a simulation study where simulation results from the proposed model of the snake robot are shown to agree well with simulation results from a more extensive model of obstacle-aided locomotion previously presented in Transeth et al. (2008b).

The chapter is organised as follows. The relation between this chapter and previous literature is briefly discussed in Sect. 10.1. Section 10.2 gives a general pre-

P. Liljebäck et al., *Snake Robots*, Advances in Industrial Control,
DOI 10.1007/978-1-4471-2996-7_10, © Springer-Verlag London 2013

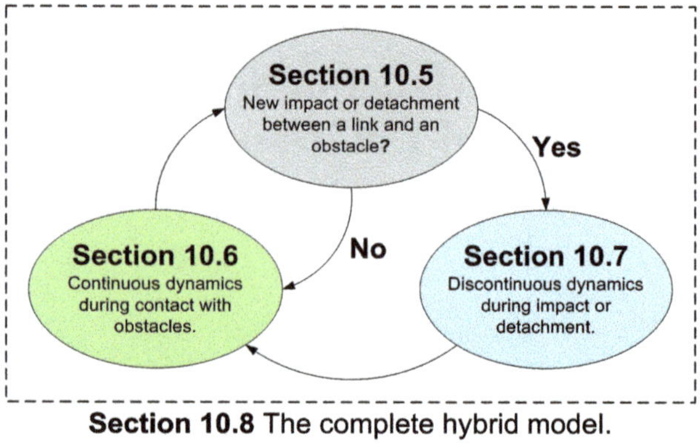

Fig. 10.1 The organisation of this chapter

sentation of the hybrid modelling framework and also the *linear complementarity problem* (LCP). The model of the snake robot without obstacles, which was presented in Chap. 2, is reformulated in a slightly different form in Sect. 10.3, and an overview of the contact modelling approach is given in Sect. 10.4. The organisation of the subsequent sections is illustrated in Fig. 10.1 to increase the readability of this chapter. In particular, Sect. 10.5 describes the mechanism for determining if a link is in contact with an obstacle, Sect. 10.6 presents the model of the continuous constrained dynamics of the robot during contact with a fixed set of obstacles, Sect. 10.7 presents the model of the discontinuous dynamics of the robot when the set of links in contact with obstacles changes, and Sect. 10.8 gives a unified presentation of the snake robot dynamics in terms of the hybrid modelling framework. Simulation results that compare the hybrid model with a model of obstacle-aided locomotion previously proposed in Transeth et al. (2008b) are presented Sect. 10.9. Finally, the chapter is summarised in Sect. 10.10.

10.1 The Relation Between This Chapter and Previous Literature

The only known works that consider the dynamics of snake robots in environments with obstacles (i.e. where obstacle contact forces are considered) are presented in Bayraktaroglu and Blazevic (2005), Date and Takita (2007), Tanev et al. (2005), Transeth et al. (2008b). In Bayraktaroglu and Blazevic (2005), a dynamic simulation software called *WorkingModel* is used to simulate a planar snake robot interacting with circular obstacles. Contact forces are calculated from a spring-damper approximation. A similar approach is employed in Tanev et al. (2005), where the simulation software *Open Dynamics Engine* (ODE) is used to model a snake robot interacting with various forms of obstacles. Date and Takita (2007) use the multi-body dynamics simulation software *Autolev* to study the motion of a snake robot during contact with a single peg, where the contact with the peg is modelled as a spring-damper system. The works in Bayraktaroglu and Blazevic (2005), Date and Takita (2007),

Tanev et al. (2005) do not provide the equations underlying the dynamics of the snake robot due to the use of general-purpose simulation software. On the other hand, the model proposed in Transeth et al. (2008b) is, to our best knowledge, the only work which explicitly presents the equations of motion underlying the obstacle interaction dynamics of a snake robot. The model, which represents a hybrid system, is formulated within the framework of nonsmooth dynamics. A *timestepping* method is used to simulate the dynamics of the robot, which means that the system equations are discretised with a time step determined by a fixed error criterion and trajectories of the system are approximated without tracking events (i.e. obstacle impacts).

The novelty of this chapter in relation to previous literature concerns the *hybrid model* of a planar snake robot interacting with obstacles in its environment. In contrast to the hybrid model presented in Transeth et al. (2008b), which is based on a *timestepping* method that approximates trajectories of the hybrid system without tracking events, the hybrid model presented in this chapter is based on *event-tracking* (Schaft and Schumacher 2000), where discrete events are tracked. We believe that the model formulation with this approach is better suited for analysis and synthesis of model-based controllers. Modelling obstacle interaction by introducing a *unilateral velocity constraint* on each contacted link of the snake robot is a novel approach. In particular, the conventional approach for modelling the obstacle interaction would be to assume that the obstacle constraint force points in the normal direction of the *obstacle* (see Brogliato 1999). With the approach described in this chapter, the shape of the obstacles does not have to be considered explicitly as we instead calculate constraint forces with respect to the normal direction of the *contacted links*, which simplifies the equations of motion.

10.2 Hybrid Dynamical Systems and Complementarity Systems

This section presents the framework employed in order to formulate the *hybrid model* of the snake robot and also the *linear complementarity problem* (LCP), which is instrumental in the modelling of obstacle contact forces in this chapter.

10.2.1 Modelling of Hybrid Dynamical Systems

A *hybrid dynamical system* is a dynamical system that exhibits both continuous and discontinuous state evolution. A snake robot interacting with obstacles is a hybrid system since the impacts between the snake and the obstacles represent discrete events.

Several modelling frameworks for hybrid systems exist, some of which are presented in Schaft and Schumacher (2000). In this book, we have chosen to employ the modelling framework described in Goebel et al. (2009) since this framework captures a wide variety of hybrid phenomena, and it also facilitates stability analysis of hybrid systems.

In accordance with Goebel et al. (2009), a hybrid system has a state vector $\mathbf{x} \in \mathbb{R}^n$ that can both *flow* (evolve continuously) and *jump* (evolve discontinuously). The data that determine the evolution of $\mathbf{x}$ are given by the four elements $(\mathbf{C}, \mathbf{F}, \mathbf{D}, \mathbf{G})$, where $\mathbf{C}$ denotes the *flow set*, $\mathbf{F}$ denotes the *flow map*, $\mathbf{D}$ denotes the *jump set*, and $\mathbf{G}$ denotes the *jump map* of the hybrid system. Whenever the state $\mathbf{x}$ belongs to the flow set $\mathbf{C}$, it flows (or evolves continuously) according to $\mathbf{F}$. During flows, the system acts as an ordinary continuous dynamical system. However, when $\mathbf{x}$ belongs to the jump set $\mathbf{D}$, it generally jumps according to $\mathbf{G}$ to a new value $\mathbf{x}^+$ (superscripts $+$ and $-$ denote 'the next value' and 'the previous value', respectively). Hence, the general form of a hybrid dynamical system is given by

$$\begin{aligned} \dot{\mathbf{x}} &= \mathbf{F}(\mathbf{x}, \mathbf{u}) \quad \text{for all } \mathbf{x} \in \mathbf{C}, \\ \mathbf{x}^+ &= \mathbf{G}(\mathbf{x}) \quad \text{for all } \mathbf{x} \in \mathbf{D}, \end{aligned} \tag{10.1}$$

where we have also included a control input, $\mathbf{u} \in \mathbb{R}^m$.

Existence and *uniqueness* of solutions is a very important issue when modelling hybrid systems. From a given initial state, $\mathbf{x}_0$, a hybrid system may have a single solution, several solutions, or no solution at all. For a general hybrid system, there are no easily verifiable necessary and sufficient conditions for existence and uniqueness of solutions. However, such conditions exist for special classes of hybrid systems, such as for *complementarity systems* (see Sect. 10.2.2).

A hybrid system is simulated by letting the state vector $\mathbf{x}$ flow according to the flow map $\mathbf{F}(\mathbf{x}, \mathbf{u})$ as long as $\mathbf{x} \in \mathbf{C}$. Whenever $\mathbf{x} \in \mathbf{D}$, the state vector jumps to the new value $\mathbf{x}^+$ according to the jump map $\mathbf{G}(\mathbf{x})$, and the simulation of the flow map is *restarted* from the new initial value given by $\mathbf{x}^+$. This approach for simulating hybrid systems is called *event-tracking* (Schaft and Schumacher 2000) since the discrete events of the model are tracked.

Remark 10.1 (A note regarding the notation in this chapter) We will employ boolean operators in the formulation of the hybrid model of the snake robot to handle the discrete nature of the obstacle contact forces. In particular, $\wedge$ denotes a logical AND operation, $\vee$ denotes a logical OR operation, $\cap$ denotes the intersection of two sets, and $\cup$ denotes the union of two sets. For ease of notation, we will also often use , to denote the logical AND operation. For example, the set $S = \{x | x > a, x < b\}$ contains all values of x that are greater than a AND less than b.

10.2.2 Complementarity Systems

A hybrid system is called a *complementarity system* if the flow of the system states is constrained by a set of *complementarity conditions* (Schaft and Schumacher 2000).

A complementarity condition between two scalar variables requires that both variables are nonnegative and that their product is always zero (i.e. one variable is always zero). In mathematical terms, the complementarity condition between two scalar variables x and y can be written $x \geq 0 \wedge y \geq 0 \wedge xy = 0$. Two vectors $\mathbf{x} \in \mathbb{R}^m$ and $\mathbf{y} \in \mathbb{R}^m$ are said to be *complementary* if, for all i, the pair of variables (x_i, y_i) is subject to a complementarity condition. We will see in Sect. 10.6 that we can formulate complementarity conditions for the links of the snake robot that are in contact with an obstacle.

The constraint equations of a complementarity system can often be formulated as a *linear complementarity problem* (LCP). An LCP asks whether there exist two complementary vectors $\mathbf{x} \in \mathbb{R}^m$ and $\mathbf{y} \in \mathbb{R}^m$ such that

$$\begin{aligned} &\mathbf{y} = \mathbf{a} + \mathbf{A}\mathbf{x}, \\ &\mathbf{x} \geq \mathbf{0}, \qquad \mathbf{y} \geq \mathbf{0}, \qquad \mathbf{x}^T\mathbf{y} = 0 \end{aligned} \tag{10.2}$$

for a given vector $\mathbf{a} \in \mathbb{R}^m$ and a matrix $\mathbf{A} \in \mathbb{R}^{m \times m}$. The constraint equations of the snake robot are given in this form in Sect. 10.6. The following result is proved in Cottle et al. (1992):

Theorem 10.1 *The LCP in* (10.2) *is uniquely solvable for all data vectors* $\mathbf{a}$ *if and only if* $\mathbf{A}$ *is a* P*-matrix.*

A P-matrix is a matrix whose *principal minors* are all positive. A principal minor of the matrix $\mathbf{A}$ is the determinant of a square submatrix of $\mathbf{A}$ consisting of the same set of rows and columns. A real *symmetrical* matrix is a P-matrix if and only if it is *positive definite*. For a real symmetrical matrix, one can therefore apply the standard criteria for positive definiteness in order to check if the matrix is a P-matrix. If the matrix is not symmetrical, one can e.g. apply the recursive algorithm proposed in Tsatsomeros and Li (2000), which is $O(2^m)$, in order to check if the matrix is a P-matrix.

Several algorithms exist for solving the LCP in (10.2). A famous approach is the so-called *Lemke's algorithm* (see e.g. Cottle et al. 1992), which basically uses trial and error to find the non-zero elements of $\mathbf{x}$ and $\mathbf{y}$, but with clever rules for changing the non-zero elements between trials.

10.3 The Dynamics of the Snake Robot Without Obstacles

The hybrid model will be developed by extending the unconstrained (no obstacles) model of the snake robot presented in (2.35). In this section, we restate this model with two modifications. The first modification, which concerns the ground friction model, is explained in Sect. 10.3.1. The second modification, which involves a slight reformulation of the model to a form more suitable for inclusion of obstacle contact forces, is described in Sect. 10.3.2.

10.3.1 The Ground Friction Model

Since the goal of Part II of this book is to study snake robot locomotion propelled by obstacle contact forces, we do not want the ground friction forces to contribute to the propulsion since it would then be difficult to know wether the robot is propelled by ground friction or by obstacle contact forces. Consequently, Theorem 4.4 from Chap. 4 suggests that the ground friction on the robot should be *isotropic*. We have considered viscous ground friction in Part I of this book based on the argument stated in Sect. 2.5.1, namely that the motion of the snake robot is qualitatively similar with anisotropic viscous friction as with anisotropic Coulomb friction. With isotropic ground friction, however, this argument is no longer relevant. We therefore choose to employ a Coulomb friction model in the following since, as stated in Sect. 2.5.1, a Coulomb friction model is more accurate from a physical point of view than a viscous friction model. Moreover, to further increase the accuracy of the ground friction model, we choose to also model the friction torque induced on a link due to the rotation of the link.

From the above discussion, we assume that each link of the snake robot is subjected to an isotropic Coulomb ground friction force acting on the CM of the link and also a friction torque acting about the link CM. We define the global frame friction force on link $i \in \{1, \dots, N\}$ in the form of (2.15) as

$$\mathbf{f}_{R,i} = \begin{bmatrix} f_{R,x,i} \\ f_{R,y,i} \end{bmatrix} = \begin{cases} -\mu m g \frac{\mathbf{v}_i}{|\mathbf{v}_i|} & \text{when } |\mathbf{v}_i| > 0, \\ \mathbf{0}_{2\times 1} & \text{when } |\mathbf{v}_i| = 0, \end{cases} \tag{10.3}$$

where $\mu \geq 0$ is the Coulomb friction coefficient, g is the gravitational acceleration constant, and $\mathbf{v}_i = [\dot{x}_i, \dot{y}_i]^T$ is the velocity of link i. We now define the scalar value $\widehat{v}_i \in \mathbb{R}$ given by

$$\widehat{v}_i = \begin{cases} \frac{1}{|\mathbf{v}_i|} & \text{when } |\mathbf{v}_i| > 0, \\ 0 & \text{when } |\mathbf{v}_i| = 0, \end{cases} \tag{10.4}$$

and also the diagonal matrix $\boldsymbol{\Gamma} = \text{diag}(\widehat{v}_1, \widehat{v}_2, \dots, \widehat{v}_N) \in \mathbb{R}^{N\times N}$, which enable us to express the friction forces on all links in the form of (2.16) as

$$\mathbf{f}_R = \begin{bmatrix} \mathbf{f}_{R,x} \\ \mathbf{f}_{R,y} \end{bmatrix} = -\mu m g \begin{bmatrix} \boldsymbol{\Gamma}\dot{\mathbf{X}} \\ \boldsymbol{\Gamma}\dot{\mathbf{Y}} \end{bmatrix} \in \mathbb{R}^{2N}. \tag{10.5}$$

The friction torque about the CM of link i is denoted by $\tau_{R,i}$ and is produced by the friction forces acting normal to the link during link rotation. As illustrated in Fig. 10.2, the friction force $\mathrm{d}f_{R,i}$ on an infinitesimal mass element $\mathrm{d}m$ of link i due to the link rotation $\dot{\theta}_i$ produces a friction torque $\mathrm{d}\tau_{R,i}$ about the CM of the link, which is given by

$$\mathrm{d}\tau_{R,i} = s\,\mathrm{d}f_{R,i} = s\big(-\mu g \cdot \text{sgn}(s\dot{\theta}_i) \cdot \mathrm{d}m\big), \tag{10.6}$$

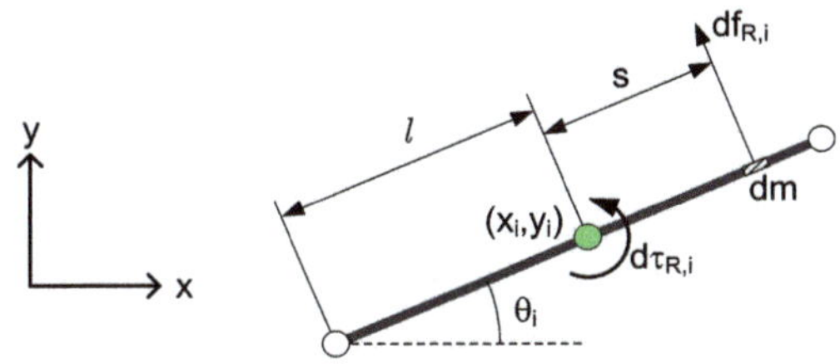

Fig. 10.2 The ground friction torque acting about the CM of each link

where s is the distance from the CM of link i to the mass element $\mathrm{d}m$. Using the relation $\mathrm{d}m = \frac{m}{2l}\,\mathrm{d}s$, we can calculate the total friction torque on link i as

$$\tau_{R,i} = \int_{-l}^{l} \mathrm{d}\tau_{R,i} = -\frac{1}{2}\mu m g l \cdot \mathrm{sgn}(\dot{\theta}_i). \tag{10.7}$$

The global frame friction torque on all links can be expressed in matrix form as

$$\boldsymbol{\tau}_R = -\frac{1}{2}\mu m g l \cdot \mathrm{sgn}(\dot{\boldsymbol{\theta}}), \tag{10.8}$$

where $\boldsymbol{\tau}_R = [\tau_{R,1}, \ldots, \tau_{R,N}]^T \in \mathbb{R}^N$.

10.3.2 The Equations of Motion Without Obstacles

It follows directly from the equations of motion in (2.33a) and (2.33b) that the model of a snake robot influenced by isotropic Coulomb ground friction forces on a flat surface can be written as

$$\mathbf{M}_\theta \ddot{\boldsymbol{\theta}} = l\mathbf{S}_\theta \mathbf{K}\mathbf{f}_{R,x} - l\mathbf{C}_\theta \mathbf{K}\mathbf{f}_{R,y} - \mathbf{W}\dot{\boldsymbol{\theta}}^2 + \boldsymbol{\tau}_R + \mathbf{D}^T\mathbf{u}, \tag{10.9}$$

$$Nm\ddot{\mathbf{p}} = Nm\begin{bmatrix}\ddot{p}_x\\ \ddot{p}_y\end{bmatrix} = \begin{bmatrix}\mathbf{e}^T\mathbf{f}_{R,x}\\ \mathbf{e}^T\mathbf{f}_{R,y}\end{bmatrix} = \mathbf{E}^T\mathbf{f}_R, \tag{10.10}$$

where $\boldsymbol{\theta} \in \mathbb{R}^N$ and $\mathbf{p} \in \mathbb{R}^2$ represent the $N+2$ generalised coordinates of the system, $\mathbf{f}_R$ contains the Coulomb friction forces defined in (10.5), $\boldsymbol{\tau}_R$ contains the friction torques defined in (10.8), and

$$\mathbf{M}_\theta = J\mathbf{I}_N + ml^2\mathbf{S}_\theta\mathbf{V}\mathbf{S}_\theta + ml^2\mathbf{C}_\theta\mathbf{V}\mathbf{C}_\theta, \tag{10.11}$$

$$\mathbf{W} = ml^2\mathbf{S}_\theta\mathbf{V}\mathbf{C}_\theta - ml^2\mathbf{C}_\theta\mathbf{V}\mathbf{S}_\theta, \tag{10.12}$$

$$\mathbf{V} = \mathbf{A}^T\left(\mathbf{D}\mathbf{D}^T\right)^{-1}\mathbf{A}, \tag{10.13}$$

$$\mathbf{K} = \mathbf{A}^T\left(\mathbf{D}\mathbf{D}^T\right)^{-1}\mathbf{D}. \tag{10.14}$$

By introducing the configuration vector of the system

$$\mathbf{q} = \begin{bmatrix}\boldsymbol{\theta}\\ \mathbf{p}\end{bmatrix} \in \mathbb{R}^{N+2}, \tag{10.15}$$

the model of the snake robot can be written compactly as

$$\mathbf{M}(\mathbf{q})\ddot{\mathbf{q}} = \mathbf{f}_u(\mathbf{q}, \dot{\mathbf{q}}, \mathbf{u}), \tag{10.16}$$

where the subscript 'u' denotes *unconstrained* since the motion is not constrained by obstacles, and where

$$\mathbf{M}(\mathbf{q}) = \begin{bmatrix} \mathbf{M}_\theta & \mathbf{0}_{N\times 1} & \mathbf{0}_{N\times 1} \\ \mathbf{0}_{1\times N} & Nm & 0 \\ \mathbf{0}_{1\times N} & 0 & Nm \end{bmatrix} \in \mathbb{R}^{(N+2)\times(N+2)}, \tag{10.17}$$

$$\mathbf{f}_u(\mathbf{q}, \dot{\mathbf{q}}, \mathbf{u}) = \begin{bmatrix} l\mathbf{S}_\theta\mathbf{K} & -l\mathbf{C}_\theta\mathbf{K} \\ \mathbf{e}^T & \mathbf{0}_{1\times N} \\ \mathbf{0}_{1\times N} & \mathbf{e}^T \end{bmatrix} \mathbf{f}_R + \begin{bmatrix} -\mathbf{W}\dot{\boldsymbol{\theta}}^2 + \boldsymbol{\tau}_R + \mathbf{D}^T\mathbf{u} \\ 0 \\ 0 \end{bmatrix}. \tag{10.18}$$

10.4 Overview of the Contact Modelling Approach

The unconstrained dynamics of the snake robot given by (10.16) will be extended in the following sections in order to include contact forces from external obstacles in the environment of the robot (see section organisation in Fig. 10.1). We now describe the contact modelling approach in more detail.

The planar environment of the snake robot consists of an arbitrary number of external obstacles with *circular* shape. We consider circular obstacles to simplify the process of detecting overlap between a link and an obstacle, as described in Sect. 10.5. Note that this assumption is not very restrictive since most objects can locally be approximated by circular shapes. The friction coefficient between the snake robot and any obstacle is denoted by $\mu_o \geq 0$. Furthermore, we assume that the shortest distance between the edges of any two obstacles is greater than the link length $2l$ to prevent contact on both sides of a link.

The interaction between a snake robot link and an obstacle is modelled by introducing a *unilateral velocity constraint* for the link when it comes into contact with the obstacle. The constraint is *unilateral* (acts in one lateral direction only) since the constraint shall allow sideways motion of the link *away* from the obstacle but prevent any sideways motion *towards* (and thereby into) the obstacle. As illustrated in Fig. 10.3, the interaction model is somewhat similar to assuming that each contacted link is equipped with tangentially mounted passive wheels with no-slip conditions. There is, however, an important difference between the two situations illustrated in Fig. 10.3 since the constraints introduced by the passive wheels act in both lateral directions of the links (i.e. they are *bilateral*), while the constraints introduced by the obstacles are *unilateral*. Section 10.6.1 presents the equations describing the unilateral constraints for all contacted links.

Remark 10.2 It was noted at the very end of Sect. 4.4.2 that links with anisotropic ground friction properties (where the normal direction friction coefficient is much

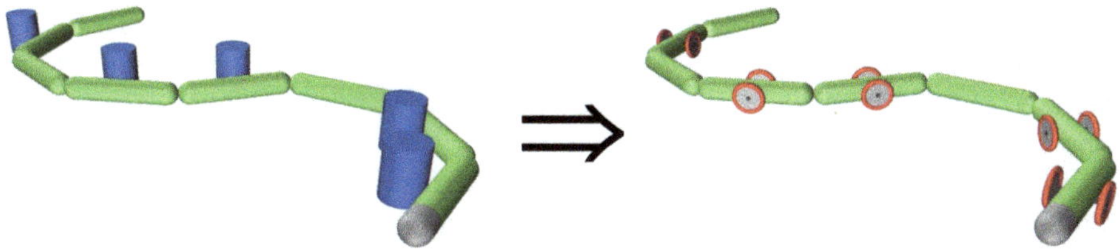

Fig. 10.3 The interaction model is similar to assuming that each contacted link is equipped with passive wheels with no-slip conditions

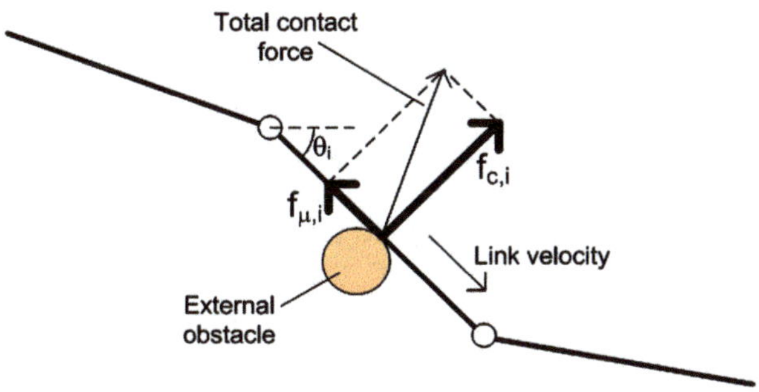

Fig. 10.4 The obstacle contact force on link i consists of the normal direction constraint force, $f_{c,i}$, which acts normal to the link, and the friction force, $f_{\mu,i}$, which acts in the tangential link direction

larger than the tangential direction coefficient) are similar to links equipped with tangentially mounted passive wheels with no-slip conditions. The comparison in Fig. 10.3 is therefore particularly interesting since it implies that the phenomenon that propels a snake robot forward due to the anisotropic ground friction properties of the links is similar to the phenomenon that propels a snake robot forward due to the interaction of the links with external obstacles. We will use this observation in Chap. 12 as an argument for continuing to consider lateral undulation motion also in environments with obstacles.

The obstacle contact force on link $i \in \{1, \dots, N\}$ consists of two orthogonal components, which are illustrated in Fig. 10.4. The first component is the *constraint force*, $\mathbf{f}_{c,i} \in \mathbb{R}^2$, acting in the normal direction of link i and away from the obstacle (parallel to the local y axis of link i). The second component is the obstacle *friction force*, $\mathbf{f}_{\mu,i} \in \mathbb{R}^2$, acting in the tangential direction of link i and in the opposite direction of the tangential link velocity (parallel to the local x axis of link i).

We assume that an obstacle contact force (i.e. the *constraint force* and the *friction force* from the obstacle) acts on the CM of a link only. Furthermore, we disregard any contact torque about the CM of the link. This simplifies the equations of motion considerably and does not have any significant influence on the overall motion of the robot when the length of the links is small.

During sustained contact with a fixed set of obstacles, we will show in Sect. 10.6 how the problem of calculating the resulting obstacle contact forces can be formulated as an LCP, which was introduced in Sect. 10.2.2. By solving the LCP, we calculate the forces on the CM of the contacted links that are needed to satisfy the unilateral velocity constraints imposed on each contacted link. These contact forces are then added to the equations of motion in (10.16) in order to cancel out the applied forces acting against the constraints. This represents the dynamics of the snake robot during continuous constrained motion.

When a link (that was previously not in contact with an obstacle) comes into contact with an obstacle, an *impact* occurs. We assume all impacts to be completely *inelastic*, meaning that the normal direction velocity of the link is completely absorbed during the impact. Furthermore, we assume all impacts to be *instantaneous*. During an impact, the contacted link is subjected to an *impulsive* constraint force in the normal direction of the link which instantaneously changes the normal direction link velocity in order to prevent it from continuing into the obstacle. This represents the discontinuous impact dynamics of the snake robot. We assume that the configuration $\mathbf{q}$ of the snake robot, which was defined in (10.15), is unaltered during an impact. We also assume that the obstacle friction forces cannot display impulsive behaviour, which means that we will disregard obstacle friction forces during an impact. Similar to the calculation of the continuous constraint forces, we will show in Sect. 10.7 that the problem of calculating the impulsive constraint forces during an impact can be formulated as an LCP. It will be seen in Sect. 10.8 that an impact triggers a *jump* in the state of the hybrid snake robot model.

The above description of the contact modelling approach is summarised by the following set of assumptions:

Assumption 10.1 *All obstacles have a circular shape.*

Assumption 10.2 *The distance between the edges of any two obstacles is greater than the link length* $2l$.

Assumption 10.3 *The friction coefficient between the snake robot and any obstacle is* $\mu_o \geq 0$.

Assumption 10.4 *An obstacle contact force acts on the CM of a link only. The link length is small so that contact torques about the link CM are negligible.*

Assumption 10.5 *Impacts between the snake robot and the obstacles are completely inelastic.*

Assumption 10.6 *All impacts are instantaneous in time, and all impact forces are impulsive.*

Assumption 10.7 *During an impact, the configuration* $\mathbf{q}$ *of the snake robot remains unaltered, while the velocity* $\dot{\mathbf{q}}$ *will generally experience a jump.*

Assumption 10.8 *Obstacle friction forces are negligible during an impact.*

Remark 10.3 The common approach when modelling mechanical systems with unilateral constraints is to calculate the direction of a constraint force with respect to the normal direction of the *constraint surface* (see Brogliato 1999), i.e. the normal direction of the obstacles in this case. With the approach taken in this chapter, the shape of the obstacles does not have to be considered explicitly as we instead calculate the constraint forces with respect to the normal direction of the *contacted links*. This simplifies the equations of motion. Note that these two approaches produce similar constraint directions when the end point of a link is not in contact with an obstacle. To verify this, consider a snake robot link in contact with a circular obstacle. Since the link is tangent to the obstacle, the normal direction of the link and the obstacle must be equal. When the end point of a link, i.e. a joint, is in contact with an obstacle, however, both links attached to the joint are in contact with the obstacle. The approach in this chapter will then produce a normal constraint force on both links attached to the joint.

Remark 10.4 A consequence of modelling obstacle contact by a unilateral force on the contacted link is that there is nothing preventing the foremost link (the head) of the snake robot from penetrating an obstacle head-on along its tangential direction. Furthermore, a consequence of Assumption 10.4 is that a link in theory can rotate 'into' an obstacle while its CM has zero normal direction velocity. These two consequences are results of the goal of keeping the mathematical model as simple as possible but are not critical in practice. In particular, head-on collisions with the head of the snake robot can be avoided through the control strategy, e.g. by assuming that the head is equipped with distance sensors that enable the robot to actively avoid head-on collisions with obstacles. Link rotations 'into' an obstacle may only occur to a very small extent during obstacle-aided locomotion since this is mostly a forward gliding type of motion.

10.5 Detection of Obstacle Impacts and Detachments

The planar environment of the snake robot consists of k circular obstacles indexed by $j \in \{1, \dots, k\}$. The global coordinates of the centre of obstacle j is denoted by (x_{O_j}, y_{O_j}). The set $\mathbf{O}_j$ of points occupied by obstacle j is given by

$$\mathbf{O}_j = \{(x, y) | (x - x_{O_j})^2 + (y - y_{O_j})^2 \le R_{O_j}^2\}, \tag{10.19}$$

where R_{O_j} is the radius of obstacle j. By studying Fig. 2.1, it is easily seen that the set $\mathbf{L}_i$ of points occupied by link $i \in \{1, \dots, N\}$ is given by

$$\mathbf{L}_i = \{(x, y) | x = x_i + s \cos\theta_i, y = y_i + s \sin\theta_i, s \in [-l, l]\}, \tag{10.20}$$

where (x_i, y_i) and θ_i are the CM coordinates and angle of link i, respectively. A collision between link i and obstacle j occurs whenever $\mathbf{L}_i \cap \mathbf{O}_j \neq \emptyset$, where $\emptyset$ denotes an empty set. There is no collision if $\mathbf{L}_i \cap \mathbf{O}_j = \emptyset$.

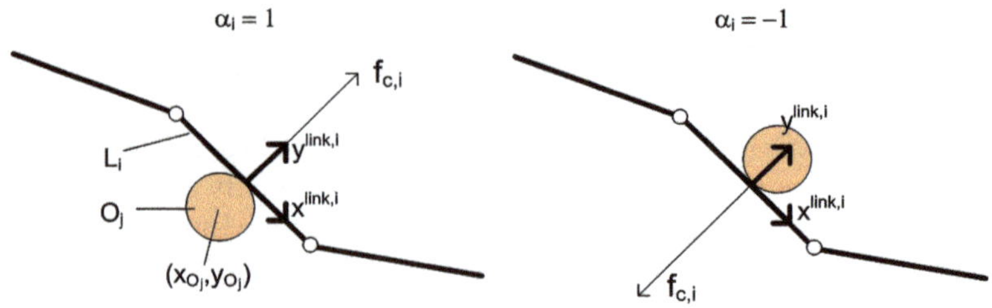

Fig. 10.5 The value of the contact parameter of link i is $\alpha_i = 1$ when the constraint force points along the positive link y axis, and $\alpha_i = -1$ when the constraint force points along the negative link y axis. We set $\alpha_i = 0$ when there is no contact

We now introduce a vector of *contact parameters*, $\boldsymbol{\alpha} = [\alpha_1, \ldots, \alpha_N]^T \in \mathbb{R}^N$. The contact parameter of link i, denoted by $\alpha_i \in \{-1, 0, 1\}$, is a discrete state value that determines if the link is in contact with an obstacle and also on which side of the link there is contact. As shown in Fig. 10.5, $\alpha_i = 1$ when the obstacle constraint force points along the positive link y axis, while $\alpha_i = -1$ when the obstacle constraint force points along the negative link y axis. We set $\alpha_i = 0$ when link i is *not* in contact with an obstacle.Whenever link i impacts an obstacle, we update the contact parameter α_i according to $\alpha_i = \widetilde{\alpha}(i)$, where the scalar function $\widetilde{\alpha}(i)$ is defined by

$$\widetilde{\alpha}(i) = -\operatorname{sgn}\Big([0, 1]\big(\mathbf{R}_{\text{link},i}^{\text{global}}\big)^T \Big(\min_{j\in\{1,\ldots,k\}} \mathbf{r}_{L_i,O_j}\Big)\Big), \tag{10.21}$$

where $\mathbf{r}_{L_i,O_j} \in \mathbb{R}^2$ is the vector from link i to obstacle j, and $\mathbf{R}_{\text{link},i}^{\text{global}}$ was defined in (2.3) as the rotation matrix from the global frame to the local frame of link i. In (10.21), the global frame vector from link i to the *closest* obstacle is first found by use of the min operator, which is assumed to find the vector with the smallest *Euclidean norm*. Subsequently, this vector is transformed to the frame of link i using the rotation matrix $\mathbf{R}_{\text{link},i}^{\text{global}}$. Finally, the y component of this vector is extracted using the vector $[0, 1]$. The sign of the y component determines on which side of the link the obstacle is located.

10.6 The Continuous Dynamics of the Snake Robot During Constrained Motion

In this section, we present the model of the continuous constrained dynamics of the snake robot during sustained contact with a fixed set of obstacles. The section consists of three parts. In particular, Sect. 10.6.1 presents the equations describing the unilateral velocity constraints for the links that are in contact with an obstacle, while Sect. 10.6.2 presents the resulting equations of motion of the snake robot under the assumption that the obstacles are frictionless ($\mu_o = 0$). The equations of motion where obstacle friction is present ($\mu_o > 0$) are derived in Sect. 10.6.3. Note that the discontinuous dynamics occurring when the set of contacted links changes is treated in Sect. 10.7.

10.6.1 The Unilateral Constraints from the Obstacles

With reference to Fig. 10.5, the unilateral velocity constraint imposed on link i during contact with an obstacle may be compactly expressed as

$$\alpha_i v_{n,i} \geq 0, \tag{10.22}$$

where α_i is the contact parameter of link i, and $v_{n,i}$ is the normal direction velocity of link i, i.e. the velocity of the CM of link i in the direction of the local link y axis. This constraint prevents sideways link motion towards (and thereby into) the obstacle. Using (2.3), we can write the velocity constraint in the global frame as

$$\alpha_i(-\dot{x}_i \sin\theta_i + \dot{y}_i \cos\theta_i) \geq 0, \tag{10.23}$$

where $(\dot{x}_i, \dot{y}_i)$ and θ_i are the CM velocity and angle of link i, respectively. Hence, using notation from Sect. 2.4, we may express the unilateral velocity constraints for all links in matrix form as

$$\text{diag}(\boldsymbol{\alpha})(-\mathbf{S}_\theta \dot{\mathbf{X}} + \mathbf{C}_\theta \dot{\mathbf{Y}}) \geq \mathbf{0}. \tag{10.24}$$

By inserting (2.12a) and (2.12b) into (10.24) and rearranging we get

$$\text{diag}(\boldsymbol{\alpha})\mathbf{C}(\mathbf{q})\dot{\mathbf{q}} \geq \mathbf{0}, \tag{10.25}$$

where $\mathbf{q} \in \mathbb{R}^{N+2}$ was defined in (10.15), and $\mathbf{C}(\mathbf{q}) \in \mathbb{R}^{N\times(N+2)}$ is given by

$$\mathbf{C}(\mathbf{q}) = \left[-l\left(\mathbf{S}_\theta \mathbf{K}^T \mathbf{S}_\theta + \mathbf{C}_\theta \mathbf{K}^T \mathbf{C}_\theta\right), -\sin\boldsymbol{\theta}, \cos\boldsymbol{\theta}\right]. \tag{10.26}$$

We denote the number of contacted links by $m \in \{0, 1, \dots, N\}$. In order to easily select the velocity constraints from (10.25) that correspond to contacted links, we define a *selection matrix* $\mathbf{S}_c(\boldsymbol{\alpha}) \in \mathbb{R}^{m\times N}$, which simply contains the m rows from the matrix $\text{diag}(\boldsymbol{\alpha}) \in \mathbb{R}^{N\times N}$ that contain a non-zero element. With this selection matrix, we can write the velocity constraints for all links that are in contact with an obstacle as

$$\overline{\mathbf{C}}(\mathbf{q}, \boldsymbol{\alpha})\dot{\mathbf{q}} \geq \mathbf{0}, \tag{10.27}$$

where $\overline{\mathbf{C}}(\mathbf{q}, \boldsymbol{\alpha}) = \mathbf{S}_c(\boldsymbol{\alpha})\mathbf{C}(\mathbf{q}) \in \mathbb{R}^{m\times(N+2)}$.

The calculation of the obstacle contact forces in the next subsection requires the time derivative of the matrix $\overline{\mathbf{C}}(\mathbf{q}, \boldsymbol{\alpha})$, which is given by

$$\dot{\overline{\mathbf{C}}}(\mathbf{q}, \boldsymbol{\alpha}) = \mathbf{S}_c(\boldsymbol{\alpha})\left[l(\mathbf{S}_\theta \widetilde{\mathbf{K}} \mathbf{C}_\theta - \mathbf{C}_\theta \widetilde{\mathbf{K}} \mathbf{S}_\theta), -\mathbf{C}_\theta \dot{\boldsymbol{\theta}}, -\mathbf{S}_\theta \dot{\boldsymbol{\theta}}\right], \tag{10.28}$$

where $\widetilde{\mathbf{K}} = \text{diag}(\dot{\boldsymbol{\theta}})\mathbf{K}^T - \mathbf{K}^T \text{diag}(\dot{\boldsymbol{\theta}})$. Note that this derivative is only valid over intervals where the set of contacted links remains constant. The derivative of $\overline{\mathbf{C}}(\mathbf{q}, \boldsymbol{\alpha})$ is not defined for time instants where an element of $\boldsymbol{\alpha}$ is changed.

The following rank property of the constraint matrix $\overline{\mathbf{C}}$ is important in order to uniquely determine the contact forces acting on the snake robot (see Sect. 10.6.2).

Property 10.1 The matrix $\overline{\mathbf{C}}$ has full rank ($\text{rank}(\overline{\mathbf{C}}) = m$) for all $(\mathbf{q}, \boldsymbol{\alpha})$.

Remark 10.5 Due to the complexity of the elements in $\overline{\mathbf{C}}$, it is difficult to present a purely mathematical proof that Property 10.1 holds. However, we can argue from a physical perspective that this property must hold. Assume that $\text{rank}(\overline{\mathbf{C}}) < m$. This implies linear dependence between some of the m rows of $\overline{\mathbf{C}}$, i.e. there must exist a row of $\overline{\mathbf{C}}$, denoted $\overline{\mathbf{C}}_i$, such that

$$\overline{\mathbf{C}}_i = \sum_{j \in \{1,\dots,m\} \setminus \{i\}} k_j \overline{\mathbf{C}}_j, \tag{10.29}$$

where $k_j \in \mathbb{R}$. The scalar $\overline{\mathbf{C}}_i \dot{\mathbf{q}}$ is the magnitude of the normal direction velocity of link i, denoted by $|v_{n,i}|$. Multiplying (10.29) by $\dot{\mathbf{q}}$ therefore gives

$$|v_{n,i}| = \overline{\mathbf{C}}_i \dot{\mathbf{q}} = \sum_{j \in \{1,\dots,m\} \setminus \{i\}} k_j |v_{n,j}|, \tag{10.30}$$

which states that the normal direction velocity of link i can be written as a linear combination of the normal direction velocities of all other contacted links of the snake robot. From a physical perspective, such a dependence could never occur unless all links are parallel since the couplings between the link velocities are given in terms of both normal *and* tangential link velocities. In particular, (2.7a) and (2.7b) in Sect. 2.4 implies that the velocity of link i can be written in terms of the velocities of link $i - 1$ and link $i + 1$. Unless links $i - 1$, i, and $i + 1$ are parallel, this is a relationship involving both the normal *and* tangential velocities of link $i - 1$ and link $i + 1$. This contradicts (10.30) since the relationship in (10.30) only contains normal direction velocities. This leaves the case of parallel links ($\theta_1 = \theta_2 = \cdots = \theta_N$) as the only way for (10.30) to be true. A straightforward calculation of $\overline{\mathbf{C}}$ in e.g. *Matlab Symbolic Toolbox* shows that $\overline{\mathbf{C}}$ always has full rank when the link angles are equal. We can therefore conclude that $\overline{\mathbf{C}}$ never drops rank.

This subsection can be summarised as follows. At any given time instant, the snake robot is in contact with m obstacles. The interaction between the robot and these m obstacles is modelled by imposing the unilateral velocity constraints in (10.27) on the m contacted links.

10.6.2 The Constrained Dynamics of the Snake Robot Without Obstacle Friction

We will now use the unilateral velocity constraints in (10.27) to derive the resulting equations of motion of the snake robot. We assume that the m contact points between the links and the obstacles have *already* been established, i.e. we consider the continuous contact dynamics of the snake robot over a time interval where the set of contacted links remains fixed. We first consider the frictionless case in this subsection, followed by contact forces *with* friction in the next subsection.

Let us first assume that the m velocity constraints on the snake robot in (10.27) are *bilateral*, i.e. that they are given by

$$\overline{\mathbf{C}}\dot{\mathbf{q}} = \mathbf{0}. \tag{10.31}$$

These are called *Pfaffian* constraints and are modelled by adding a term to the equations of motion in (10.16) as follows (see Goldstein et al. 2002):

$$\mathbf{M}\ddot{\mathbf{q}} = \mathbf{f}_u + \overline{\mathbf{C}}^T \boldsymbol{\lambda}. \tag{10.32}$$

The term $\overline{\mathbf{C}}^T \boldsymbol{\lambda}$ ensures compliance with the imposed velocity constraints, where $\boldsymbol{\lambda} \in \mathbb{R}^m$ is a vector of scalars known as *Lagrange multipliers* (Goldstein et al. 2002). The Lagrange multipliers are important because multiplier λ_j equals the magnitude of the constraint force that ensures compliance with the jth constraint. This means that if the jth constraint in (10.31) corresponds to the velocity constraint on link i, then λ_j equals the magnitude of the constraint force $\mathbf{f}_{c,i}$ acting on link i.

We now argue that (10.32) also represent the equations of motion of the snake robot when the velocity constraints are *unilateral*, as in (10.27). This is quite obvious since the influence of a *unilateral* constraint on the snake robot when this constraint is active (i.e. when the unilateral constraint is preventing sideways motion of a link) must necessarily be identical to the influence that the corresponding *bilateral* constraint would have. In other words, the nature of the constraint (i.e. unilateral or bilateral) is not apparent when the constraint is active since it is only active in one direction at a time. The only difference between the bilateral and unilateral case concerns the calculation of the constraint forces given by $\boldsymbol{\lambda}$. In the bilateral case, one may calculate $\boldsymbol{\lambda}$ directly by differentiating (10.31) with respect to time, inserting (10.32), and solving for $\boldsymbol{\lambda}$. In the unilateral case, however, the constraint forces must be calculated so that they comply with the directional requirements of the constraints, i.e. we require that $\boldsymbol{\lambda} \geq \mathbf{0}$. This means that the constraint forces can only point *away* from the obstacles.

We will now handle this directional requirement by employing the theory of *linear complementarity problems* (LCPs) introduced in Sect. 10.2.2. This approach is based on the work in Lötstedt (1982). The key observation is that the normal direction velocity of a contacted link and the corresponding constraint force are subjected to a *complementarity condition*. If the normal direction velocity is non-zero (i.e. the link is moving away from the obstacle), then the corresponding constraint force must be zero. Likewise, the normal direction velocity must be zero if the corresponding constraint force is non-zero. This complementarity condition also applies to the normal direction *acceleration* of a contacted link and the corresponding constraint force.

From the above discussion, the equations of motion of the snake robot that include *unilateral* constraint forces from *frictionless* obstacles are given by

$$\mathbf{M}\ddot{\mathbf{q}} = \mathbf{f}_u + \overline{\mathbf{C}}^T \boldsymbol{\lambda}, \tag{10.33}$$

$$\overline{\mathbf{C}}\dot{\mathbf{q}} \geq \mathbf{0}, \qquad \boldsymbol{\lambda} \geq \mathbf{0}, \qquad \boldsymbol{\lambda}^T \overline{\mathbf{C}}\dot{\mathbf{q}} = 0. \tag{10.34}$$

The vector $\overline{\mathbf{C}}\dot{\mathbf{q}} \in \mathbb{R}^m$ contains the normal direction velocity of each contacted link in the direction *away* from each obstacle. The normal direction *acceleration* of each contacted link in the direction *away* from each obstacle, denoted by $\overline{\mathbf{a}}_n \in \mathbb{R}^m$, is given by

$$\overline{\mathbf{a}}_n = \frac{\mathrm{d}}{\mathrm{d}t}(\overline{\mathbf{C}}\dot{\mathbf{q}}) = \overline{\mathbf{C}}\ddot{\mathbf{q}} + \dot{\overline{\mathbf{C}}}\dot{\mathbf{q}} \geq \mathbf{0}. \tag{10.35}$$

By solving (10.33) for $\ddot{\mathbf{q}}$ and inserting into (10.35), we finally arrive at the following model of the continuous contact dynamics of the snake robot with frictionless obstacles:

$$\mathbf{M}\ddot{\mathbf{q}} = \mathbf{f}_u + \overline{\mathbf{C}}^T \boldsymbol{\lambda}, \tag{10.36}$$

$$\begin{aligned} \overline{\mathbf{a}}_n &= \overline{\mathbf{C}}\mathbf{M}^{-1}\mathbf{f}_u + \dot{\overline{\mathbf{C}}}\dot{\mathbf{q}} + \overline{\mathbf{C}}\mathbf{M}^{-1}\overline{\mathbf{C}}^T \boldsymbol{\lambda}, \\ \overline{\mathbf{a}}_n &\geq \mathbf{0}, \qquad \boldsymbol{\lambda} \geq \mathbf{0}, \qquad \boldsymbol{\lambda}^T \overline{\mathbf{a}}_n = 0. \end{aligned} \tag{10.37}$$

Equation (10.37) is in the form of the general LCP given in (10.2) with $\mathbf{A} = \overline{\mathbf{C}}\mathbf{M}^{-1}\overline{\mathbf{C}}^T$ and $\mathbf{a} = \overline{\mathbf{C}}\mathbf{M}^{-1}\mathbf{f}_u + \dot{\overline{\mathbf{C}}}\dot{\mathbf{q}}$. In order to calculate the dynamics of the snake robot at any given time instant, this LCP must be solved for the unknowns $\overline{\mathbf{a}}_n$ and $\boldsymbol{\lambda}$ subject to the complementarity conditions. The calculated $\boldsymbol{\lambda}$ gives the constraint forces from the obstacles and is plugged into (10.36) in order to calculate $\ddot{\mathbf{q}}$.

In order to determine the existence and uniqueness properties of the LCP in (10.37), we will need the following result, which is proved in Bernstein (2009, Proposition 8.1.2, item *xiii*):

Proposition 10.1 *Let* $\mathbf{M} \in \mathbb{R}^{N \times N}$ *be a* symmetrical *and* positive definite *matrix* ($\mathbf{M} > 0$), *and* $\mathbf{C} \in \mathbb{R}^{m \times N}$ *be a matrix of full rank* (rank$(\mathbf{C}) = m$). *Then* $\mathbf{C}\mathbf{M}\mathbf{C}^T > 0$.

We can now state the following result concerning the existence and uniqueness properties of the LCP in (10.37):

Theorem 10.2 *The LCP in* (10.37) *always has a unique solution* $(\overline{\mathbf{a}}_n, \boldsymbol{\lambda})$.

Proof From Theorem 10.1 we see that the proof is complete if we can show that $\mathbf{A} = \overline{\mathbf{C}}\mathbf{M}^{-1}\overline{\mathbf{C}}^T$ is a P-matrix. Since $\overline{\mathbf{C}}$ has full rank (by Property 10.1) and $\mathbf{M} = \mathbf{M}^T > 0$ (the inertia matrix is always symmetrical and positive definite), we have from Proposition 10.1 that $\mathbf{A} = \overline{\mathbf{C}}\mathbf{M}^{-1}\overline{\mathbf{C}}^T > 0$. Since $\mathbf{A}$ is symmetrical and positive definite, it must also be a P-matrix. This completes the proof. □

Remark 10.6 The LCP in (10.37) can be regarded as the problem of, at a given time instant, determining which obstacle contacts that will persist onto the next time instant and which will not. A link contact will persist onto the next time instant if the corresponding value of $\boldsymbol{\lambda}$ is non-zero. If, however, the value of $\bar{\mathbf{a}}_n$ for a link contact is non-zero, then the link will detach from the obstacle.

10.6.3 The Constrained Dynamics of the Snake Robot with Obstacle Friction

We employ a Coulomb friction model in order to describe the gliding friction force between the links and the obstacles. In accordance with Fig. 10.4, we define the obstacle friction force on link i as

$$\mathbf{f}_{\mu,i} = \begin{bmatrix} \mathbf{f}_{\mu,x,i} \\ \mathbf{f}_{\mu,y,i} \end{bmatrix} = -\mu_o \begin{bmatrix} \cos\theta_i \\ \sin\theta_i \end{bmatrix} \operatorname{sgn}(v_{t,i}) |\mathbf{f}_{c,i}|, \tag{10.38}$$

where $\mu_o \geq 0$ is the Coulomb friction coefficient of the obstacles, $|\mathbf{f}_{c,i}|$ is the magnitude of the obstacle constraint force acting on link i, and $v_{t,i}$ is the tangential direction velocity of link i, i.e. the velocity of the CM of link i in the direction of the local link x axis. The obstacle friction forces on all the links can be expressed as

$$\mathbf{f}_{\mu} = \begin{bmatrix} \mathbf{f}_{\mu,x} \\ \mathbf{f}_{\mu,y} \end{bmatrix} = -\mu_o \begin{bmatrix} \mathbf{C}_\theta \\ \mathbf{S}_\theta \end{bmatrix} \operatorname{diag}\big(\operatorname{sgn}(\mathbf{v}_t)\big) |\mathbf{f}_c|, \tag{10.39}$$

where $\mathbf{f}_{\mu,x} = [\mathbf{f}_{\mu,x,1}, \dots, \mathbf{f}_{\mu,x,N}]^T \in \mathbb{R}^N$ and $\mathbf{f}_{\mu,y} = [\mathbf{f}_{\mu,y,1}, \dots, \mathbf{f}_{\mu,y,N}]^T \in \mathbb{R}^N$ contain the obstacle friction forces on the links in the global x and y directions, respectively, $|\mathbf{f}_c| = [|\mathbf{f}_{c,1}|, \dots, |\mathbf{f}_{c,N}|]^T \in \mathbb{R}^N$ contains the magnitude of the constraint force on each link, and $\mathbf{v}_t = [v_{t,1}, \dots, v_{t,N}]^T \in \mathbb{R}^N$ contains the tangential link velocities in the local x direction of each link.

The mapping between the friction forces and the acceleration of the configuration vector $\ddot{\mathbf{q}}$ is identical to the mapping between the ground friction forces $\mathbf{f}_R$ and $\ddot{\mathbf{q}}$ given in (10.18) since $\mathbf{f}_R$ and $\mathbf{f}_\mu$ both act on the CM of the links. By using the easily verifiable relation $|\mathbf{f}_c| = |\mathbf{S}_c(\boldsymbol{\alpha})|^T \boldsymbol{\lambda}$, where $\mathbf{S}_c(\boldsymbol{\alpha})$ is the selection matrix introduced in (10.27), and $\boldsymbol{\lambda}$ is the vector of Lagrange multipliers introduced in (10.32), we may write the accelerations due to the obstacle friction forces, temporarily denoted $\ddot{\mathbf{q}}_{f_\mu}$, as

$$\ddot{\mathbf{q}}_{f_\mu} = -\mu_o \boldsymbol{\Lambda} \boldsymbol{\lambda}, \tag{10.40}$$

where $\boldsymbol{\Lambda} \in \mathbb{R}^{(N+2)\times m}$ is given by

$$\boldsymbol{\Lambda}(\mathbf{q}, \dot{\mathbf{q}}, \boldsymbol{\alpha}) = \begin{bmatrix} l\mathbf{S}_\theta \mathbf{K} & -l\mathbf{C}_\theta \mathbf{K} \\ \mathbf{e}^T & \mathbf{0}_{1\times N} \\ \mathbf{0}_{1\times N} & \mathbf{e}^T \end{bmatrix} \begin{bmatrix} \mathbf{C}_\theta \\ \mathbf{S}_\theta \end{bmatrix} \operatorname{diag}\big(\operatorname{sgn}(\mathbf{v}_t)\big) \big|\mathbf{S}_c(\boldsymbol{\alpha})\big|^T. \tag{10.41}$$

By adding (10.40) to the equations of motion in (10.36) and following the same approach that led to the LCP in (10.37), we get the following equations describing the continuous contact dynamics of the snake robot that include obstacle friction forces:

$$\mathbf{M}\ddot{\mathbf{q}} = \mathbf{f}_u + (\overline{\mathbf{C}}^T - \mu_o \boldsymbol{\Lambda})\boldsymbol{\lambda}, \tag{10.42}$$

$$\begin{gathered} \overline{\mathbf{a}}_n = \overline{\mathbf{C}}\mathbf{M}^{-1}\mathbf{f}_u + \dot{\overline{\mathbf{C}}}\dot{\mathbf{q}} + \overline{\mathbf{C}}\mathbf{M}^{-1}(\overline{\mathbf{C}}^T - \mu_o \boldsymbol{\Lambda})\boldsymbol{\lambda}, \\ \overline{\mathbf{a}}_n \geq \mathbf{0}, \qquad \boldsymbol{\lambda} \geq \mathbf{0}, \qquad \boldsymbol{\lambda}^T \overline{\mathbf{a}}_n = 0. \end{gathered} \tag{10.43}$$

We again identify (10.43) as an LCP of the general form given in (10.2) with $\mathbf{A} = \overline{\mathbf{C}}\mathbf{M}^{-1}(\overline{\mathbf{C}}^T - \mu_o \boldsymbol{\Lambda})$ and $\mathbf{a} = \overline{\mathbf{C}}\mathbf{M}^{-1}\mathbf{f}_u + \dot{\overline{\mathbf{C}}}\dot{\mathbf{q}}$. When obstacle friction is present ($\mu_o > 0$), we can no longer guarantee existence and uniqueness of the solution to the LCP in (10.43) since it is no longer evident that the matrix $\mathbf{A}$ is a P-matrix ($\mathbf{A}$ is no longer symmetrical, which complicates the P-matrix check). This existence and uniqueness issue is a general and well-known problem for acceleration LCPs that include Coulomb friction (see e.g. Brogliato 1999; Lötstedt 1981; Mason and Wang 1988; Song et al. 2000; Trinkle et al. 1997). Due to the complexity of determining if $\mathbf{A}$ is a P-matrix, we are unable to provide an analytical upper bound of μ_o that must be satisfied to guarantee existence and uniqueness of the solution to the LCP in (10.43). However, we can still state the following result:

Theorem 10.3 *For a given* $(\mathbf{q}, \dot{\mathbf{q}}, \boldsymbol{\alpha})$, *there exists a* $\mu_o^* > 0$ *such that the LCP in* (10.43) *has a unique solution* $(\overline{\mathbf{a}}_n, \boldsymbol{\lambda})$ *for* $\mu_o \in [0, \mu_o^*)$.

Proof Recall from Sect. 10.2.2 that $\mathbf{A} = \overline{\mathbf{C}}\mathbf{M}^{-1}(\overline{\mathbf{C}}^T - \mu_o \boldsymbol{\Lambda})$ is a P-matrix if all principal minors of $\mathbf{A}$ are positive. We know from Theorem 10.2 that the LCP in (10.43) *always* has a unique solution for $\mu_o = 0$ since $\mathbf{A}$ is a P-matrix in this case. All principal minors of $\mathbf{A}$ must therefore be positive for $\mu_o = 0$. Assume now that we increase μ_o until a principal minor of $\mathbf{A}$ becomes zero and denote the corresponding value of the friction coefficient by $\mu_o^* > 0$. It is then evident that the P-matrix property of $\mathbf{A}$ must be preserved for $\mu_o < \mu_o^*$, i.e. existence and uniqueness of the solution to the LCP in (10.43) must hold for $\mu_o < \mu_o^*$. This completes the proof. □

Remark 10.7 During our numerical treatments of the LCP in (10.43) so far, we have not yet encountered a single instance where $\mathbf{A}$ has failed to be a P-matrix. We therefore conjecture that μ_o must have an unrealistically high value in order for $\mathbf{A}$ to no longer be a P-matrix and that the LCP in (10.43) will always be uniquely solvable during our simulations of the snake robot.

Remark 10.8 As implied by Theorem 10.1, the matrix $\mathbf{A}$ being a P-matrix is only required to guarantee existence and uniqueness of the solution to a LCP for *all* data vectors $\mathbf{a}$. The LCP in (10.43) may therefore have a unique solution for a given $(\mathbf{q}, \dot{\mathbf{q}}, \boldsymbol{\alpha})$ even if $\mathbf{A}$ is not a P-matrix (Cottle et al. 1992).

10.7 The Discontinuous Dynamics of the Snake Robot During Obstacle Impacts and Detachments

In this section, we present the model of the discontinuous dynamics of the snake robot occurring when the set of links in contact with an obstacle changes. We first present the dynamics during obstacle impacts in Sect. 10.7.1, followed by the dynamics during obstacle detachments in Sect. 10.7.2.

10.7.1 The Discontinuous Dynamics of the Snake Robot During Obstacle Impacts

By Assumption 10.5, an *inelastic impact* occurs when a link (that was previously not in contact with an obstacle) comes into contact with an obstacle. By Assumption 10.6, the impact is instantaneous in time, and the resulting impact forces are impulsive, resulting in a discontinuous *jump* in the velocity of the snake robot. Following an approach in Brogliato (1999), we model the impact as

$$\mathbf{M}(\mathbf{q}^+)\dot{\mathbf{q}}^+ - \mathbf{M}(\mathbf{q}^-)\dot{\mathbf{q}}^- = \mathbf{F}_{\text{impulse}}, \tag{10.44}$$

where $\mathbf{F}_{\text{impulse}} \in \mathbb{R}^{N+2}$ denotes the generalised impulsive impact forces, and $\mathbf{q}^-$, $\dot{\mathbf{q}}^-$, $\mathbf{q}^+$, and $\dot{\mathbf{q}}^+$ denote the generalised coordinates and velocities immediately *before* and *after* the impact, respectively. This superscript notation is commonly used when modelling hybrid systems, as described in Sect. 10.2.1. By Assumption 10.7, the configuration of the snake robot is unaltered during an impact ($\mathbf{q}^+ = \mathbf{q}^-$). This means that $\mathbf{M}(\mathbf{q}^-) = \mathbf{M}(\mathbf{q}^+) = \mathbf{M}(\mathbf{q})$. By Assumption 10.8, the impact forces are frictionless. Following the same argumentation that led to the expression in (10.36), we can now rewrite (10.44) as

$$\mathbf{M}(\mathbf{q})(\dot{\mathbf{q}}^+ - \dot{\mathbf{q}}^-) = \overline{\mathbf{C}}^T(\mathbf{q}, \boldsymbol{\alpha}^+)\boldsymbol{\lambda}, \tag{10.45}$$

where $\boldsymbol{\lambda} \in \mathbb{R}^m$ is a vector of *impulsive constraint forces*. Note that the constraint matrix $\overline{\mathbf{C}}$, which was defined in (10.27), depends on the contact parameter vector *after* the impact, i.e. $\boldsymbol{\alpha}^+$. This is because the contact parameter of the impacted link is zero immediately *before* the impact (i.e. if link i impacts an obstacle, then $\alpha_i^- = 0$). In order to include this link in the impact dynamics, we must calculate $\overline{\mathbf{C}}$ based on the value of α_i *after* the impact, i.e. $\alpha_i^+ = \widetilde{\alpha}(i)$, where $\widetilde{\alpha}(i)$ is given by (10.21). Note also that $\overline{\mathbf{C}}(\mathbf{q}, \boldsymbol{\alpha}^-) \in \mathbb{R}^{(m-1)\times(N+2)}$, while $\overline{\mathbf{C}}(\mathbf{q}, \boldsymbol{\alpha}^+) \in \mathbb{R}^{m\times(N+2)}$.

We will now calculate the impulsive constraint forces $\boldsymbol{\lambda}$ and the post-impact velocity $\dot{\mathbf{q}}^+$ by following an approach presented in Schaft and Schumacher (2000). The post-impact velocity and the impulsive constraint forces are naturally subjected to the same complementarity conditions as given in (10.34). We therefore have that

$$\overline{\mathbf{C}}(\mathbf{q}, \boldsymbol{\alpha}^+)\dot{\mathbf{q}}^+ \geq \mathbf{0}, \qquad \boldsymbol{\lambda} \geq \mathbf{0}, \qquad \boldsymbol{\lambda}^T\overline{\mathbf{C}}(\mathbf{q}, \boldsymbol{\alpha}^+)\dot{\mathbf{q}}^+ = 0. \tag{10.46}$$

Solving (10.45) for $\dot{\mathbf{q}}^+$ and premultiplying by $\overline{\mathbf{C}}(\mathbf{q}, \boldsymbol{\alpha}^+)$ give

$$\overline{\mathbf{C}}\dot{\mathbf{q}}^+ = \overline{\mathbf{C}}\dot{\mathbf{q}}^- + \overline{\mathbf{C}}\mathbf{M}^{-1}\overline{\mathbf{C}}^T\boldsymbol{\lambda}. \tag{10.47}$$

Denoting the normal direction velocities of each of the contacted links (in the direction *away* from each obstacle) by the vector $\overline{\mathbf{v}}_n \in \mathbb{R}^m$, we may combine (10.46) and (10.47) into the following LCP describing the impact dynamics of the snake robot:

$$\begin{aligned} &\overline{\mathbf{v}}_n^+ = \overline{\mathbf{v}}_n^- + \overline{\mathbf{C}}\mathbf{M}^{-1}\overline{\mathbf{C}}^T\boldsymbol{\lambda}, \\ &\overline{\mathbf{v}}_n^+ \geq \mathbf{0}, \qquad \boldsymbol{\lambda} \geq \mathbf{0}, \qquad \boldsymbol{\lambda}^T\overline{\mathbf{v}}_n^+ = 0. \end{aligned} \tag{10.48}$$

The LCP in (10.48) is in the general form of the LCP given in (10.2) with $\mathbf{A} = \overline{\mathbf{C}}\mathbf{M}^{-1}\overline{\mathbf{C}}^T$ and $\mathbf{a} = \overline{\mathbf{v}}_n^-$, and must be solved for the unknowns $\overline{\mathbf{v}}_n^+$ and $\boldsymbol{\lambda}$. Subsequently, the post-impact velocity is found by solving (10.45) for $\dot{\mathbf{q}}^+$ and inserting the calculated $\boldsymbol{\lambda}$. The following result concerns the existence and uniqueness properties of the LCP in (10.48):

Theorem 10.4 *The LCP in* (10.48) *always has a unique solution* $(\overline{\mathbf{v}}_n^+, \boldsymbol{\lambda})$.

Proof The proof is identical to the proof of Theorem 10.2. □

This subsection is now summarised. The discontinuous impact dynamics of the snake robot when link i impacts an obstacle and the state immediately before the impact is $(\mathbf{q}^-, \dot{\mathbf{q}}^-, \boldsymbol{\alpha}^-)$ is given by

$$\mathbf{q}^+ = \mathbf{q}^-, \tag{10.49a}$$

$$\dot{\mathbf{q}}^+ = \dot{\mathbf{q}}^- + \mathbf{M}^{-1}(\mathbf{q})\overline{\mathbf{C}}^T(\mathbf{q}, \boldsymbol{\alpha}^+)\boldsymbol{\lambda}, \tag{10.49b}$$

$$\alpha_j^+ = \begin{cases} \widetilde{\alpha}(i) & \text{when } j = i, \\ \alpha_j^- & \text{when } j \neq i, \end{cases} \tag{10.49c}$$

where $j \in \{1, \dots, N\}$, $\widetilde{\alpha}(i)$ is given by (10.21), and $\boldsymbol{\lambda}$ is calculated from the LCP in (10.48).

10.7.2 The Discontinuous Dynamics of the Snake Robot During Obstacle Detachments

If, at any time instant, the normal direction acceleration vector $\overline{\mathbf{a}}_n$ in the solution to the LCP in (10.43) contains a non-zero element, then the link corresponding to this non-zero element will accelerate away from the obstacle, meaning that the link will *detach* from the obstacle. The detachment dynamics is trivial compared to the impact dynamics since it only involves setting the contact parameter of the detached link to zero. In other words, whenever $\alpha_i \neq 0$ and the collision detection mechanism described in Sect. 10.5 detects that link i no longer overlaps with an obstacle, then the state of the snake robot is updated according to the discontinuous detachment dynamics given by

$$\mathbf{q}^+ = \mathbf{q}^-, \tag{10.50a}$$

$$\dot{\mathbf{q}}^+ = \dot{\mathbf{q}}^-, \tag{10.50b}$$

$$\alpha_j^+ = \begin{cases} 0 & \text{when } j = i, \\ \alpha_j^- & \text{when } j \neq i, \end{cases} \tag{10.50c}$$

where $j \in \{1, \dots, N\}$.

10.8 The Complete Hybrid Model of the Snake Robot in an Obstacle Environment

In this section, we employ the framework of *hybrid dynamical systems*, which was described in Sect. 10.2.1, in order to encapsulate the *continuous* dynamics from Sect. 10.6 and the *discontinuous* dynamics from Sect. 10.7 into a single *hybrid* model, or a *hybrid plant*. We will denote the hybrid model as a *plant* to distinguish it from a hybrid *controller* that we will present in Chap. 12. In the following subsections, we first present the *jump set* $\mathbf{D}_p$, *jump map* $\mathbf{G}_p$, *flow set* $\mathbf{C}_p$, and *flow map* $\mathbf{F}_p$ of the plant, followed by a summary of the hybrid plant in the last subsection. We define the state vector of the hybrid plant as

$$\mathbf{x} = \begin{bmatrix} \mathbf{q} \\ \dot{\mathbf{q}} \\ \boldsymbol{\alpha} \end{bmatrix} \in \mathbb{R}^{3N+4}. \tag{10.51}$$

10.8.1 The Jump Set

A *jump* in the state vector $\mathbf{x}$ of the plant occurs when a link impacts an obstacle (jump in $\dot{\mathbf{q}}$ and $\boldsymbol{\alpha}$) or when a link detaches from an obstacle (jump in $\boldsymbol{\alpha}$). By employing the notation from Sect. 10.5, the *jump set* corresponding to an *impact* between link i and an obstacle may be expressed as

$$\mathbf{D}_{L_i}^{\text{Impact}} = \{\mathbf{x} | \mathbf{L}_i \cap \mathbf{O}_j \neq \emptyset, j \in \{1, \dots, k\}, \widetilde{\alpha}(i)\mathbf{C}_i\dot{\mathbf{q}} < 0\}, \tag{10.52}$$

where $\widetilde{\alpha}(i)$ is given by (10.21), and $\mathbf{C}_i$ denotes the ith row of the matrix $\mathbf{C}(\mathbf{q})$ in (10.26). We use $\widetilde{\alpha}(i)$ instead of α_i in (10.52) because $\alpha_i = 0$ before the impact has taken place. The *jump set* corresponding to link i *detaching* from an obstacle may be expressed as

$$\mathbf{D}_{L_i}^{\text{Detach}} = \{\mathbf{x}|\mathbf{L}_i \cap \mathbf{O}_j = \emptyset, j \in \{1,\ldots,k\}, \alpha_i \neq 0\}. \tag{10.53}$$

The jump sets comprising the *impacts* and the *detachments* of all the links, respectively, are given by

$$\mathbf{D}^{\text{Impact}} = \bigcup_{i\in\{1,\ldots,N\}} \mathbf{D}_{L_i}^{\text{Impact}}, \qquad \mathbf{D}^{\text{Detach}} = \bigcup_{i\in\{1,\ldots,N\}} \mathbf{D}_{L_i}^{\text{Detach}}. \tag{10.54}$$

The complete jump set of the hybrid plant may now be compactly expressed as

$$\mathbf{D}_p = \mathbf{D}^{\text{Impact}} \cup \mathbf{D}^{\text{Detach}}. \tag{10.55}$$

10.8.2 The Jump Map

The *jump map* corresponding to the *impact* between link i and an obstacle is presented in (10.49a)–(10.49c), while the *jump map* corresponding to link i *detaching* from an obstacle is presented in (10.50a)–(10.50c). By combining (10.49a)–(10.49c) and (10.50a)–(10.50c), the complete jump map of the plant can be expressed as

$$\mathbf{x}^+ = \mathbf{G}_p(\mathbf{x}) = \begin{bmatrix} \mathbf{q}^+ \\ \dot{\mathbf{q}}^+ \\ \boldsymbol{\alpha}^+ \end{bmatrix} \quad \text{for all } \mathbf{x} \in \mathbf{D}_p, \tag{10.56}$$

where

$$\mathbf{q}^+ = \mathbf{q}^-, \tag{10.57a}$$

$$\dot{\mathbf{q}}^+ = \begin{cases} \dot{\mathbf{q}}^- + \mathbf{M}^{-1}(\mathbf{q})\overline{\mathbf{C}}^T(\mathbf{q},\boldsymbol{\alpha}^+)\boldsymbol{\lambda} & \text{when } \mathbf{x} \in \mathbf{D}^{\text{Impact}}, \\ \dot{\mathbf{q}}^- & \text{when } \mathbf{x} \notin \mathbf{D}^{\text{Impact}}, \end{cases} \tag{10.57b}$$

$$\alpha_i^+ = \begin{cases} \widetilde{\alpha}(i) & \text{when } \mathbf{x} \in \mathbf{D}_{L_i}^{\text{Impact}}, \\ 0 & \text{when } \mathbf{x} \in \mathbf{D}_{L_i}^{\text{Detach}}, \\ \alpha_i^- & \text{when } \mathbf{x} \notin (\mathbf{D}_{L_i}^{\text{Impact}} \cup \mathbf{D}_{L_i}^{\text{Detach}}). \end{cases} \tag{10.57c}$$

The value of $\widetilde{\alpha}(i)$ is given by (10.21), and $\boldsymbol{\lambda}$ is calculated from the LCP in (10.48).

10.8.3 The Flow Set

We define the *flow set* of the plant so that the state vector $\mathbf{x}$ always flows as long as the *jump set* is empty. The flow set is therefore given as

$$\mathbf{C}_p = \{\mathbf{x}|\mathbf{x} \notin \mathbf{D}_p\}. \tag{10.58}$$

10.8.4 The Flow Map

The *flow map* of $\mathbf{q}$ is simply $\dot{\mathbf{q}}$, and the *flow map* of $\dot{\mathbf{q}}$ is given by (10.42). The contact vector $\boldsymbol{\alpha}$ remains unchanged between jumps of $\mathbf{x}$, which means that the *flow map* of $\boldsymbol{\alpha}$ is the zero vector. The complete flow map of the plant is given by

$$\dot{\mathbf{x}} = \mathbf{F}_p(\mathbf{x}, \mathbf{u}) = \begin{bmatrix} \dot{\mathbf{q}} \\ \ddot{\mathbf{q}} \\ \mathbf{0}_{N\times 1} \end{bmatrix} \quad \text{for all } \mathbf{x} \in \mathbf{C}_p, \tag{10.59}$$

where

$$\ddot{\mathbf{q}} = \mathbf{M}^{-1}\big(\mathbf{f}_u(\mathbf{q}, \dot{\mathbf{q}}, \mathbf{u}) + \big(\overline{\mathbf{C}}^T - \mu_o \boldsymbol{\Lambda}\big)\boldsymbol{\lambda}\big), \tag{10.60}$$

and $\boldsymbol{\lambda}$ is calculated from the LCP in (10.43).

10.8.5 Summary of the Complete Hybrid Plant

In accordance with Sect. 10.2.1, the complete hybrid model of the plant is written

$$\begin{aligned} \dot{\mathbf{x}} &= \mathbf{F}_p(\mathbf{x}, \mathbf{u}) \quad \text{for all } \mathbf{x} \in \mathbf{C}_p, \\ \mathbf{x}^+ &= \mathbf{G}_p(\mathbf{x}) \quad \text{for all } \mathbf{x} \in \mathbf{D}_p. \end{aligned} \tag{10.61}$$

The evolution of the state vector of the plant has the following existence and uniqueness properties:

Theorem 10.5 *Given a control input* $\mathbf{u}$, *the evolution of the state* $\mathbf{x}$ *of the hybrid plant in* (10.61) *from any initial state can always be uniquely determined when the obstacles are frictionless* ($\mu_o = 0$). *With obstacle friction, there exists a* $\mu_o^* > 0$ *such that existence and uniqueness of the evolution of* $\mathbf{x}$ *is guaranteed for* $\mu_o \in [0, \mu_o^*)$, *but not guaranteed for* $\mu_o \geq \mu_o^*$.

Proof From (10.58), the *flow* and *jump set* are mutually exclusive, so we can always uniquely determine whether $\mathbf{x}$ should flow or jump. By Theorem 10.4, the *jump map* of $\mathbf{x}$ is always unique. By Theorem 10.2, the *flow map* of $\mathbf{x}$ is always unique with frictionless obstacles. By Theorem 10.3, there exists a $\mu_o^* > 0$ such that the *flow map* of $\mathbf{x}$ is always unique when $\mu_o \in [0, \mu_o^*)$. This completes the proof. □

10.9 Simulation Study: Comparison of the Hybrid Model with Previous Experimental and Simulation Results

In order to investigate the validity of the hybrid plant proposed in (10.61), we have compared simulation results from this hybrid plant with experimental and simulation results presented in Transeth et al. (2008b). The simulator of the hybrid plant in

(10.61) was implemented in *Matlab R2008b* on a laptop running *Windows XP*. The continuous dynamics of the plant was calculated with the *ode45* solver in Matlab with a relative and absolute error tolerance of 10^{-3}.

The work Transeth et al. (2008b), which was described in Sect. 1.3, presents experimental results of obstacle-aided locomotion and also simulation results from a model developed based on the framework of nonsmooth dynamics. The joints of both the physical and simulated snake robots in Transeth et al. (2008b) were controlled according to the gait pattern lateral undulation that was defined in (4.32). In particular, joint i of the snake robot was controlled according to the reference

$$\phi_{i,\text{ref}} = \alpha \sin\bigl(\omega t + (i-1)\delta\bigr) \tag{10.62}$$

with $i \in \{1, \dots, N-1\}$, $\alpha = 40^\circ$, $\omega = 80^\circ/\text{s}$, and $\delta = 50^\circ$. Furthermore, the parameters characterising the snake robot in Transeth et al. (2008b) were $N = 11$, $l = 0.061$ m, $m = 0.682$ kg, and $J = 0.0013$ kgm^2. Obstacles of diameter 25 cm were placed in accordance with the expected motion of the robot (see Transeth et al. 2008b for details). We implemented the plant in (10.61) in accordance with the above parameters and set the control input $\mathbf{u}$ according to the joint controller in (4.33) to control the joints of the robot according to (10.62) with $k_p = 20$ and $k_d = 5$.

The experimental and simulation results from Transeth et al. (2008b) are reprinted in Fig. 10.6, while the corresponding simulation results from the hybrid plant proposed in this chapter are presented in Fig. 10.7. We see that there is a close resemblance between the results from Transeth et al. (2008b) and the simulation results based on the plant in (10.61). The y direction amplitudes in Figs. 10.6(b) and 10.7(b) are slightly different because the model in Transeth et al. (2008b) also considers the width of each link, while the model in this chapter assumes the width to be infinitesimal. In our opinion, the simulation results support the conjecture that, despite its simplifying assumptions, the model proposed in this chapter captures the essential part of the dynamics of a snake robot interacting with obstacles.

10.10 Chapter Summary

This chapter is summarised as follows:

- We have proposed an isotropic Coulomb ground friction model that also considers the ground friction torque on the links. The unconstrained (no obstacles) model of the snake robot with this new friction model is given in (10.16).
- We have extended the unconstrained model in (10.16) to include contact forces from external obstacles in the environment around the robot. In particular:
  - Under Assumptions 10.1–10.8 presented in Sect. 10.4, the motion of the snake robot is constrained according to the velocity constraints in (10.27) when one or several snake robot links are in contact with obstacles.

Fig. 10.6 A reprint of Figs. 14 and 15 from Transeth et al. (2008b), which show experimentally measured and simulated lateral undulation in an environment with obstacles

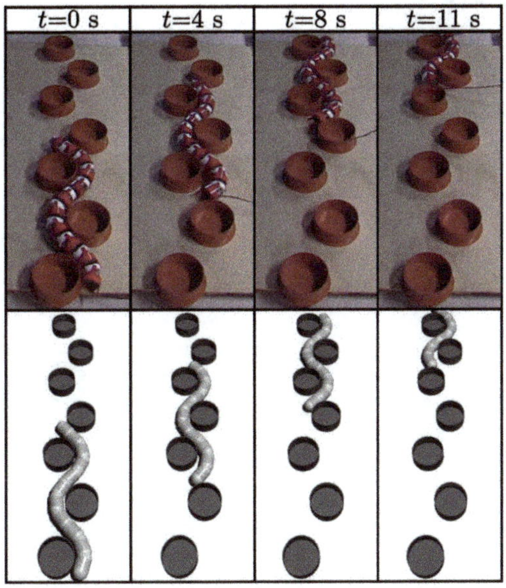

(a) The experimental (top) and simulated (bottom) robot motion.

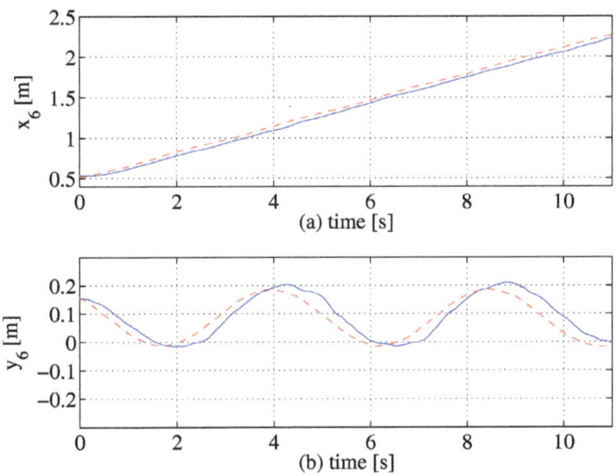

(b) The experimentally measured (solid line) and simulated (dashed line) position of link 6.

- The continuous constrained dynamics of the snake robot during sustained contact with a fixed set of obstacles is described by (10.36) when the obstacles are frictionless and by (10.42) when obstacle friction is present.
- Calculation of the constrained dynamics *without* obstacle friction requires us to solve the LCP in (10.37). By Theorem 10.2, this LCP always has a unique solution.

Fig. 10.7 Simulation of lateral undulation in an environment with obstacles based on the hybrid model presented in this chapter. The simulated scenario is similar to the scenario shown in Fig. 10.6

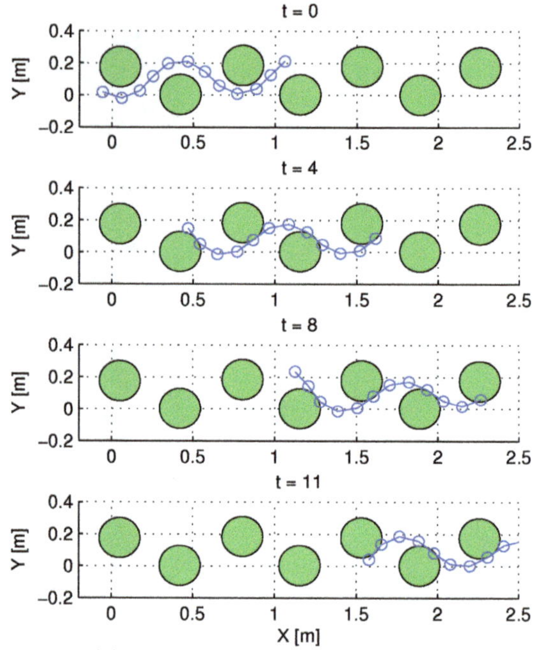

(a) The simulated motion of the robot.

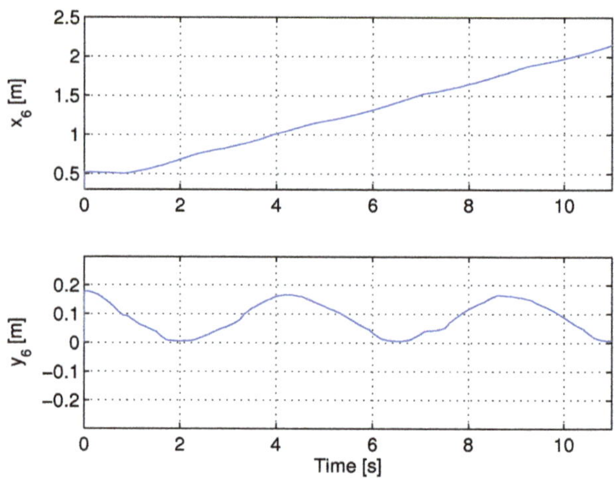

(b) The simulated position of link 6.

– Calculation of the constrained dynamics *with* obstacle friction requires us to solve the LCP in (10.43). By Theorem 10.3, there always exists a non-zero scalar $\mu_o^* > 0$ such that this LCP has a unique solution when the obstacle friction coefficient μ_o satisfies $0 \le \mu_o < \mu_o^*$. We were unable to derive an analytical expression for μ_o^*.

- The discontinuous *impact* dynamics of the snake robot occurring when a link (that was previously not in contact with an obstacle) comes into contact with an obstacle is described by (10.49a)–(10.49c), which requires us to solve the LCP in (10.48). By Theorem 10.4, this LCP always has a unique solution.
- The discontinuous *detachment* dynamics of the snake robot occurring when a link detaches from an obstacle is described by (10.50a)–(10.50c).
- The complete *hybrid model* of a planar snake robot in an environment with obstacles is given by (10.61).

Chapter 11
Development of a Mechanical Snake Robot for Obstacle-Aided Locomotion

In this chapter, we describe the development of the snake robot *Kulko*, which is shown in Fig. 11.1. While the snake robot *Wheeko*, described in Chap. 3, was used as the experimental platform in Part I of this book, Kulko serves as the experimental platform in Part II. The joint modules of Kulko are covered by contact force sensors to allow the robot to sense its environment, and spherical shells that give the robot a smooth outer surface, thereby allowing slithering (gliding) motion in uneven and cluttered environments. In the following, we will detail the design and implementation of the robot and present experimental results that validate the function of the contact force measurement system. At the end of this chapter, we will also propose an alternative strategy for contact force sensing, which has the advantage that the sensor system can be well protected *inside* the snake robot.

The chapter is organised as follows. The relation between this chapter and previous literature is briefly discussed in Sect. 11.1. We provide an overview of the snake robot design in Sect. 11.2, followed by a description of the exterior gliding surface, the force measurement system, and the power and control system in Sects. 11.3, 11.4, and 11.5, respectively. Section 11.6 presents an experimental investigation of the contact force measurement system, while Sect. 11.7 describes the setup of the motion control experiments that have been carried out with the robot. An alternative approach to contact force sensing is outlined in Sect. 11.8. Finally, the chapter is summarised in Sect. 11.9.

11.1 The Relation Between This Chapter and Previous Literature

Previous snake robot design efforts have given very limited attention to the exterior gliding surface of such robots, and to methods for enabling snake robots to sense their environment. To our best knowledge, snake robots equipped with an explicit contact force sensing mechanism have only been considered previously in Hirose (1993), Chen et al. (2008), Bayraktaroglu (2008), Liljebäck et al. (2006), Fjerdingen

P. Liljebäck et al., *Snake Robots*, Advances in Industrial Control,
DOI 10.1007/978-1-4471-2996-7_11, © Springer-Verlag London 2013

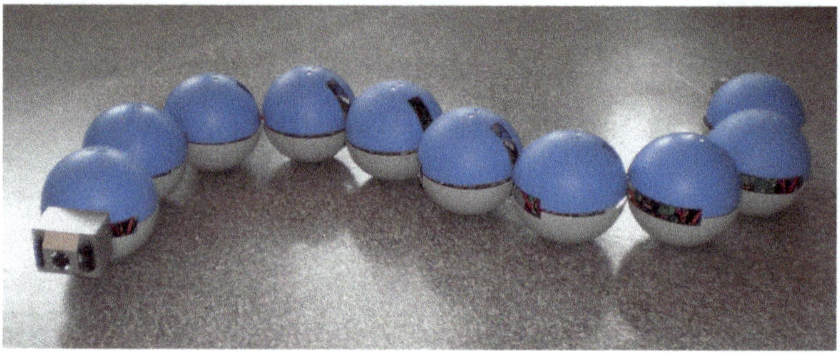

Fig. 11.1 The snake robot *Kulko* developed for locomotion in uneven and cluttered environments

et al. (2008), and Taal et al. (2009). The works Hirose (1993), Chen et al. (2008), and Taal et al. (2009) consider force sensing for wheeled snake robots, while Bayraktaroglu (2008), Liljebäck et al. (2006), and Fjerdingen et al. (2008) consider force sensing for wheel-less snake robots. The force sensing mechanisms in Hirose (1993) and Bayraktaroglu (2008) are based on discrete contact switches, while the force sensing mechanisms in Chen et al. (2008), Liljebäck et al. (2006), Fjerdingen et al. (2008), and Taal et al. (2009) are based on analogue force measurements. In other words, wheel-less snake robots with the ability to measure the amplitude of external contact forces have been presented in Liljebäck et al. (2006) and Fjerdingen et al. (2008). These two snake robots are, however, only able to assess the magnitude of external forces applied at certain areas of their joint modules.

The novelty of this chapter in relation to previous literature is therefore the design of a spherical-shaped joint mechanism for a snake robot that (1) allows the joint modules to be covered by shells, thereby giving the robot a smooth outer surface independently of how the joints are flexed and that (2) allows contact force sensors to be installed underneath the shells, thereby enabling the robot to sense its environment. To our best knowledge, Kulko is the first reported snake robot that can measure the magnitude of external forces applied along its body. Another novel feature of this chapter is an alternative strategy for contact force sensing (see Sect. 11.8), which has the advantage that the sensor system can be well protected *inside* the snake robot.

11.2 Overview of the Snake Robot Design

As we stated in Hypothesis 9.1, snake robot locomotion in a cluttered environment requires that the snake robot can *sense* its environment, which can be achieved by equipping the robot with contact force sensing capabilities along its body. A force sensing system for a snake robot is challenging since the robot is articulated. In particular, the force sensing capabilities of the robot should be maintained independently of how the joints are flexed, which represents a significant design challenge.

Enabling a snake robot to glide forward in a cluttered environment requires that the body of the robot is sufficiently smooth, i.e. free of obstructive features. In particular, irregularities along the body may potentially induce large friction forces on the robot that obstruct the gliding motion. Obtaining a sufficiently smooth surface combined with contact force sensing along the articulated body is challenging.

The idea behind the design of Kulko, which was conceived with the above challenges in mind, is to encapsulate each joint module by a spherical shell that gives the joint a smooth outer surface independently of how the joint is flexed. Contact force sensing is thereby achieved by mounting force sensors underneath each spherical shell. As shown in Fig. 11.1, the complete snake robot consists of a serial connection of 10 identical ball-shaped joint modules. The smooth exterior surface and the force sensing capabilities of the robot are maintained independently of how the joints are flexed.

Since the scope of this book is *planar* snake robot locomotion (as explained in Sect. 1.4), Kulko was primarily developed to study obstacle-aided locomotion on a *horizontal* surface with vertical obstacles, which corresponds to the environment captured by the hybrid model proposed in Chap. 10. As a result, the contact force sensor system of Kulko was implemented to primarily measure horizontal contact forces, even though the design can be modified to measure contact forces of arbitrary direction.

In order to study obstacle-aided locomotion, we required Kulko to be propelled solely by contact forces from external obstacles. To this end, we deliberately designed the snake robot to have *isotropic* ground friction properties by making the shells completely smooth. Had we designed the robot with *anisotropic* friction properties (e.g. by making shells with grooves to produce larger friction forces in the normal direction of the body), the efficiency of the motion would have been improved according to the analysis in Sect. 4.4. However, we then would have been unable to know if the robot was propelled by obstacle contact forces or by the propulsive ground friction forces produced due to the anisotropic friction property.

In the following sections, we describe the various components of Kulko in more detail. The joint actuation mechanism of Kulko is identical to the actuation mechanism of Wheeko and is therefore described in Chap. 3.

11.3 The Exterior Gliding Surface

The smooth exterior gliding surface of Kulko is obtained by covering each joint module by two hemispherical shells, as shown in Fig. 11.2. Each hemispherical shell is 1.5-mm thick, weighs 42 g, and has an outer diameter of 140 mm. The shells were moulded from a plastic material.

As illustrated to the left in Fig. 11.3, four aluminium plates (indicated with a red colour) are bent around the joint in order to support the shells and also to allow for contact force measurements. Each shell is attached to the joint mechanism by two screws, as shown to the right in Fig. 11.3. The locations of the attachment

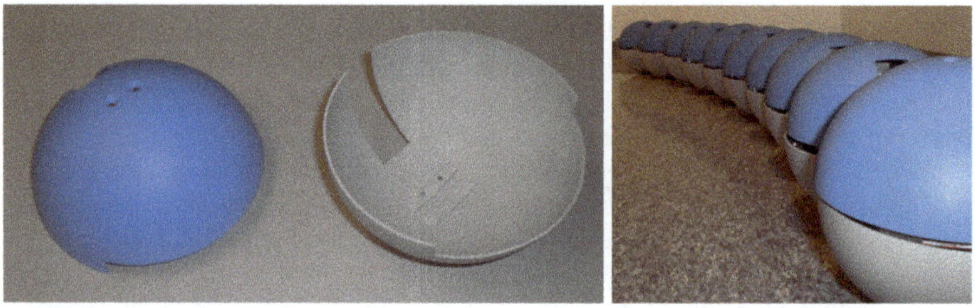

Fig. 11.2 *Left*: The upper and lower hemispherical shell of a joint module. *Right*: The smooth gliding surface along the snake robot

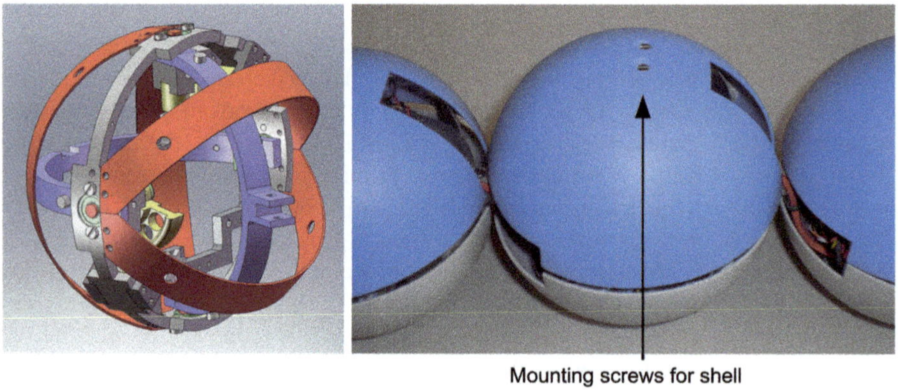

Fig. 11.3 *Left*: The four curved aluminium plates (with *red* colour) used for mounting force sensors. *Right*: The pair of screws for attaching the shell to the joint mechanism

screws define the *top* and *bottom* of the snake robot, respectively. The splice between the two hemispherical shells lies in the horizontal plane. The shells have a slit on each side corresponding to the range of motion of the connection points to the two neighbouring joints.

Note that an even more smooth exterior surface can be obtained by installing a thin hollow cylinder of e.g. a plastic material between each joint module. This approach is illustrated in Fig. 11.4 but has not been implemented on Kulko since it was not necessary for the experiments reported in this book.

11.4 The Contact Force Measurement System

11.4.1 Assumptions Underlying the Sensor System

As explained in Sect. 11.2, the main goal of Kulko is to demonstrate obstacle-aided locomotion on *horizontal* surfaces with vertical obstacles. The design described in

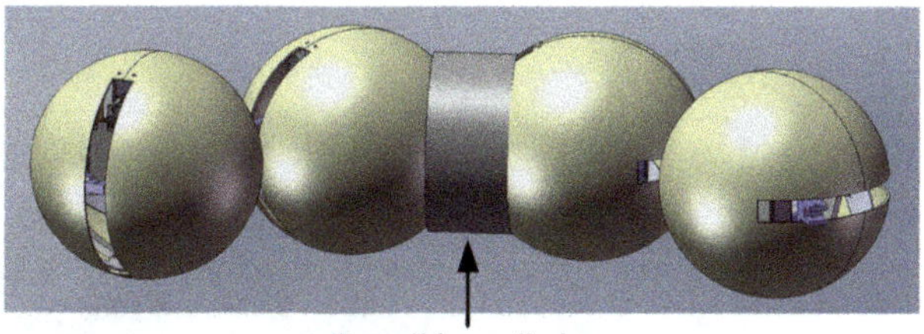

Fig. 11.4 Installing a thin *cylinder* between each joint module will further smoothen the exterior surface of the snake robot

the following therefore assumes that all contact forces are applied at the sides of the joint and not at the top or bottom. This assumption affects the placement of the contact force sensors. Note that the design can be modified to measure contact forces of arbitrary direction. However, these modifications are not considered in this book.

We require the contact force measurement system to provide information about contact forces with respect to the *macroscopic* shape of the snake robot. Information about the specific location of an applied contact force within a single joint module is not believed to be of significant interest during obstacle-aided locomotion since the location of a force within a single joint module only has a minor effect on the motion compared to the location of the force with respect to the overall shape of the robot. This means that the sensor system is only required to determine the *magnitude* of a contact force and also at which *side* of a joint module it is applied, but not the specific location where the force is applied on the outer shell. It should be noted that information about the force location within a joint module could be extracted from the force measurements by relating the magnitude of the measured forces to the relative placement of each sensor.

Since the location of the contact force with respect to the shell is not determined by the sensor system, it will not be possible to determine the exact direction of the contact forces. However, we conjecture that it will be adequate to approximate the direction of any contact force as being *normal to the macroscopic shape* of the snake robot at the location where the force is applied. This approach is in line with the contact modelling approach of the hybrid model presented in Chap. 10.

11.4.2 The Sensor System Setup

A set of force sensing resistors (FSRs) are used to measure the external contact forces applied to each joint module. An FSR is a polymer thick film device that exhibits a decrease in electrical resistance when the force applied to the active surface area of the sensor increases. Due to effects such as hysteresis, an FSR is not suited for precision measurements. However, we conjecture that obstacle-aided locomotion with a snake robot does not require very precise force measurements, which, combined with their low cost and ease of use, make FSRs suitable as force sensors for snake robots.

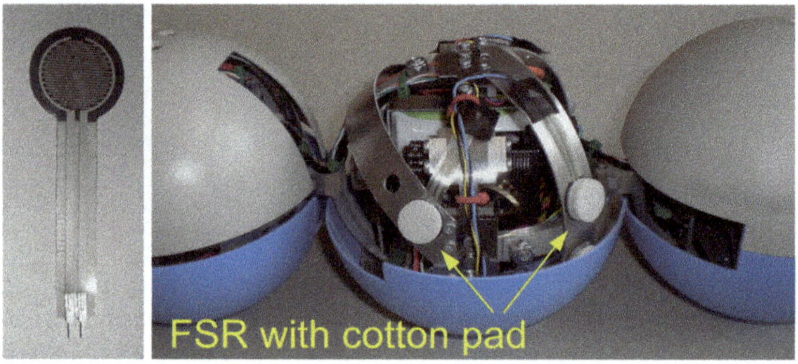

Fig. 11.5 *Left*: FSR (force sensing resistor) used to measure contact forces. *Right*: FSRs covered by *cotton pads* mounted to a joint module

Fig. 11.6 The voltage divider circuit used to measure the resistance through the FSR

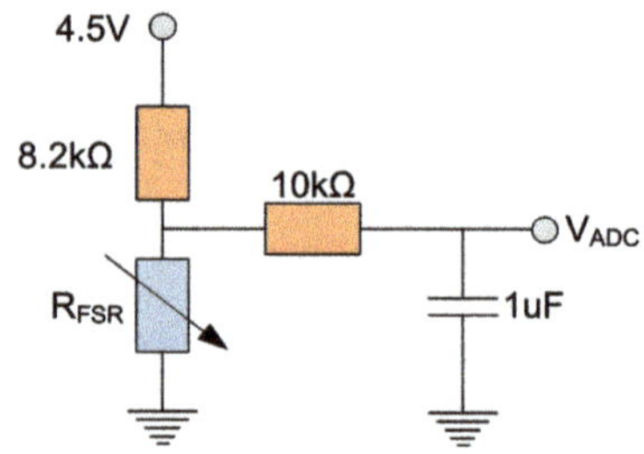

The FSR chosen for Kulko has a diameter (active sensor area) of 13 mm and is shown to the left in Fig. 11.5. The right of Fig. 11.5 shows the placement of the FSRs on the curved aluminium plates covering each joint. A small cotton pad (3-mm thick) is placed over each FSR in order to distribute the applied force across the entire active area of the sensor. Four FSRs are placed at each side of a joint module in order to be able to measure horizontal contact forces, as explained in Sect. 11.4.1. There are, in other words, eight FSRs mounted to each joint module. The exact placement of the FSRs around the joint is not critical since, as explained in the next subsection, the magnitude of the contact force is estimated by simply summing the contact forces measured by each FSR.

Note that the hemispherical shells enclosing the sensors are not completely rigid, i.e. the shells are, to some extent, deformable. This, combined with the deformability of the cotton pads placed over each FSR, means that there is compliance between the sensors and the location of an applied force.

The controller board for the joint, which is described in Sect. 11.5, contains a set of identical voltage divider circuits for measuring the resistance through the FSRs. The circuit diagram for the voltage divider circuit is shown in Fig. 11.6. The voltage V_{ADC}, where ADC denotes *analog-to-digital converter*, is the FSR measurement signal and is given as a function of the variable resistance R_{FSR} across the FSR.

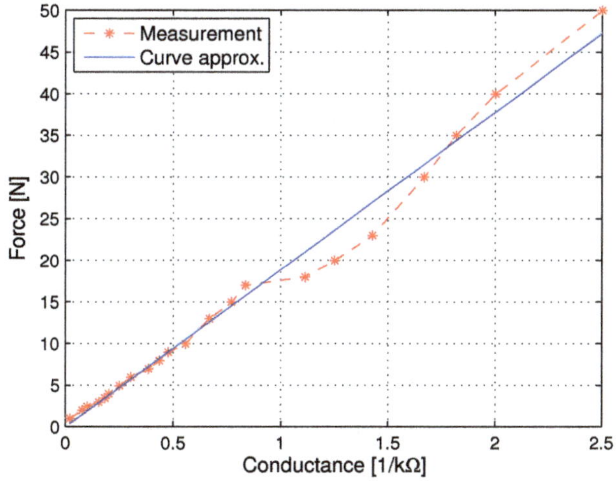

Fig. 11.7 The measured conductance (1/R) of the FSR as a function of applied force (measurements indicated by '*'). The *solid line* shows the linear curve approximation to these measurements

11.4.3 Calculation of Contact Forces

The force vs. resistance characteristic of an FSR is extremely nonlinear. However, as shown in Fig. 11.7, there is a near linear relationship between the conductance (1/resistance) of an FSR and the force applied to it. The measurements in the figure are indicated by '*' and were carried out by placing an FSR on a digital scale. We used the scale to measure the force applied to the FSR while simultaneously measuring the electrical resistance through the FSR. A linear curve approximation to these measurements is plotted with a solid line in Fig. 11.7. The linear curve approximates the relationship between the force F_{FSR} applied to an FSR as a function of its conductance G_{FSR} and resistance R_{FSR}. Based on the measurements, the expression for this linear curve was estimated as

$$F_{\mathrm{FSR}} = 18.9 \cdot G_{\mathrm{FSR}} = \frac{18.9}{R_{\mathrm{FSR}}}. \tag{11.1}$$

A simple mapping may now be derived between the FSR measurement voltage in Fig. 11.6, V_{ADC}, and the estimated applied force, F_{FSR}. The measurement voltage is given by

$$V_{\mathrm{ADC}} = \frac{R_{\mathrm{FSR}}}{R_{\mathrm{FSR}} + 8.2} 4.5. \tag{11.2}$$

Solving (11.1) for R_{FSR}, inserting into (11.2), and solving for F_{FSR} give

$$F_{\mathrm{FSR}} = \frac{4.5 - V_{\mathrm{ADC}}}{8.2 V_{\mathrm{ADC}}} 18.9. \tag{11.3}$$

As explained in Sect. 11.4.2, each side of the joint mechanism is equipped with four FSRs. Since the spherical shell covering the joint mechanism is only in contact with the internal structure of the joint through the FSR measuring points, the magnitude of an external contact force applied to the joint may be estimated by simply summing the forces measured at each FSR. Note that the attachment of the shells causes the shells to induce a constant pressure on the force sensors even when there are no external forces acting on the shells. This produces a constant force offset which we subtract from the force measurements and which we calculate as the average force during the first second after the snake robot is powered up. We denote the four FSR measurements on the left side of the joint by $F_{\text{FSR,left},1}, \dots, F_{\text{FSR,left},4}$, the measurements on the right side of the joint by $F_{\text{FSR,right},1}, \dots, F_{\text{FSR,right},4}$, and the force offset on the left and right side of the joint by $F_{\text{left,offset}}$ and $F_{\text{right,offset}}$, respectively. Consequently, we can estimate the total external forces, F_{left} and F_{right}, applied to the left and right side of the joint, respectively, as

$$F_{\text{left}} = \left(\sum_{i=1}^{4} F_{\text{FSR,left},i}\right) - F_{\text{left,offset}}, \tag{11.4a}$$

$$F_{\text{right}} = \left(\sum_{i=1}^{4} F_{\text{FSR,right},i}\right) - F_{\text{right,offset}}. \tag{11.4b}$$

11.5 The Power and Control System

Motion control and supply of power to the components of each joint module are handled by three custom-designed circuit boards installed in each joint module. These circuit boards are shown in Fig. 11.8 and are described in the following subsections.

11.5.1 The Power System

Figure 11.9 illustrates the flow of power to the various components of a joint module. Each joint is powered by two serially connected Lithium Ion batteries from *A123Systems* of type ANR26650M1. The batteries produce a supply voltage of about 6.6 V at a capacity of 2.3 Ah. The batteries were chosen due to their ability to deliver high currents (rated at 70-A continuous discharge current) and also their short charge time (rated at 15-min charge time at 10-A charge current). In particular, the two Hitec servo motors driving each joint (see Sect. 3.2) draw high current pulses each time the motor direction changes rapidly. Had the power system not been able to deliver such high currents, then the system voltage would drop and reset the microcontroller card each time the motor direction changes rapidly.

Each battery is charged by an individual battery charger card shown at the top right in Fig. 11.8. The charging is automatically initiated by applying an external voltage to the external power connectors located at the tail of the snake robot (see the right of Fig. 11.10). Each battery is connected in series with a relay, which is controlled by a power off button also located at the tail of the snake robot.

Fig. 11.8 The three custom-designed circuit boards located in each joint module. *Top left*: Microcontroller card that controls the joint mechanism. *Top right*: Battery charger card. *Bottom*: Motor power supply card

Fig. 11.9 The flow of power to the components of a joint module

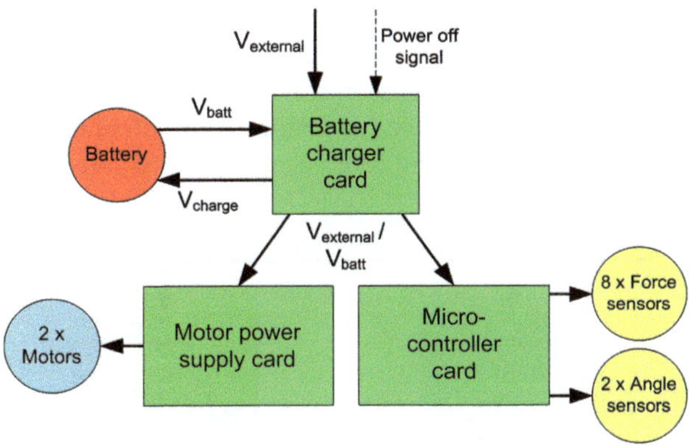

The motor power supply card, shown at the bottom of Fig. 11.8, supplies power to the two servo motors driving each joint. This card converts the voltage supplied by the battery charger card to the motor voltage (6 V). The voltage supplied by the battery charger card is either the battery voltage (when external power is disconnected) or the external voltage.

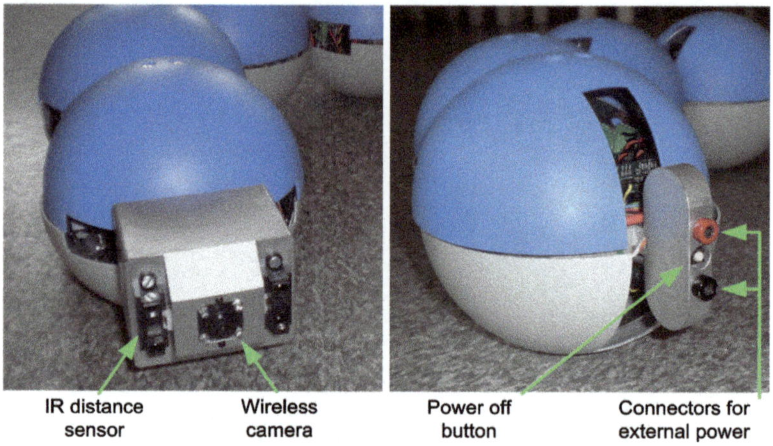

Fig. 11.10 *Left*: The head of the snake robot. *Right*: The tail of the snake robot

11.5.2 The Control System

The data flow between the components of the snake robot is illustrated in Fig. 11.11. Motion control of each joint module is handled by the microcontroller card shown to the top left in Fig. 11.8, which is based on the *Atmel* microcontroller AT90CAN128. This card continuously reads angular measurements from the two magnetic encoders (see Fig. 3.3 in Sect. 3.2) and also contact force sensor data from the FSRs (see Fig. 11.5). This card also generates PWM pulses that control the two servo motors driving each joint module. The card has a CAN bus interface for communicating with the other modules of the snake robot.

The brain (or head) of the snake robot, which is shown to the left in Fig. 11.10, contains the same microcontroller card that controls the motion of the joints. The brain card is responsible for sending joint reference angles to all joint modules over the CAN bus. The joint reference angles are calculated on an external computer in

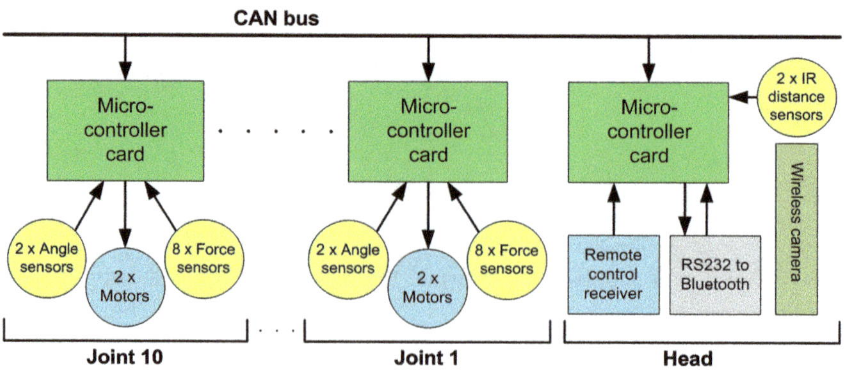

Fig. 11.11 The data flow between the modules of the snake robot

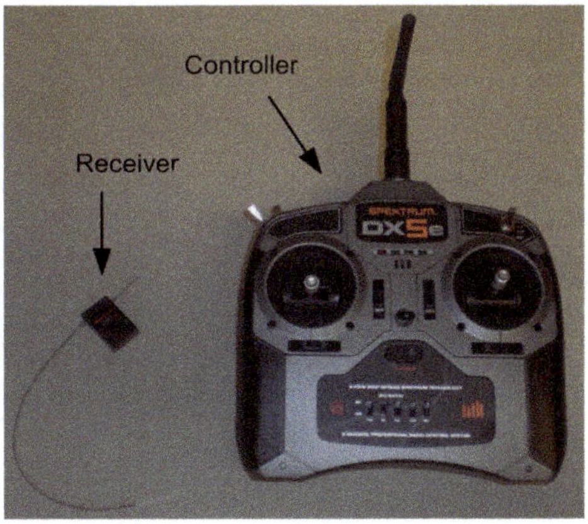

Fig. 11.12 The *remote controller* and the *receiver* used for manually controlling the snake robot

accordance with any desired control strategy and sent to the brain card via a wireless connection based on Bluetooth. The refresh rate for the two reference angles of each joint module is about 20 Hz.

For simple demonstration purposes (not for experimental purposes), the snake robot can also be manually controlled with a commercially available radio transmitter (the DX5e developed by *Spektrum*), which is shown in Fig. 11.12. The receiver of the radio controller is connected to the brain card, which calculates joint reference angles based on the input from the radio controller. The mapping from radio control input to the resulting joint reference angles will not be detailed here as it is not relevant to the experiments reported in this book.

As shown to the left in Fig. 11.10, the head of the snake robot is equipped with a small wireless camera and two IR distance sensors (*Sharp* GP2D120). These sensors have not been employed in the experiments reported in this book, but are intended for future experiments where the goal is to prevent the head from colliding with obstacles in its path.

11.6 The Performance of the Snake Robot

In this section, we present experimental results that validate the function of the contact force measurement system. We also show the ability of the snake robot to display different motion patterns.

11.6.1 Experimental Validation of the Contact Force Measurement System

The force measurement system of Kulko was experimentally investigated as shown in Fig. 11.13. In particular, we placed the robot against a wall (to prevent it from

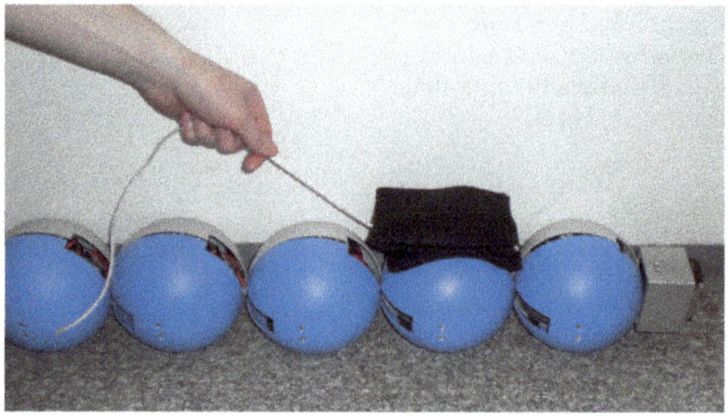

Fig. 11.13 The experimental investigation of the contact force measurement system of Kulko

rolling over) so that the left side of the robot was facing upwards. Loads with different weights were then dragged backwards along the part of the snake robot facing upwards (i.e. the left side), while each joint module estimated the applied contact forces according to (11.4a) and (11.4b) at a sampling frequency of 10 Hz. The joint modules reported the measured forces over the CAN bus to the brain module, which redirected these measurements to an external computer over the wireless Bluetooth connection.

Three different loads weighing 1350 g, 2750 g, and 4300 g, respectively, were dragged from the head and backwards along the snake robot. The resulting force measurements at joint 4, joint 5, and joint 6 (joint 1 is the foremost module) are shown in Figs. 11.14(a)–(c), respectively. In theory, the amplitude of each force curve should be 13.5 N, 27.5 N, and 43 N, respectively, for the three different loads. Despite some deviations, the measured forces agree well with the weight of the loads. As described in Sect. 11.4.2, an FSR is not suitable for precision measurements, and consequently, some deviations were expected. However, we conjecture that obstacle-aided locomotion with a snake robot primarily requires the ability to detect a contact force and also, to some extent, assess the magnitude of this force. The experimental results indicate that the proposed sensor setup is able to meet these requirements.

Note that an ideal sensor system would produce a linear horizontal curve corresponding to the weight of the load being dragged along the snake body. Since the plots of the measured forces are instead given as peaks, it is clear that the sensor system does not measure forces *between* the joints very well. However, we do not consider this to be a critical issue in order to demonstrate obstacle-aided locomotion, especially not if the obstacles are large compared to the size of each joint module.

11.6.2 Demonstration of Motion Patterns

Some of the motion capabilities of Kulko are demonstrated in Fig. 11.15. In Fig. 11.15(a), the snake robot conducts sidewinding across a flat surface. This is a sideways motion produced by propagating both horizontal and vertical body waves

Fig. 11.14 Forces measured by joints 4–6 when three different loads were dragged along the snake robot

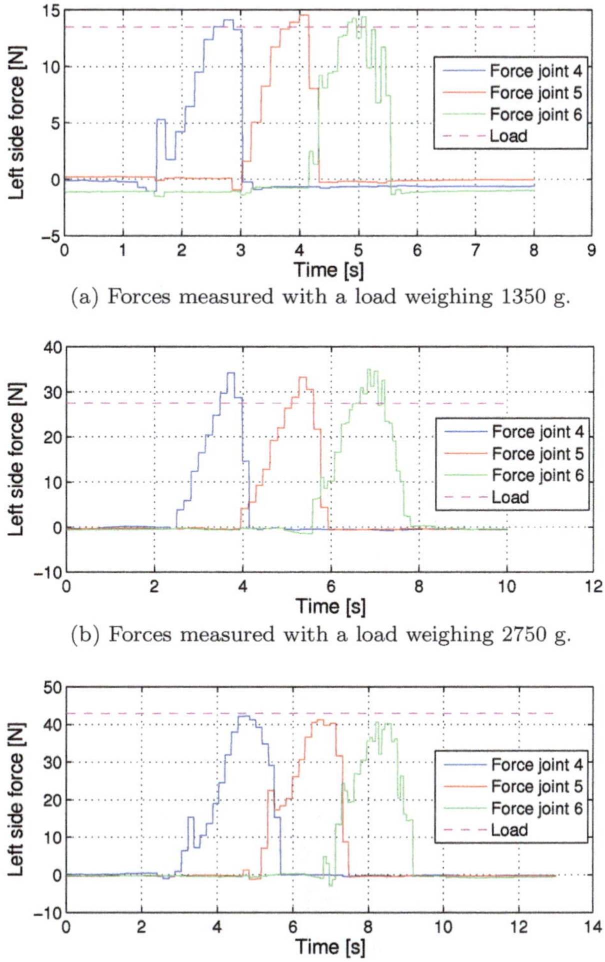

(a) Forces measured with a load weighing 1350 g.

(b) Forces measured with a load weighing 2750 g.

(c) Forces measured with a load weighing 4300 g.

backwards along the snake (see e.g. Transeth et al. 2008a). In Fig. 11.15(b), the snake robot conducts lateral rolling, which is a rolling motion produced by continuously creating a U-shape with the snake body that tips over to one side (see e.g. Mori and Hirose 2002).

In summary, we conjecture that the motion capabilities of the snake robot are satisfactory and adequate in order to demonstrate obstacle-aided locomotion.

11.7 The Experimental Setup of the Snake Robot

The motion control experiments carried out with Kulko are reported in Chaps. 12 and 13. These experiments were carried out on a black horizontal surface measur-

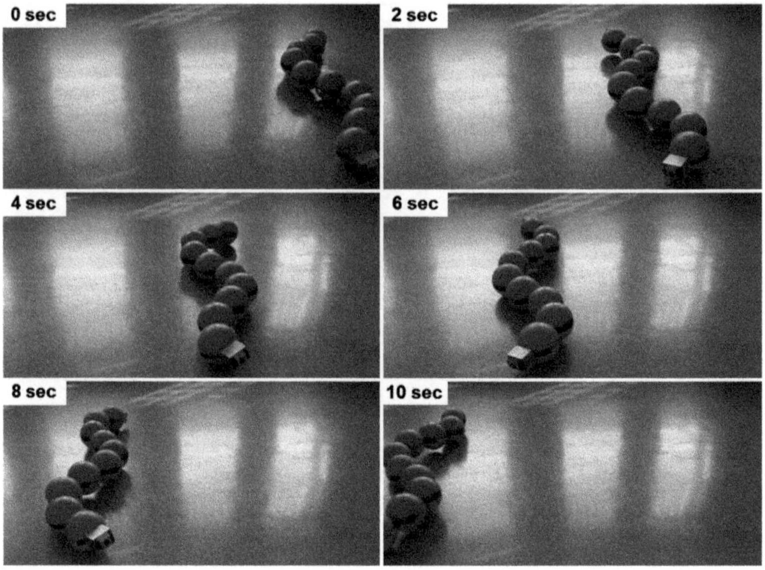

(a) Kulko sidewinding across the floor.

(b) Kulko conducting lateral rolling across the floor.

Fig. 11.15 Demonstration of the motion capabilities of Kulko

ing about 100 cm in width and 200 cm in length. As shown in Fig. 11.16, circular obstacles were placed around the robot. The location of each obstacle could easily be changed by means of a grid of mounting holes in the floor. The friction coefficient between the floor and the robot, denoted by μ, and the friction coefficient between the obstacles and the robot, denoted by μ_o, were found experimentally

Fig. 11.16 The experimental setup. Three cameras mounted in the ceiling measured the position of the snake robot in a course with *obstacles*

to lie between 0.2 and 0.3. Note that the physical environment of the snake robot corresponds to the environment captured by the hybrid model proposed in Chap. 10.

The horizontal position of the snake robot during the experiments was measured by mounting black circular markers to the snake robot, as shown to the right in Fig. 11.16, and then tracking these markers by use of the camera system shown to the left in Fig. 11.16. This camera system is identical to the camera system used in the experimental setup of Wheeko and is described in more detail in Sect. 3.5.

11.8 An Alternative Approach for Measuring External Contact Forces

We end this chapter by proposing an alternative strategy for environment sensing based on force measurements. With respect to Kulko, this alternative approach leads to a simpler instrumentation system. The idea behind this approach was, however, conceived only a short time before the writing of this book. We are therefore unable to present an experimental validation of the approach in this book.

The idea behind this alternative approach is simply to calculate the external forces on each link of the robot based on measurements of the joint constraint forces that occur at the connection between the links. As illustrated in Fig. 11.17, a major advantage of this approach is that external forces on each link can be determined based on force measurements conducted *inside* the robot. In particular, the top of Fig. 11.17 illustrates force sensing based on direct measurement of the external forces acting on each link, which basically requires that the sensing area covers the link completely. Kulko is based on this force sensing principle. Force sensing based on measured joint constraint forces, on the other hand, only requires instrumentation at the locations of the joints, and all the instrumentation can be well protected

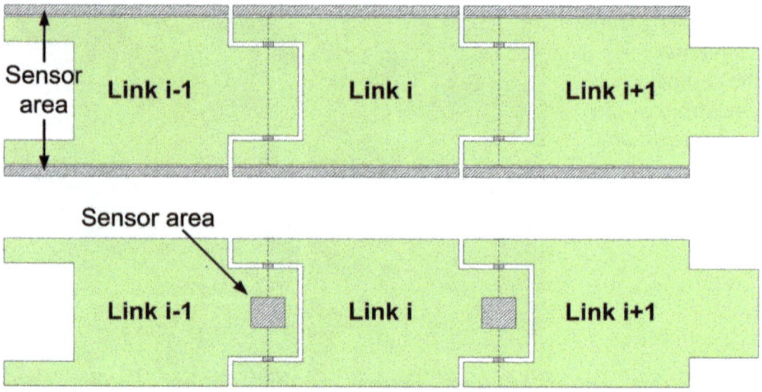

Fig. 11.17 *Top*: *Sensor area* required for direct measurement of external forces. *Bottom*: *Sensor area* required for calculating external forces based on internally measured joint constraint forces

inside the joint modules. This alternative approach, which is illustrated in the bottom of Fig. 11.17, simplifies the development of the physical coverage of a snake robot since the coverage can be developed more or less independently of the contact force sensor system. Techniques for measuring the joint constraint forces, of which there are many, remain a topic of future work.

To verify that the external forces on the snake robot indeed can be determined from the joint constraint forces, we investigate the force balance for the links. The force balance of a link which is not in contact with an obstacle is given by (2.26a) and (2.26b). During contact with an obstacle, the link is additionally influenced by an obstacle constraint force and an obstacle friction force, as described in Sect. 10.4. The force balance of link $i \in \{1, \dots, N\}$ can therefore be written as

$$m\ddot{\mathbf{p}}_i = \mathbf{f}_{R,i} + \mathbf{f}_{c,i} + \mathbf{f}_{\mu,i} + \mathbf{h}_i - \mathbf{h}_{i-1}, \tag{11.5}$$

where m and $\ddot{\mathbf{p}}_i = [\ddot{x}_i, \ddot{y}_i]^T \in \mathbb{R}^2$ are the mass and translational acceleration, respectively, of link i, $\mathbf{f}_{R,i} \in \mathbb{R}^2$ is the ground friction force, $\mathbf{f}_{c,i} \in \mathbb{R}^2$ is the constraint force from any external obstacle, $\mathbf{f}_{\mu,i} \in \mathbb{R}^2$ is the obstacle friction force, $\mathbf{h}_i \in \mathbb{R}^2$ is the joint constraint force from link $i+1$, and $-\mathbf{h}_{i-1} \in \mathbb{R}^2$ is the joint constraint force from link $i-1$. With the alternative force sensing approach, the joint constraint forces $\mathbf{h}_i$ and $\mathbf{h}_{i-1}$ are measured. The sum of the external forces on link i, which we denote by $\mathbf{f}_{\text{ext},i} \in \mathbb{R}^2$, can thereby be calculated as

$$\mathbf{f}_{\text{ext},i} = \mathbf{f}_{R,i} + \mathbf{f}_{c,i} + \mathbf{f}_{\mu,i} = m\ddot{\mathbf{p}}_i - \mathbf{h}_i + \mathbf{h}_{i-1}, \tag{11.6}$$

which is given solely from the measured joint constraint forces when the velocity of the link is zero or constant so that $\ddot{\mathbf{p}}_i = \mathbf{0}$. Moreover, since snake locomotion is usually a smooth gliding form of locomotion with slowly varying link velocities, we conjecture that (11.6), with $\ddot{\mathbf{p}}_i$ set to zero, also in general will provide a good approximation of $\mathbf{f}_{\text{ext},i}$. Alternatively, the estimate of $\mathbf{f}_{\text{ext},i}$ can be improved by also measuring $\ddot{\mathbf{p}}_i$, which is easily achieved by installing a small acceleration sensor inside each link.

11.9 Chapter Summary

This chapter is summarised as follows:

- We have presented the design of the snake robot *Kulko*, which was developed for the purpose of experiments related to obstacle-aided locomotion in uneven and cluttered environments.
- The robot consists of 10 identical joint modules covered by contact force sensors (to allow the robot to sense its environment) and spherical shells that give the robot a smooth outer surface (to allow slithering motion in cluttered environments).
- The internal structure of Kulko is identical to the internal structure of the snake robot *Wheeko*, which was described in Chap. 3.
- We have presented experimental results that validate the function of the contact force measurement system and also demonstrated some of the motion capabilities of the robot.
- We have proposed an alternative strategy for contact force sensing which is based on measuring the joint constraint forces at the connection between the links. The advantage of this approach is that the sensor system can be well protected inside the snake robot.

Chapter 12
Hybrid Control of Obstacle-Aided Locomotion

In direct accordance with Hypothesis 9.1, we propose in this chapter a control strategy that enables a snake robot to propel its body forward by active use of the interaction with obstacles in its environment. This form of propulsion is called obstacle-aided locomotion and was introduced in Chap. 9. Obstacle-aided locomotion represents an interesting control problem for which previous research is very limited. The literature review presented in Sect. 1.3 clearly shows that a large majority of control strategies proposed for snake robots so far assume that the environment of the robot is flat. In fact, the works in Bayraktaroglu and Blazevic (2005), Bayraktaroglu (2008), Hirose (1993) present, to our best knowledge, the only known control strategies related to obstacle-aided locomotion. We believe control strategies for snake robots that consider environment interaction are important since the main advantage of these mechanisms are their potential ability to move in uneven and cluttered environments.

The difference in complexity between flat surface locomotion, which was considered in Part I of this book, and obstacle-aided locomotion is significant. Unlike flat surface locomotion, where we know that periodic body waves will propel a snake robot forward under anisotropic ground friction conditions, there exists no clear intuition as to how we can control a snake robot so that it is propelled forward by obstacle contact forces. One obvious and major challenge is that we do not know in advance how, when, and where the snake robot will make contact with its environment. A second major challenge is to develop a general strategy for adjusting the shape of the robot so that forward propulsion is achieved in any given contact situation.

Our proposed solution to this problem is simple and, in many ways, obvious. In particular, we are seeking a form of locomotion where obstacle contact forces are what propel the snake robot forward. Moreover, since each obstacle contact force consists of an obstacle constraint force acting normal to the link and an obstacle friction force opposing the link motion in the tangential link direction (see Sect. 10.4), forward propulsion must necessarily be achieved mainly due to the normal direction constraint forces on the links. From this observation we state the following hypothesis:

P. Liljebäck et al., *Snake Robots*, Advances in Industrial Control,
DOI 10.1007/978-1-4471-2996-7_12,

Hypothesis 12.1 *Obstacle-aided snake robot locomotion can be achieved by producing body shape changes where the links in contact with obstacles are rotated so that the components of the contact forces in the desired direction of motion are increased.*

In order to investigate this fundamental control principle, we will in this chapter introduce the concepts of *jam detection* and *jam resolution*. A snake robot which moves in a cluttered environment without taking the environment interaction into account is likely to become *jammed* between the obstacles in its path. We will show that a control strategy based on the control principle in Hypothesis 12.1 is efficient for resolving such jams and maintaining the propulsion of the snake robot. Since a jam of the robot can be regarded as a discrete event, we will employ a hybrid formulation in the control strategy similar to the formulation of the hybrid model in Chap. 10, i.e. we will propose a *hybrid controller*. The performance of the controller will be illustrated with simulation results and with experimental results based on the snake robot Kulko. In Chap. 13, the control principle in Hypothesis 12.1 will be employed to propose a more general control law that combines environment adaptation with path following capabilities in a cluttered environment.

The chapter is organised as follows. The relation between this chapter and previous literature is discussed in Sect. 12.1. A general description of hybrid controllers is presented in Sect. 12.2. The objective and the basic assumptions underlying the hybrid controller are presented in Sects. 12.3 and 12.4, respectively. Section 12.5 presents the hybrid controller, while the closed-loop system (i.e. the hybrid model with the hybrid controller) is summarised in Sect. 12.6. Sections 12.7 and 12.8 present, respectively, simulation results and experimental results that illustrate the performance of the controller. Finally, the chapter is summarised in Sect. 12.9.

12.1 The Relation Between This Chapter and Previous Literature

Only a few works in previous literature consider control strategies for snake robots where the surface is no longer assumed to be flat (i.e. in environments with obstacles). To our best knowledge, the works in Bayraktaroglu and Blazevic (2005), Bayraktaroglu (2008), Hirose (1993) are the only works in previous literature which present control strategies for snake robots that employ explicit contact sensing in the feedback loop. In Hirose (1993), a strategy for lateral inhibition is proposed where the shape of a snake robot is modified based on contact force sensing along the snake body in order to avoid obstacles. The work in Bayraktaroglu and Blazevic (2005) proposes an inverse dynamics approach by formulating and numerically solving an optimisation problem in order to, for a given set of obstacle contacts, calculate the contact forces required to propel the robot in a desired direction. A strategy for calculating the actual torque inputs to the joints from the desired contacts was, however, not presented. A kinematic approach is proposed in Bayraktaroglu (2008), where a curve fitting procedure is used to determine an appropriate shape of the robot with respect to the detected obstacles. Subsequently, this shape is propagated backwards along the snake body under the assumption that this will push the robot forward.

Sensing the environment of a snake robot must not necessarily involve contact force sensing since the environment can be indirectly sensed through the joint angle measurements and/or the actuator torques. This approach is considered in Date and Takita (2007), where the joint torques of a snake robot are specified solely in terms of the measured joint angles to achieve motion through a winding corridor, in Andruska and Peterson (2008), which presents a control strategy that uses motor current measurements to adjust the shape of a snake robot moving through an elastically deformable channel, and in Kuwada et al. (2008), where the deviations of the joint angles from their setpoints are used to adapt the body shape of a snake robot moving inside pipe structures.

The main novel result of this chapter in relation to previous literature concerns a *hybrid controller* for obstacle-aided locomotion aimed at resolving situations where the snake robot is jammed between obstacles. Included in this result is the control principle proposed in Hypothesis 12.1. We strongly believe that this control principle is applicable also in the general case of three-dimensional snake robot locomotion. The concept of detecting and resolving snake robot jams has, to our best knowledge, not been treated in previous literature, but is a genuine challenge during snake robot locomotion in cluttered environments. To our knowledge, this is the first published control strategy for a snake robot involving feedback and explicit use of measured contact forces to achieve propulsion. Note that the work Hirose (1993) also considers snake locomotion based on measured contact forces. However, the contact forces in Hirose (1993) are employed to *avoid* obstacles, whereas the contact forces in this chapter are employed to push the snake robot forward. The experimental investigation of the controller by use of the snake robot Kulko is also considered a novel result of this chapter. To our best knowledge, this is the first reported experiment where a snake robot is propelled forward based on measurements of the amplitude of contact forces along the body of the robot. The works Bayraktaroglu (2008), Hirose (1993) also report experiments where a snake robot is propelled by obstacle contact forces. However, the control strategies in these works do not consider the amplitude of the contact forces since discrete contact switches on the robots are used to detect the obstacles.

12.2 Preliminary Note on Hybrid Controllers

Consider any continuous or hybrid plant (i.e. a model of a dynamical system) with state vector $\mathbf{x} \in \mathbb{R}^n$ and control input $\mathbf{u} \in \mathbb{R}^m$. If the controller that generates the control input $\mathbf{u}$ for the system consists of an algorithm with discrete-valued states, then we denote this a *hybrid controller* (Goebel et al. 2009). A hybrid controller is a hybrid system with state $\boldsymbol{\eta} \in \mathbb{R}^p$ (which can contain e.g. logic states, timers, and counters) that evolves as a function of both the controller state η and the plant state $\mathbf{x}$. The control input is generally calculated according to a function $\mathbf{u} = \kappa(\mathbf{x}, \eta)$. We can describe a hybrid controller by the hybrid modelling framework introduced in Sect. 10.2.1. Sometimes hybrid controllers are used to control plants that are continuous-time systems (see e.g. Goebel et al. 2009 for various examples). In this chapter, however, we will propose a hybrid controller for a snake robot described by the hybrid model proposed in Chap. 10, i.e. we will consider a hybrid controller for a hybrid plant.

12.3 Control Objective

A major challenge during obstacle-aided locomotion is to prevent the snake robot from being *jammed* between the obstacles. In a jammed situation, the *propulsive* components (i.e. the force components in the desired direction of motion) of the contact forces from the obstacles are too small to overcome the friction forces on the robot, and hence the forward motion stops. In this jammed situation, the obstacle contact forces will also prevent a number of the snake robot joints from moving to their reference angle. The goal of this chapter is to employ the control principle in Hypothesis 12.1 to develop a strategy for detecting and resolving situations where the motion of the robot is jammed. To this end, we choose the control goal to be locomotion along the global x axis with a positive and non-zero forward velocity. The control problem is thereby to design a feedback control law for the joint torques $\mathbf{u} \in \mathbb{R}^{N-1}$ such that the following control objective is reached:

$$\dot{p}_x(t) > 0. \tag{12.1}$$

Note that the controller only targets the propulsion of the snake robot without considering the direction of the motion. Directional control in an obstacle environment is considered in Chap. 13. Another important limitation of the control strategy is the underlying requirement of sufficiently many obstacles in reach of the snake robot. Without obstacles, there are no push points that the robot can use for propulsion. At the same time, the control strategy will not work if there are too many obstacles so that the path of the robot is blocked. These are not critical issues, however, since the main purpose of the controller is to demonstrate how a snake robot can utilise contact forces from objects in its environment to achieve propulsion.

12.4 Notation and Basic Assumptions

With reference to the obstacle constraint force $\mathbf{f}_{c,i}$ defined in Sect. 10.4, the controller will make use of the following quantities:

Definition 12.1 (Measured contact force) The *measured* contact force on link $i \in \{1, \ldots, N\}$ is denoted by $\rho_i \in \mathbb{R}$ and is defined as the component of the constraint

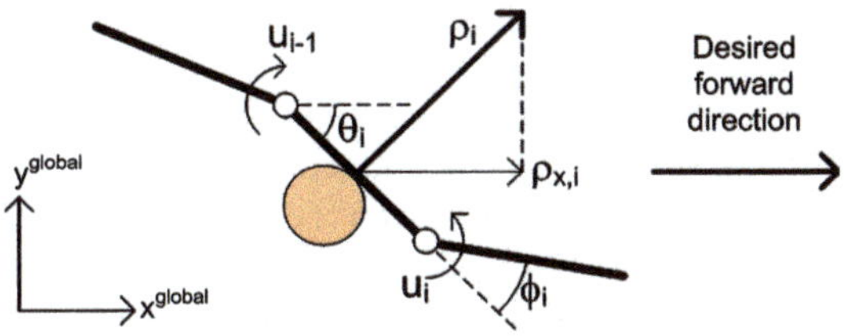

Fig. 12.1 The propulsive component $\rho_{x,i}$ of the contact force on link i

force vector $\mathbf{f}_{c,i}$ along the y axis of the local frame of link i (see illustration of the local link frame in Fig. 2.1), i.e. as

$$\rho_i = [-\sin\theta_i, \cos\theta_i]\mathbf{f}_{c,i}. \tag{12.2}$$

Moreover, the force measurements of all links are assembled in the vector $\boldsymbol{\rho} = [\rho_1, \dots, \rho_N]^T \in \mathbb{R}^N$.

Remark 12.1 Since, as described in Sect. 10.4, the constraint force vector $\mathbf{f}_{c,i}$ always acts along the local y axis of link i, we have that $|\rho_i| = |\mathbf{f}_{c,i}|$.

Definition 12.2 (Propulsive component) The *propulsive component* of the contact force on link $i \in \{1, \dots, N\}$ is defined as the component of the constraint force vector $\mathbf{f}_{c,i}$ along the desired *forward* direction of motion.

Since the desired forward direction of the controller considered in this chapter is along the global x axis, the propulsive component of a contact force is given as the component of the constraint force vector along the global x axis. As illustrated in Fig. 12.1, this component is denoted by $\rho_{x,i} \in \mathbb{R}$ and is easily calculated as

$$\rho_{x,i} = -\rho_i \sin\theta_i. \tag{12.3}$$

Remark 12.2 Chapter 13 considers directional control of snake robots in an obstacle environment where the desired forward direction of motion at any time is along the current heading of the robot in order to maintain the forward velocity. The propulsive component of a contact force in Chap. 13 is therefore given as the component along the current heading instead of along the global x axis.

We will base the controller on the following assumptions:

Assumption 12.1 *The initial heading of the snake robot is along the global x axis.*

Assumption 12.2 *The control system has access to measurements of the joint angles $\boldsymbol{\phi}$, the joint angle velocities $\dot{\boldsymbol{\phi}}$, the contact forces $\boldsymbol{\rho}$, and at least one of the absolute link angles θ_i for some $i \in \{1, \dots, N\}$.*

Note that the remaining link angles can be calculated from $\boldsymbol{\phi}$ and θ_i.

12.5 The Hybrid Controller for Obstacle-Aided Locomotion

The control strategy proposed in the following is a *hybrid controller* consisting of a *leader-follower* scheme and a *jam resolution* scheme, and also a supervisory mechanism for switching between these two schemes, denoted the *jam detection* scheme. The *leader-follower* scheme is carried out as long as the robot is able to move without being jammed between the obstacles. If the *jam detection* scheme detects a jam, then the *jam resolution* scheme is carried out in order to effectively 'unlock' the jammed joints.

The *leader-follower* scheme, the *jam detection* scheme, the *jam resolution* scheme, and the joint angle controller are presented in Sects. 12.5.1, 12.5.2, 12.5.3, and 12.5.4, respectively, without considering the hybrid nature of the controller. In Sect. 12.5.5, the complete hybrid controller is summarised and formulated in terms of the framework described in Sect. 12.2.

12.5.1 The Leader-Follower Scheme

In Part I of this book, we employed cyclic body wave motion as a means for propelling a snake robot forward. As we are now considering environments with obstacles, we will continue to base the motion of the snake robot on such cyclic body wave motion. The reason for this choice is based on Remark 10.2 in Sect. 10.4, where we noted that the phenomenon that propels a snake robot forward due to the anisotropic ground friction properties of the links is similar to the phenomenon that propels a snake robot forward due to the interaction of the links with external obstacles. Since cyclic body wave motion produces forward propulsion under anisotropic ground friction conditions, we can thereby expect this to also be the case in environments with obstacles.

In the leader-follower scheme, we will therefore produce oscillatory body wave motion similar to the lateral undulation gait defined in (4.32). However, we do not specify the body shape motion in terms of (4.32) since the shape of the robot then would be completely predefined, which makes no sense when the environment is not known in advance. Instead, we define the oscillatory body shape motion based on the observation that each part of a biological snake conducting lateral undulation follows the path traced out by the head (see e.g. Gray 1946), which suggests that we should choose the head joint angle (the angle of the foremost joint), ϕ_{N-1}, as the reference angle for all subsequent joints. This approach is called *leader-following* since all joints follow the motion of the head, i.e. the leader. In contrast to the predetermined lateral undulation motion in (4.32), environment adaptation is, at least to some extent, inherently present in the leader-following motion since all joints follow the *actual* angle of the head joint.

In order to generate a leader-follower based control reference to the joints, the head joint angle ϕ_{N-1} is propagated backwards along the snake body at a constant

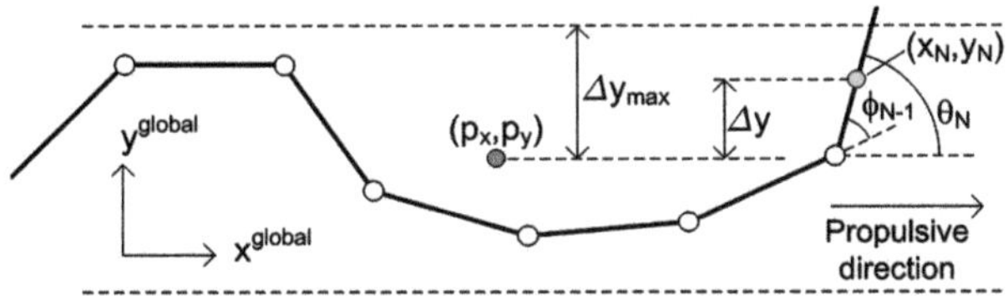

Fig. 12.2 The control strategy for the head of the snake robot

predefined propagation velocity v_{ref} and used as the reference angle for all subsequent joints according to

$$\phi_{i,\text{ref}}(t) = \phi_{N-1}\big(t - (N - i - 1)\Delta t\big), \tag{12.4}$$

where $i \in \{1, \ldots, N-2\}$, and where the time offset Δt between two consecutive joints with intermediate distance $2l$ is found as $\Delta t = 2l/v_{ref}$.

In order to achieve the sinusoidal motion characteristic of lateral undulation, we alternate between moving the head in the leftward and rightward direction with respect to the global positive x axis. This may be achieved by choosing the reference angle for the head link, $\theta_{N,\text{ref}}$, equal to a suitable positive constant θ_{left} when the head should move leftward and a negative constant θ_{right} when the head should move rightward. The criterion for switching between these two reference directions is defined to be the instant when the amplitude of the head motion perpendicular to the desired forward direction becomes greater than some predefined amplitude. In mathematical terms, we switch the head direction of motion when the distance Δy between the position of the head along the global y axis, y_N, and the y axis coordinate of the CM of the snake robot, p_y, becomes greater than some predefined amplitude, $\Delta y_{\max}$. This criterion is illustrated in Fig. 12.2.The distance Δy is easily calculated as a function of measured state values only. In particular, by inspecting the last row of the matrix equation in (2.11b), it can easily be verified that Δy is given by

$$\Delta y = y_N - p_y = \sum_{j=1}^{N-1} \frac{l}{N} j(\sin\theta_j + \sin\theta_{j+1}). \tag{12.5}$$

The reference angle of the head link is in other words set according to the rule

$$\begin{array}{lll} \text{Leftward motion:} & \theta_{N,\text{ref}} = \theta_{\text{left}} & \text{until } \Delta y > \Delta y_{\max}, \\ \text{Rightward motion:} & \theta_{N,\text{ref}} = \theta_{\text{right}} & \text{until } \Delta y < -\Delta y_{\max}, \end{array} \tag{12.6}$$

which means that $\theta_{N,\text{ref}} \in \{\theta_{\text{left}}, \theta_{\text{right}}\}$. Since the head joint (i.e. joint $N-1$) is at the front of the snake robot, the rotation of the head joint mainly affects the angle of the head link (i.e. link N) and not the angle of the subsequent links. From the relation $\phi_{N-1} = \theta_{N-1} - \theta_N$ (see (2.1)) we can therefore track the head link reference angle in (12.6) by controlling the head joint angle ϕ_{N-1} according to the reference

$$\phi_{N-1,\text{ref}} = \theta_{N-1} - \theta_{N,\text{ref}}. \tag{12.7}$$

Note that $\phi_{N-1,\text{ref}}$ will experience a jump each time $\theta_{N,\text{ref}}$ switches. However, the actuator torque applied at joint $N-1$ will still be bounded since the derivative of $\phi_{N-1,\text{ref}}$ with respect to time is not included in the joint controller presented below in Sect. 12.5.4.

To summarise, the reference angles for all the joints of the snake robot in this leader-follower scheme are

$$\phi_{N-1,\text{ref}}(t) = \theta_{N-1}(t) - \theta_{N,\text{ref}}, \tag{12.8a}$$

$$\phi_{i,\text{ref}}(t) = \phi_{N-1}\bigl(t - (N-i-1)\Delta t\bigr), \tag{12.8b}$$

where $i \in \{1, \ldots, N-2\}$. The design parameters θ_{left}, θ_{right}, Δy_{max}, and v_{ref} were introduced in order to calculate these reference angles.

Remark 12.3 The implementation of (12.8a) and (12.8b) requires a buffer which keeps track of the angle history of the head joint, $\phi_{N-1}(t)$.

12.5.2 The Jam Detection Scheme

We define a single joint of the snake robot to be *jammed* if the deviation between the joint angle and its reference angle exceeds a certain limit, $\Delta\phi_{\text{max}}$. It is reasonable to assume that a jam of a single joint will resolve by itself. However, two jammed joints could be caused by a situation where the obstacle contact forces cause the jammed joints to act 'against' each other. This situation may not always resolve by itself. The entire snake robot is therefore defined to be *jammed* if two or more joints are jammed. If the robot is jammed over a continuous period longer than $t_{\text{jam,max}}$, the leader-follower scheme is stopped in order to carry out the jam resolution scheme. We let the robot execute jam resolution for a predefined amount of time $t_{\text{resolution,max}}$ since it is difficult to come up with a specific criterion for when a jam has been resolved. Subsequently, the leader-follower scheme continues.

In summary, the design parameters $\Delta\phi_{\text{max}}$, $t_{\text{jam,max}}$, and $t_{\text{resolution,max}}$ determine the switching between the *leader-follower* scheme and the *jam resolution* scheme.

12.5.3 The Jam Resolution Scheme

As explained in the introduction, the main purpose of the controller proposed in this chapter is to investigate the control principle proposed in Hypothesis 12.1. To this end, we define the jam resolution scheme in accordance with this control principle; namely, we rotate the links affected by contact forces so that the propulsive component of each contact force increases. In a jammed situation, the propulsive components of the contact forces from the obstacles are too small to overcome the friction forces opposing the motion of the robot. Rotating the contacted links (and thereby the direction of the contact forces) to increase the total propulsive contact force should therefore resolve the jammed situation.

The propulsive component of the contact force on link $i \in \{1, \dots, N\}$, denoted by $\rho_{x,i}$, was defined in (12.3). The change of the propulsive force due to a change of the link angle is found by differentiating (12.3) with respect to θ_i, which gives

$$\frac{\partial \rho_{x,i}}{\partial \theta_i} = -\rho_i \cos\theta_i. \tag{12.9}$$

We see that changing a link angle near perpendicular to the direction of motion (large θ_i) has a greater effect on the propulsive force than a similar change of a link angle near parallel to the direction of motion (small θ_i). During jam resolution, we therefore prioritise to rotate links with a high propulsive force gradient with respect to the link angle, which suggests that the link angles should be changed according to

$$\Delta\theta_{i,\text{ref}} = k_\theta \frac{\partial \rho_{x,i}}{\partial \theta_i} = -k_\theta \rho_i \cos\theta_i, \tag{12.10}$$

where $k_\theta > 0$ is a controller gain.

Let us now derive how the joint angles ϕ_{i-1} and ϕ_i at each side of link i should be changed to comply with (12.10). We choose that the contact force on link i only should affect the angle of link i, so that $\Delta\theta_{i-1,\text{ref}} = \Delta\theta_{i+1,\text{ref}} = 0$. Since we have from (2.1) that $\phi_i = \theta_i - \theta_{i+1}$, we can immediately write the desired change of the joint angles ϕ_{i-1} and ϕ_i due to the contact force on link i as

$$\Delta\phi_{i-1,\text{ref}} = \Delta\theta_{i-1,\text{ref}} - \Delta\theta_{i,\text{ref}} = k_\theta \rho_i \cos\theta_i, \tag{12.11}$$

$$\Delta\phi_{i,\text{ref}} = \Delta\theta_{i,\text{ref}} - \Delta\theta_{i+1,\text{ref}} = -k_\theta \rho_i \cos\theta_i. \tag{12.12}$$

By combining the desired change of joint angle ϕ_i due to the measured contact forces on the link at each side of the joint, i.e. the contribution from both ρ_i and ρ_{i+1}, we get that the angle of joint $i \in \{1, \dots, N-1\}$ should be changed during jam resolution as

$$\Delta\phi_{i,\text{ref}} = k_\theta(-\rho_i \cos\theta_i + \rho_{i+1} \cos\theta_{i+1}). \tag{12.13}$$

We now explain two important controller design choices. First of all, during jam resolution, we leave the head joint angle ϕ_{N-1} unchanged to maintain a smooth head angle. We thereby avoid that any jam resolution motion of the head link propagates backwards to all other links once the leader-follower scheme resumes. This would create undesirable body shapes. Second, we use the contact forces that were measured at the instant the jam resolution scheme was initiated as feedback so that the force measurements used in the feedback loop are *constant* during jam resolution. This ensures a steady rotation of the contacted links in accordance with the contact forces that produced the jam. If the force measurements had been updated *during* jam resolution, then jam resolution would very quickly be aborted for most of the contacted links because the link rotation carried out during jam resolution generally causes the links to detach from the obstacles. We denote the measured contact forces on all links at the instant the jam resolution scheme was initiated by $\boldsymbol{\rho}_{\text{jam}} = [\rho_{\text{jam},1}, \ldots, \rho_{\text{jam},N}]^T \in \mathbb{R}^N$.

From the above discussion, the reference angles for all the joints of the snake robot in the jam resolution scheme are

$$\phi_{N-1,\text{ref}} = \phi_{N-1}, \tag{12.14a}$$

$$\phi_{i,\text{ref}} = \phi_i + k_\theta(-\rho_{\text{jam},i} \cos\theta_i + \rho_{\text{jam},i+1} \cos\theta_{i+1}), \tag{12.14b}$$

where $i \in \{1, \ldots, N-2\}$, and k_θ is a design parameter.

12.5.4 The Joint Angle Controller

The *leader-follower* scheme and the *jam resolution* scheme provide the joint reference angles $\boldsymbol{\phi}_{\text{ref}} = [\phi_{1,\text{ref}}, \ldots, \phi_{N-1,\text{ref}}]^T \in \mathbb{R}^{N-1}$. To ensure that the reference angles comply with the maximum allowable deflection of the joints, which we specify as $[-\phi_{\max}, \phi_{\max}]$ for some $\phi_{\max} > 0$, we saturate the reference angle of joint $i \in \{1, \ldots, N-1\}$ according to

$$\widehat{\phi}_{i,\text{ref}} = \max\bigl(\min(\phi_{i,\text{ref}}, \phi_{\max}), -\phi_{\max}\bigr), \tag{12.15}$$

where $\max(\cdot)$ and $\min(\cdot)$ are operators that return the maximum and minimum value of their arguments, respectively. To make the joint angles $\boldsymbol{\phi} = [\phi_1, \ldots, \phi_{N-1}]^T \in \mathbb{R}^{N-1}$ track the saturated reference angles in (12.15), we employ the simple joint controller from Sect. 4.8.1 and set the actuator torque of joint $i \in \{1, \ldots, N-1\}$ according to

$$u_i = k_p(\widehat{\phi}_{i,\text{ref}} - \phi_i) - k_d\dot{\phi}_i, \tag{12.16}$$

where $k_p > 0$ and $k_d > 0$ are controller gains. A velocity reference is not included in (12.16) since the transitions between the schemes of the control strategy produce steps in the reference angles, which would lead to large and undesirable velocity references.

Remark 12.4 Compliance is an important issue during control based on force feedback. However, there is no need to explicitly consider compliance for the proposed control strategy since we do not attempt to explicitly control the contact forces on the snake robot. Note that the proportional action of the joint torque controller in (12.16) introduces compliance in the system since the dynamic properties of a proportional controller are similar to those of a mechanical spring.

12.5.5 The Complete Hybrid Controller

Based on the controller schemes presented in the preceding subsections, we now provide a formal and precise specification of the complete *hybrid controller* of the

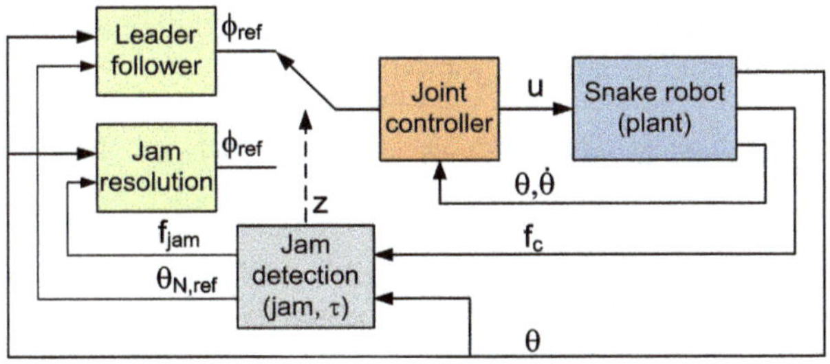

Fig. 12.3 The block diagram of the closed-loop system (hybrid plant and hybrid controller)

snake robot in terms of the modelling framework briefly described in Sect. 12.2. We begin by defining the state vector of the hybrid controller as

$$\eta = \begin{bmatrix} \theta_{N,\mathrm{ref}} \\ \boldsymbol{\rho}_{\mathrm{jam}} \\ \mathrm{jam} \\ \tau \\ z \end{bmatrix}, \tag{12.17}$$

where $\theta_{N,\mathrm{ref}} \in \{\theta_{\mathrm{left}}, \theta_{\mathrm{right}}\}$ is the current reference angle of the head link, $\boldsymbol{\rho}_{\mathrm{jam}} \in \mathbb{R}^N$ denotes the measured constraint forces on all links at the instant the jam resolution scheme is initiated, jam $\in \{0, 1\}$ is a boolean variable indicating if the robot is currently jammed (jam $= 1$ indicates a jam), $\tau \in \mathbb{R}_{\geq 0}$ is a timer variable, and finally $z \in \{0, 1\}$ is a boolean variable that decides if the robot should currently execute the *leader-follower* scheme ($z = 0$) or the *jam resolution* scheme ($z = 1$). The block diagram of the closed-loop system is illustrated in Fig. 12.3, where the state vector $\boldsymbol{\eta}$ is maintained inside the *jam detection* block. The state of the hybrid plant in (10.61) and the state of the hybrid controller is hereafter written as $(\mathbf{x}, \boldsymbol{\eta})$.

In the following subsections, we define the *jump set* $\mathbf{D}_c$, *jump map* $\mathbf{G}_c$, *flow set* $\mathbf{C}_c$, and *flow map* $\mathbf{F}_c$ of the hybrid controller, where subscript c is used to distinguish these sets from the corresponding sets of the hybrid plant in Sect. 10.8.

The Jump Set

A *jump* in the state vector $\boldsymbol{\eta}$ of the controller occurs either when the direction of the head link should change, when the jam state changes, or when the jam resolution scheme is initiated or stopped (controlled by switching the value of z).

In direct accordance with Sect. 12.5.1, the direction of the head link should change when the state of the plant and the controller, $(\mathbf{x}, \boldsymbol{\eta})$, belongs to the jump set

$$\begin{aligned} \mathbf{D}_{\mathrm{dir}} = {} & \left\{(\mathbf{x}, \boldsymbol{\eta}) | z = 0, \theta_{N,\mathrm{ref}} = \theta_{\mathrm{left}}, \Delta y > \Delta y_{\max}\right\} \\ & \cup \left\{(\mathbf{x}, \boldsymbol{\eta}) | z = 0, \theta_{N,\mathrm{ref}} = \theta_{\mathrm{right}}, \Delta y < -\Delta y_{\max}\right\}. \end{aligned} \tag{12.18}$$

In order to determine the jam state, we define the following index set corresponding to pairs of jammed joints:

$$\mathbf{I}_{\text{jam}} = \left\{(i, j)|i \neq j, |\phi_i - \phi_{i,\text{ref}}| > \Delta\phi_{\max}, |\phi_j - \phi_{j,\text{ref}}| > \Delta\phi_{\max}\right\}. \tag{12.19}$$

We consider pairs of jammed joints since the robot is defined to be jammed when two or more joints are jammed, i.e. when $\mathbf{I}_{\text{jam}} \neq \emptyset$. In accordance with Sect. 12.5.2, the jump set of the jam state is given by

$$\begin{aligned}\mathbf{D}_{\text{jam}} = &\left\{(\mathbf{x}, \boldsymbol{\eta})|z = 0, \text{jam} = 0, \mathbf{I}_{\text{jam}} \neq \emptyset\right\} \\ &\cup \left\{(\mathbf{x}, \boldsymbol{\eta})|z = 0, \text{jam} = 1, \mathbf{I}_{\text{jam}} = \emptyset\right\}.\end{aligned} \tag{12.20}$$

In accordance with Sect. 12.5.2, the switching variable z should change when $(\mathbf{x}, \boldsymbol{\eta})$ belongs to the jump set

$$\begin{aligned}\mathbf{D}_{\text{res}} = &\left\{(\mathbf{x}, \boldsymbol{\eta})|z = 0, \text{jam} = 1, \tau > t_{\text{jam,max}}\right\} \\ &\cup \left\{(\mathbf{x}, \boldsymbol{\eta})|z = 1, \tau > t_{\text{resolution,max}}\right\}.\end{aligned} \tag{12.21}$$

The complete *jump set* of the hybrid controller may now be compactly expressed as

$$\mathbf{D}_c = \mathbf{D}_{\text{dir}} \cup \mathbf{D}_{\text{jam}} \cup \mathbf{D}_{\text{res}}. \tag{12.22}$$

The Jump Map

From the description of the controller schemes in Sects. 12.5.1, 12.5.2, and 12.5.3 we can directly state the *jump map* of the hybrid controller as

$$\boldsymbol{\eta}^+ = \mathbf{G}_c(\mathbf{x}, \boldsymbol{\eta}) = \begin{bmatrix} \theta_{N,\text{ref}}^+ \\ \boldsymbol{\rho}_{\text{jam}}^+ \\ \text{jam}^+ \\ \tau^+ \\ z^+ \end{bmatrix}, \tag{12.23}$$

where

$$\theta_{N,\text{ref}}^+ = \begin{cases} \theta_{\text{left}} & \text{when } (\mathbf{x}, \boldsymbol{\eta}) \in \mathbf{D}_{\text{dir}}, \Delta y < -\Delta y_{\max}, \\ \theta_{\text{right}} & \text{when } (\mathbf{x}, \boldsymbol{\eta}) \in \mathbf{D}_{\text{dir}}, \Delta y > \Delta y_{\max}, \\ \theta_{N,\text{ref}}^- & \text{otherwise,} \end{cases} \tag{12.24a}$$

$$\boldsymbol{\rho}_{\text{jam}}^+ = \mathbf{f}_c, \tag{12.24b}$$

$$\text{jam}^+ = \begin{cases} 1 & \text{when } (\mathbf{x}, \boldsymbol{\eta}) \in \mathbf{D}_{\text{jam}}, \text{jam} = 0, \\ 0 & \text{when } ((\mathbf{x}, \boldsymbol{\eta}) \in \mathbf{D}_{\text{jam}}, \text{jam} = 1) \vee (\mathbf{x}, \boldsymbol{\eta}) \in \mathbf{D}_{\text{res}}, \\ \text{jam}^- & \text{otherwise,} \end{cases} \tag{12.24c}$$

$$\tau^+ = \begin{cases} 0 & \text{when } (\mathbf{x}, \eta) \in \mathbf{D}_{\text{jam}} \cup \mathbf{D}_{\text{res}}, \\ \tau^- & \text{otherwise}, \end{cases} \tag{12.24d}$$

$$z^+ = \begin{cases} 1 & \text{when } (\mathbf{x}, \eta) \in \mathbf{D}_{\text{res}}, z = 0, \\ 0 & \text{otherwise}. \end{cases} \tag{12.24e}$$

The Flow Set

We define the *flow set* of the hybrid controller so that the state vector η always flows as long as the jump set is empty. The flow set is therefore simply given as

$$\mathbf{C}_c = \{(\mathbf{x}, \eta) | (\mathbf{x}, \eta) \notin \mathbf{D}_c\}. \tag{12.25}$$

The Flow Map

The only variable in the state vector η that should change between jumps is the timer variable, τ. Since the time derivative of τ is 1, the *flow map* is given by

$$\dot{\eta} = \mathbf{F}_c(\mathbf{x}, \eta) = \begin{bmatrix} \dot{\theta}_{N,\text{ref}} \\ \dot{\boldsymbol{\rho}}_{\text{jam}} \\ \dot{\text{jam}} \\ \dot{\tau} \\ \dot{z} \end{bmatrix} = \begin{bmatrix} 0 \\ \mathbf{0}_{N\times 1} \\ 0 \\ 1 \\ 0 \end{bmatrix}. \tag{12.26}$$

Calculation of the Control Input for the Plant

The joint torques $\mathbf{u}$ of the snake robot are calculated as

$$\mathbf{u} = \begin{cases} \kappa_{\text{nojam}}(\mathbf{x}, \eta) & \text{when } z = 0, \\ \kappa_{\text{jam}}(\mathbf{x}, \eta) & \text{when } z = 1, \end{cases} \tag{12.27}$$

where $\kappa_{\text{nojam}} : \mathbf{C}_c \to \mathbb{R}^{N-1}$ is defined by (12.8a)–(12.8b) and (12.16), and $\kappa_{\text{jam}} : \mathbf{C}_c \to \mathbb{R}^{N-1}$ is defined by (12.14a)–(12.14b) and (12.16).

12.6 Summary of the Closed-Loop System

We are now ready to summarise the complete closed-loop system consisting of the hybrid model of the snake robot from Chap. 10 and the hybrid controller presented in Sect. 12.5.5. The block diagram of the closed-loop system is illustrated in Fig. 12.3.

In accordance with the hybrid modelling framework presented in Sect. 10.2.1, the closed-loop system is a hybrid system with state $(\mathbf{x}, \eta)$ and data $(\mathbf{C}, \mathbf{F}, \mathbf{D}, \mathbf{G})$, where $\mathbf{x}$ is the state of the hybrid plant in (10.61), and η is the state of the hybrid controller defined in (12.17). The closed-loop system flows as long as neither $\mathbf{x}$ nor η should jump. In other words, $(\mathbf{x}, \eta)$ flows as long as both $\mathbf{x} \in \mathbf{C}_p$ and $(\mathbf{x}, \eta) \in \mathbf{C}_c$, and jumps when $\mathbf{x} \in \mathbf{D}_p$ or $(\mathbf{x}, \eta) \in \mathbf{D}_c$. The closed-loop system can therefore be written as

$$\begin{aligned} \begin{bmatrix} \dot{x} \\ \dot{\eta} \end{bmatrix} &= \mathbf{F}(\mathbf{x}, \eta) \quad \text{for all } (\mathbf{x}, \eta) \in \mathbf{C}, \\ \begin{bmatrix} \mathbf{x}^+ \\ \eta^+ \end{bmatrix} &= \mathbf{G}(\mathbf{x}, \eta) \quad \text{for all } (\mathbf{x}, \eta) \in \mathbf{D}, \end{aligned} \tag{12.28}$$

where

$$\mathbf{C} = \{(\mathbf{x}, \eta) | \mathbf{x} \in \mathbf{C}_p \wedge (\mathbf{x}, \eta) \in \mathbf{C}_c\}, \tag{12.29}$$

$$\mathbf{D} = \{(\mathbf{x}, \eta) | \mathbf{x} \in \mathbf{D}_p \vee (\mathbf{x}, \eta) \in \mathbf{D}_c\}, \tag{12.30}$$

$$\mathbf{x}^+ = \begin{cases} \mathbf{G}_p(\mathbf{x}) & \text{when } \mathbf{x} \in \mathbf{D}_p, \\ \mathbf{x}^- & \text{otherwise}, \end{cases} \tag{12.31}$$

$$\eta^+ = \begin{cases} \mathbf{G}_c(\mathbf{x}, \eta) & \text{when } (\mathbf{x}, \eta) \in \mathbf{D}_c, \\ \eta^- & \text{otherwise}. \end{cases} \tag{12.32}$$

Since, by design, the evolution of η always can be uniquely determined, the control input $\mathbf{u}$ to the plant is always well defined. We can therefore conclude that the existence and uniqueness properties stated in Theorem 10.5 also apply to the closed-loop system in (12.28).

12.7 Simulation Study: The Performance of the Hybrid Controller

In this section, we investigate the performance of the hybrid controller described in Sect. 12.5 by simulating the closed-loop system summarised in (12.28).

12.7.1 Simulation Parameters

The simulator was implemented in *Matlab R2008b* on a laptop running *Windows XP*. The continuous dynamics in (12.28) was calculated with the *ode45* solver in Matlab with a relative and absolute error tolerance of 10^{-3}.

In accordance with the notation from Sect. 2.3, the parameters characterising the simulated snake robot were $N = 10$, $l = 0.07$ m, $m = 1$ kg, and $J = 0.0016$ kgm^2.

These parameters characterise the snake robot Kulko described in Chap. 11. The ground friction coefficient was set to $\mu = 0.3$, and the obstacle friction coefficient was set to $\mu_o = 0.2$. Two different obstacle environments were considered. In the first environment, the obstacles were chosen to be three rows (parallel to the x axis) of circular objects with equal radius $R_{O_j} = 10$ cm, $j \in \{1, \dots, k\}$. The centre distance between two obstacles in a row and the distance between two rows were 25 cm. The middle row was displaced with respect to the other two rows by 12.5 cm along the x axis. In the second environment, obstacles of varying radius were placed in a random fashion around the snake robot.

The various parameters of the hybrid controller were $\theta_{\text{left}} = 50°$, $\theta_{\text{right}} = -50°$, $\Delta y_{\max} = 0.14$ m, $v_{\text{ref}} = 0.2$ m/s, $\Delta\phi_{\max} = 20°$, $t_{\text{jam,max}} = 0.5$ s, $t_{\text{resolution,max}} = 0.5$ s, $k_\theta = 0.05$, $\phi_{\max} = 50°$, $k_p = 20$, and $k_d = 5$. The initial link angles and position of the snake robot were $\boldsymbol{\theta} = [7°, -32°, -57°, -46°, -8°, 33°, 53°, 45°, 12°, -23°]^T$ and $\mathbf{p} = \mathbf{0}_{2\times 1}$, respectively. The initial shape was more or less randomly chosen in order to give the robot an initial curl around the obstacles without intersecting them.

In order to ensure a unique solution to the LCP in (10.43), we verified at each timestep of the simulation that the matrix $\mathbf{A}$ of the LCP in (10.43) was a P-matrix by employing the P-matrix test algorithm presented in Tsatsomeros and Li (2000).

12.7.2 Attempting Lateral Undulation in Open-Loop in a Structured Obstacle Environment

We begin by illustrating the need for feedback of obstacle contact forces by controlling the snake robot, not according to the hybrid controller, but according to the open-loop lateral undulation gait defined in (4.32). This simulation is similar to the simulation described in Sect. 10.9. However, in the present simulation, we do *not* place the obstacles in accordance with the expected evolution of the shape and position of the snake robot as was done in Sect. 10.9. Instead, we place the obstacles according to the first obstacle environment described in the previous subsection. We chose the gait pattern parameters in (4.32) as $\alpha = 40°$, $\omega = 40°$/s, and $\delta = 40°$.

The initial ($t = 0$ s) and final ($t = 20$ s) shape and position of the snake robot are shown at the top of Fig. 12.4, where the trace of the head is indicated with a dotted line. We see that the robot was only able to crawl about 0.5 m in 20 s. Since there was no adaptation of the motion to the environment, the obstacles prevented the snake robot from assuming the predetermined body shape. This simulation illustrates that a snake robot is generally unable to locomote in a cluttered environment when the joint motion is preprogrammed.

12.7.3 Hybrid Controller in an Obstacle Environment

The next simulation shows the effectiveness of the hybrid controller proposed in this chapter when the snake robot moves for 30 s in the two obstacle environments

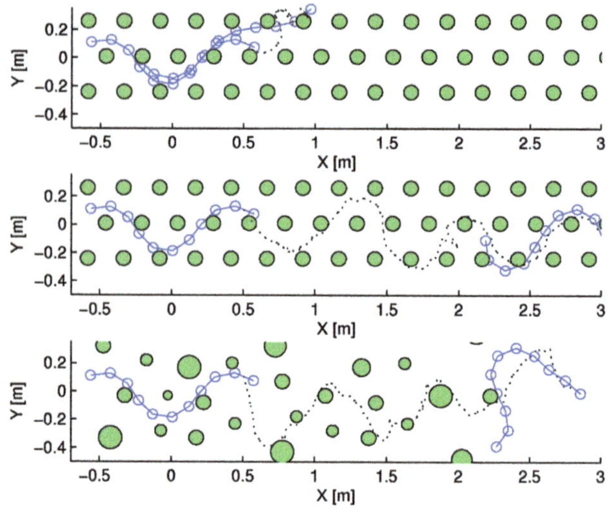

Fig. 12.4 The initial and final shape and position, and the trace of the head of the snake robot. *Top*: Lateral undulation (jam resolution disabled). *Middle*: Jam resolution in the first obstacle environment. *Bottom*: Jam resolution in the second obstacle environment

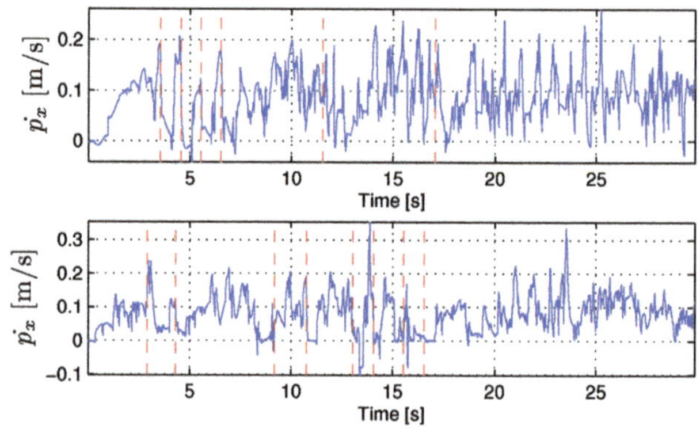

Fig. 12.5 The global x direction velocity of the snake robot in the first (*top*) and second (*bottom*) obstacle environment. *Vertical dashed lines* indicate when the jam resolution scheme is initiated

described above. The initial ($t = 0$ s) and final ($t = 30$ s) shape and position of the snake robot in these two environments are shown in the middle and at the bottom of Fig. 12.4, respectively, while a plot of the x direction velocity of the snake, $\dot{p}_x$, is shown in Fig. 12.5. Vertical dashed lines in Fig. 12.5 indicate time instants where the jam resolution scheme is initiated. After 30 s, the snake robot has managed to crawl about 2.5 m along the global x axis in both environments. Figure 12.5 shows that the velocity in both environments varies around 10 cm/s. The jam resolution scheme was initiated six and eight times, respectively, in the first and second environments, and all the jams were successfully resolved by the proposed algorithm.

In order to give an idea of the forces involved in obstacle-aided locomotion, we provide a plot of the constraint forces on the centre link (link 5), ρ_5, at the top of Fig. 12.6. The actuator torque applied to joint 5, u_5, is plotted at the bottom of this figure. We see that constraint forces above 200 N occur during the motion, and that the applied joint torque is sometimes as high as 10 Nm.

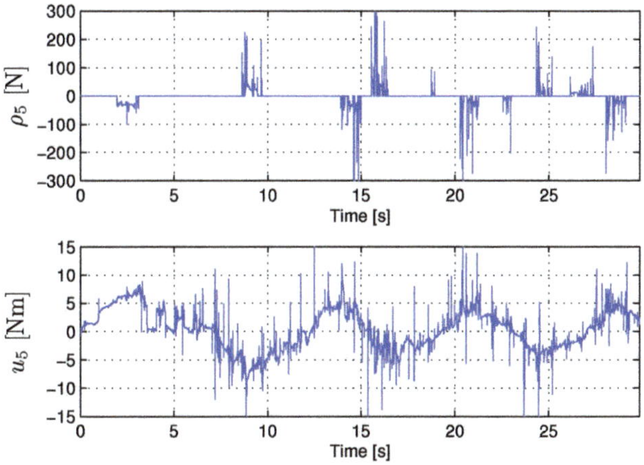

Fig. 12.6 The obstacle constraint forces on link 5 (*top*) and the actuator torques applied to joint 5 (*bottom*) in the first obstacle environment

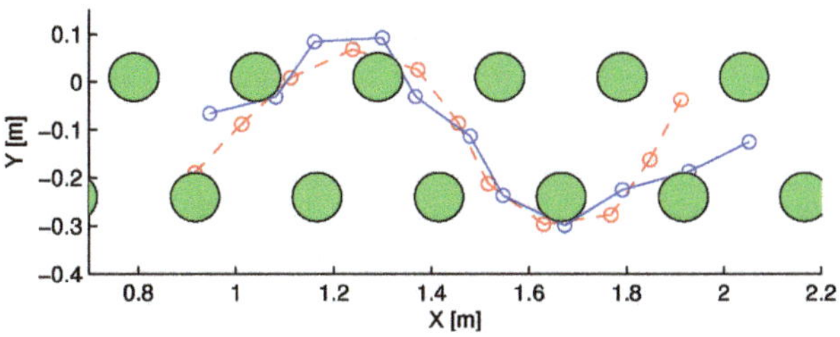

Fig. 12.7 The shape of the snake robot before (*dashed*) and after (*solid*) an instance of the jam resolution scheme

To clearly illustrate the effect of the jam resolution scheme, a plot of the snake robot before (dashed) and after (solid) a jam resolution is shown in Fig. 12.7. The figure shows the jam occurring in the first obstacle environment at time $t = 17.05$ s and ending at time $t = 17.55$ s, which is caused by contact forces acting on links 4 and 8 (link 1 is the tail). The jam is resolved by rotating link 4 clockwise and link 8 counterclockwise, thereby increasing the propulsive components of the two constraint forces enough to overcome the friction forces from the ground and the obstacles.

In summary, the simulation results illustrate how the proposed jam detection and resolution scheme can help to maintain the propulsion of a snake robot in a cluttered environment. The successful performance of the jam resolution scheme is a support of the control principle proposed in Hypothesis 12.1.

12.8 Experimental Study: The Performance of the Hybrid Controller

This section presents experimental results that illustrate the performance of the hybrid controller described in Sect. 12.5. The experiments show that the snake robot Kulko, which was described in Chap. 11, is propelled through several different obstacle courses when the robot is controlled according to the hybrid controller.

12.8.1 Experimental Setup

The experimental setup, which is described in more detail in Chap. 11, consisted of the snake robot Kulko and the camera-based position measurement system. The hybrid controller described in Sect. 12.5 was implemented on an external computer with parameters $N = 10$, $l = 0.07$ m, $\theta_{\text{left}} = 50°$, $\theta_{\text{right}} = -50°$, $\Delta y_{\max} = 14$ cm, $v_{\text{ref}} = 5$ cm/s, $\Delta\phi_{\max} = 20°$, $t_{\text{jam,max}} = 1$ s, $t_{\text{resolution,max}} = 1$ s, and $k_\theta = 0.05$. The joint torque controller given by (12.16) was not implemented since accurate torque control is not supported by the servo motors installed in the snake robot. The joint angles were instead controlled according to a proportional controller implemented in the microcontroller of each joint module.

Three different obstacle environments were considered. The first obstacle environment contained five obstacles with x coordinates (−123.9, −89.6, −48.4, −8.2, −0.6) cm, y coordinates $(20.2, -15.7, 13.2, -23.5, 24.8)$ cm, and diameters $(30, 20, 30, 30, 20)$ cm, respectively. The second obstacle environment contained four obstacles with x coordinates $(-90.9, -35.5, 5.1, 31.7)$ cm, y coordinates $(-20.3, 4.2, -28.9, 15.9)$ cm, and diameters $(30, 30, 30, 30)$ cm, respectively. The third and final obstacle environment contained four obstacles with x coordinates $(-93.1, -79.4, -17.4, 14.6)$ cm, y coordinates $(-61.7, -6.3, -18.9, 24.3)$ cm, and diameters $(30, 30, 30, 30)$ cm, respectively.

Three trials were carried out in each obstacle environment. The initial link angles in the first, second, and third environments were approximately $\boldsymbol{\theta} = [49°, 43°, 6°, 14°, -19°, -20°, -3°, 16°, 11°, 1°]^T$, $\boldsymbol{\theta} = [58°, 47°, 25°, -14°, -35°, -27°, -12°, 3°, 28°, 25°]^T$, and $\boldsymbol{\theta} = [-4°, -4°, -2°, 15°, 30°, 65°, 40°, 29°, -3°, -22°]^T$, respectively. The initial position of the head link was ($x_N = 0$, $y_N = 0$), and the initial reference angle for the head link was $\theta_{N,\text{ref}} = \theta_{\text{right}}$ in all three environments.

12.8.2 Experimental Results

The experimental results from the three obstacle environments are shown in Figs. 12.8 and 12.9, Figs. 12.10 and 12.11, and Figs. 12.12 and 12.13, respectively. Three trials were carried out in each obstacle environment. The measured position of the head link along the forward direction (the global x axis) during the three trials in each environment are shown in Figs. 12.8(a), 12.10(a), and 12.12(a), respectively, and the control scheme executed during each trial (i.e. *leader-following* or *jam resolution*) is shown in Figs. 12.8(b), 12.10(b), and 12.12(b). Furthermore, the sideways position (along the global y axis) and the orientation of the head link are shown in Figs. 12.8(c)–(d), 12.10(c)–(d), and 12.12(c)–(d). In order to give an idea of the forces needed to propel the robot forward, the measured contact forces on joint modules 3–8 (module 1 is the tail) during the first trial in each environment are shown in Figs. 12.8(e)–(f), 12.10(e)–(f), and 12.12(e)–(f), respectively. The motion of the snake robot during the first trial in each environment is visualised in Figs. 12.9, 12.11, and 12.13, respectively.

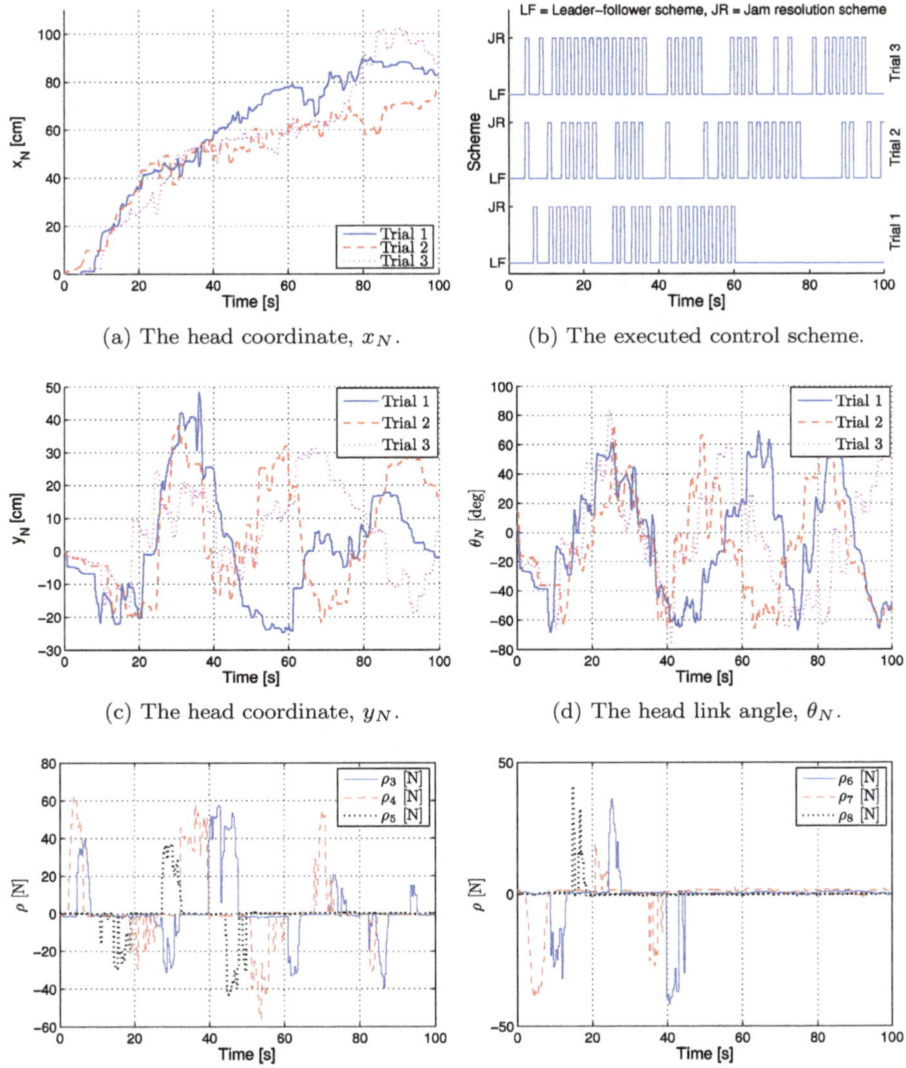

(a) The head coordinate, x_N.

(b) The executed control scheme.

(c) The head coordinate, y_N.

(d) The head link angle, θ_N.

(e) Contact forces, ρ_3–ρ_5 (Trial 1).

(f) Contact forces, ρ_6–ρ_8 (Trial 1).

Fig. 12.8 Experimental results of obstacle-aided locomotion in the first obstacle environment

As seen by the plots of the head position along the forward direction in Figs. 12.8(a), 12.10(a), and 12.12(a), the overall forward propulsion of the snake robot was maintained throughout the trials in all three environments. This was also the main goal of the experiments. In other words, using the same controller with the same set of controller parameters, the snake robot was able to move through three different obstacle environments. The plots of the sideways position and orientation of the head link suggest that the reference angles from the leader-follower scheme

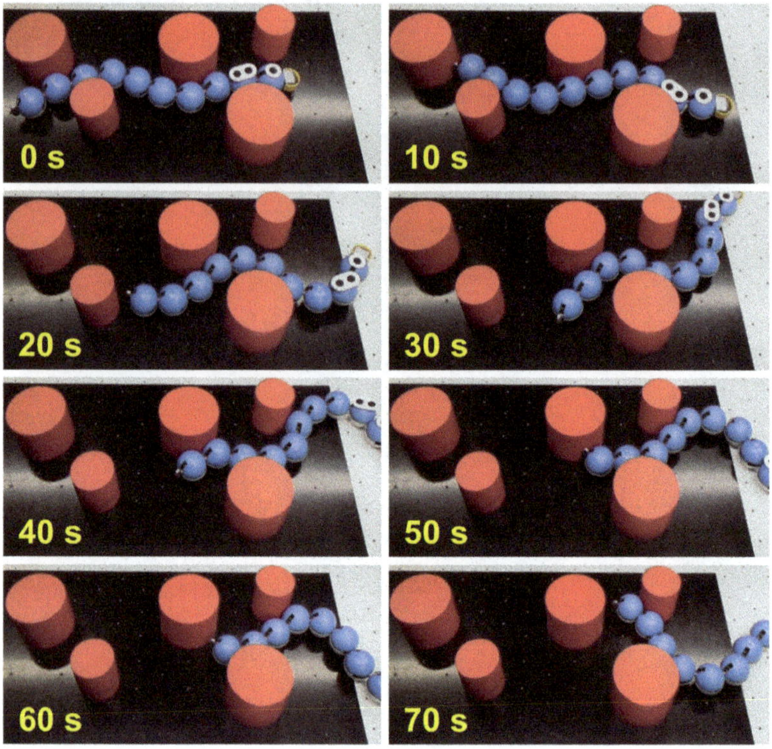

Fig. 12.9 The motion of the snake robot in the first obstacle environment (Trial 1)

were rather different in the individual trials in each environment. However, there is a fairly good repeatability in the forward direction plots from the trials in each environment, which is indicative of the robustness and environment adaptability properties of the proposed controller.

The forward direction plots show that the forward speed of the robot was relatively slow in all trials. This limited speed was due to the limited torque of the joints of the snake robot compared to the rather large ground and obstacle friction forces opposing the motion. In particular, the snake robot is rather heavy (about 10 kg) compared to its maximum actuator torque (about 4 Nm). To cope with the limited strength of the physical snake robot, the propagation velocity of the head joint angle in the leader-follower scheme was set to a rather small value during the experiments, namely $v_{\text{ref}} = 5$ cm/s. Since v_{ref} determines the propagation velocity of the body waves produced during the locomotion, a small value of v_{ref} will naturally lead to a small forward speed of the robot. In other words, we claim that the limited speed during the experiments was caused by limitations of the physical snake robot and is not a general property of the proposed control strategy. Had the experiments been carried out using a snake robot with a larger actuator strength to weight ratio, then the controller parameters could have been adjusted to increase the forward speed significantly.

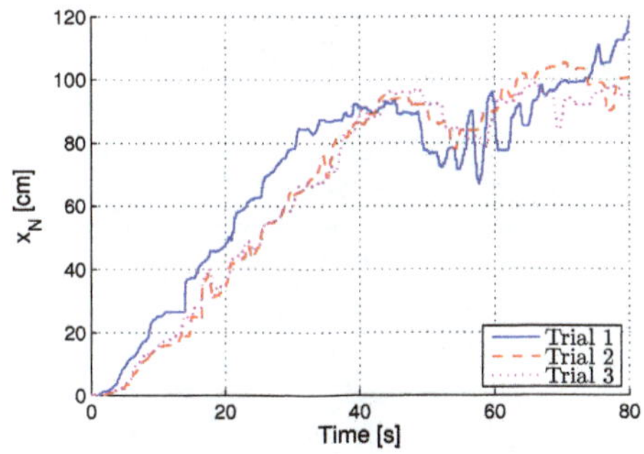

(a) The head coordinate, x_N.

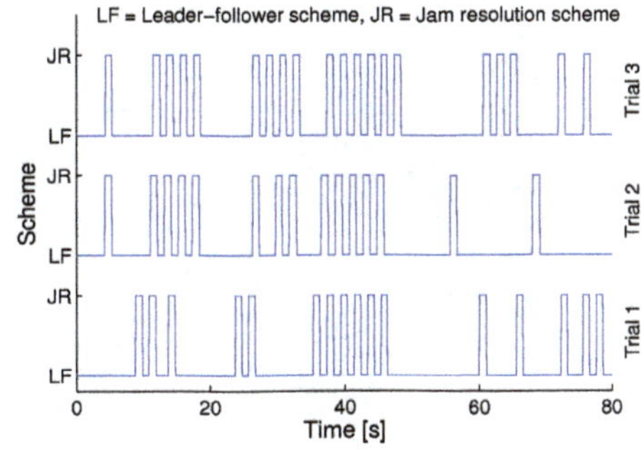

(b) The executed control scheme.

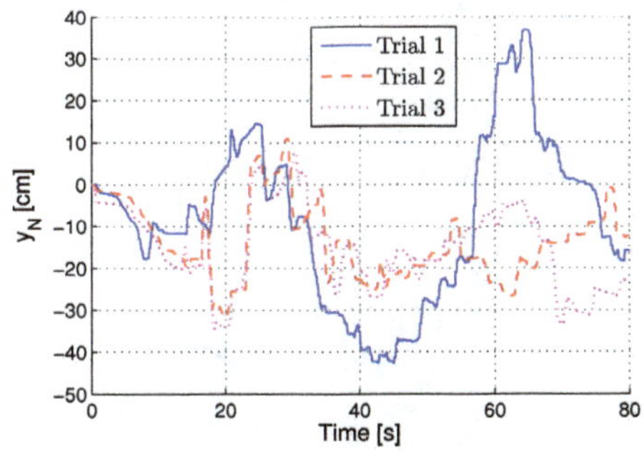

(c) The head coordinate, y_N.

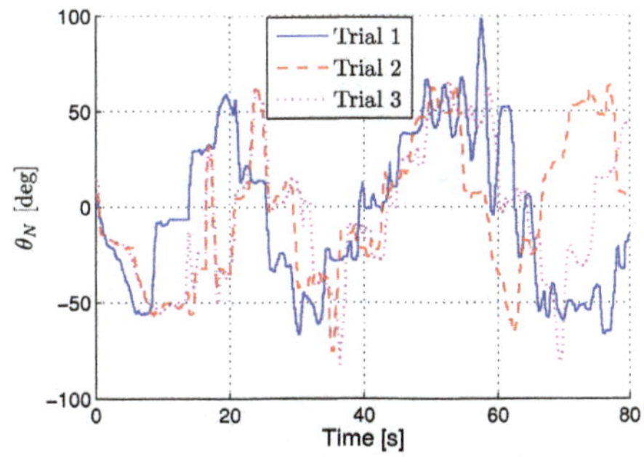

(d) The head link angle, θ_N.

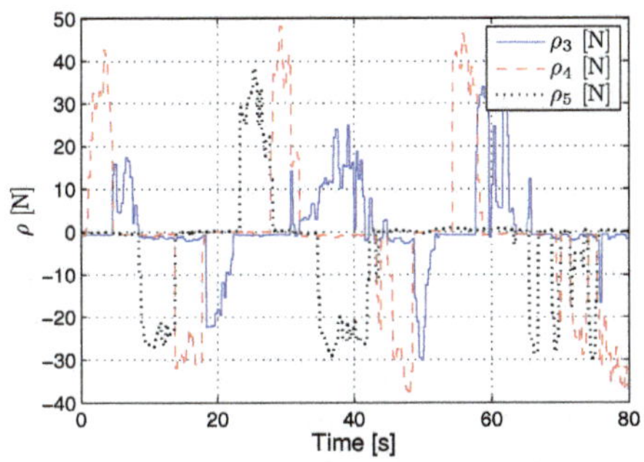

(e) Contact forces, ρ_3–ρ_5 (Trial 1).

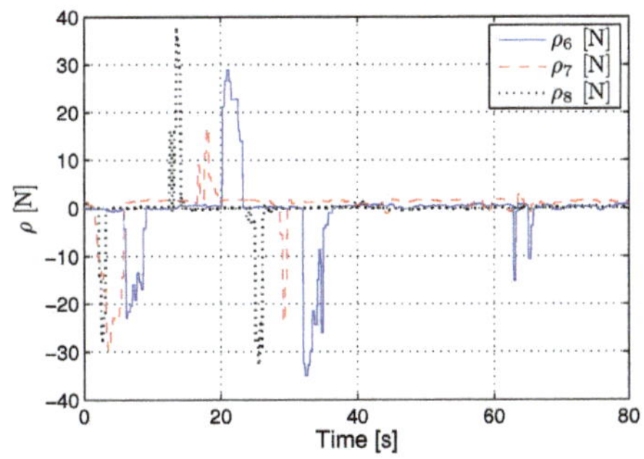

(f) Contact forces, ρ_6–ρ_8 (Trial 1).

Fig. 12.10 Experimental results of obstacle-aided locomotion in the second obstacle environment

There is a clear tendency in the forward direction plots from all three environments that the forward velocity of the robot starts to decrease after about 50 s. The reason for this decrease in velocity is seen from the visualisations in Figs. 12.9, 12.11, and 12.13, which show that the robot used about 50 s to pass through each obstacle course. Since the flat surface outside the obstacle courses contained no push-points that the robot could use for propulsion, the forward velocity decreased as the robot moved out of each obstacle course. The proposed control strategy is,

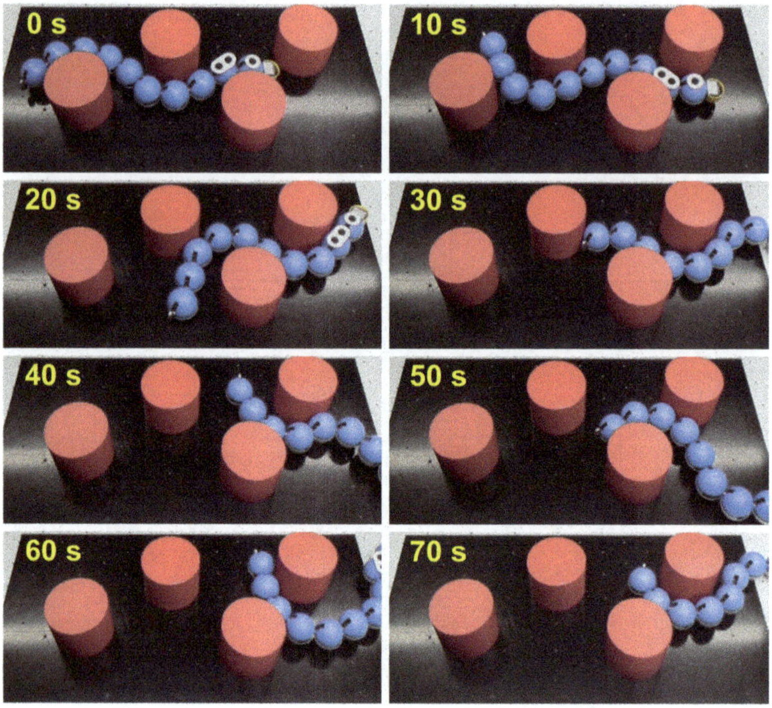

Fig. 12.11 The motion of the snake robot in the second obstacle environment (Trial 1)

in other words, dependent on external objects in order to propel a snake robot with isotropic ground friction properties forward.

As seen in Figs. 12.8(b), 12.10(b), and 12.12(b), the snake robot was jammed and executed jam resolution several times during each trial. In order to visualise the behaviour of the snake robot during jam resolution, Fig. 12.14 shows the snake robot in the first obstacle environment at two time instants when it was jammed ($t = 6$ s and $t = 21$ s) and after jam resolution had been carried out ($t = 7$ s and $t = 22$ s). It is clearly seen from the figure that the jam resolution scheme increased the angles of the jammed joints with respect to the forward direction, thereby increasing the propulsive components of the subsequent obstacle contact forces at these locations. This behaviour is also the intended purpose of the jam resolution scheme, as described in Sect. 12.5.3.

The high number of jams that occurred during the trials in the three obstacle environments is an interesting observation since it suggests that the *jam* state of the snake robot should be treated as a *continuous* rather than a discrete state. Furthermore, the high number of jams suggests that leader-following should not be conducted in open loop, but rather combined with continuous use of the measured contact forces. In particular, the jam resolution scheme was active a large number of times during the experiments because the leader-follower scheme did not consider the environment interaction, which caused the robot to become jammed over and

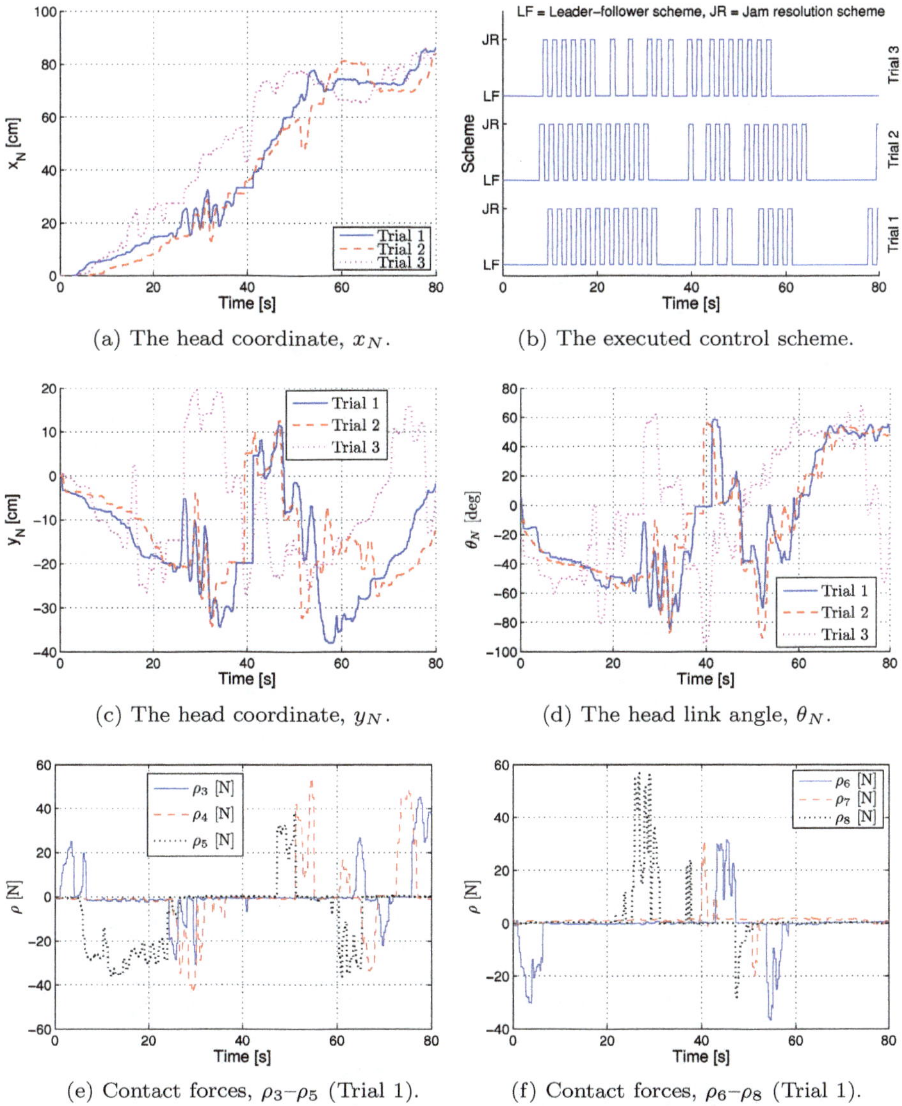

(a) The head coordinate, x_N.

(b) The executed control scheme.

(c) The head coordinate, y_N.

(d) The head link angle, θ_N.

(e) Contact forces, ρ_3–ρ_5 (Trial 1).

(f) Contact forces, ρ_6–ρ_8 (Trial 1).

Fig. 12.12 Experimental results of obstacle-aided locomotion in the third obstacle environment

over. Based on this observation, we will in Chap. 13 propose a continuous control law for obstacle-aided locomotion where jam resolution is a continuous action that is performed in parallel with the cyclic wave motion of the snake robot.

In summary, the experimental results indicate that the proposed control strategy is robust with respect to maintaining the overall forward propulsion of a snake robot in various obstacle environments. The ability of the robot to resolve jams clearly suggests that rotating links in contact with obstacles to increase the propulsive force

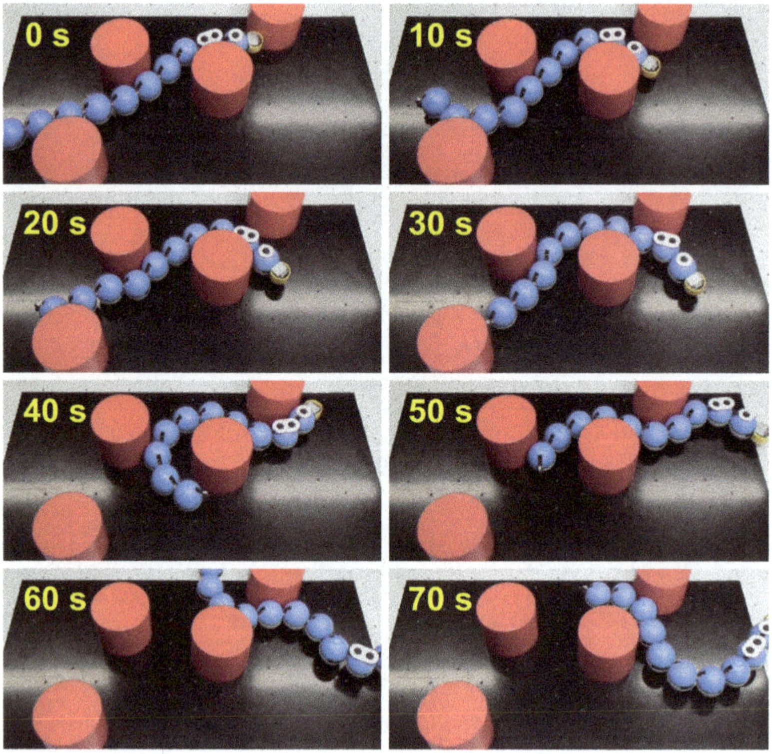

Fig. 12.13 The motion of the snake robot in the third obstacle environment (Trial 1)

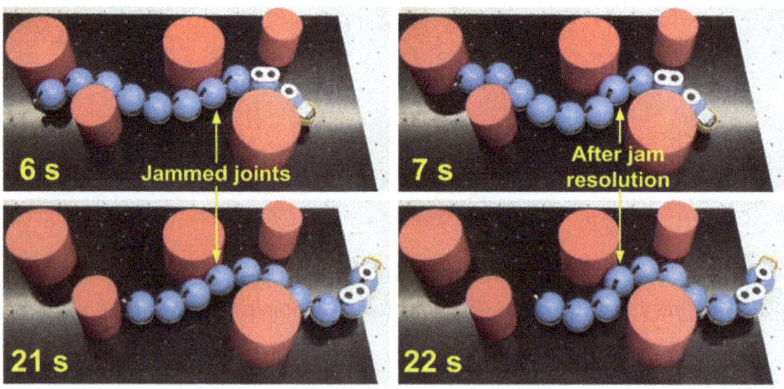

Fig. 12.14 The shape of the snake robot before and after jam resolution in the first obstacle environment

on the robot is a control principle that should be pursued in further work on obstacle-aided locomotion. We therefore claim that the experimental results support Hypothesis 12.1.

12.9 Chapter Summary

This chapter is summarised as follows:

- We have proposed a fundamental control principle (see Hypothesis 12.1), where we claim that obstacle-aided snake robot locomotion is achieved by producing body shape changes where the links in contact with obstacles are rotated to increase the propulsive forces on the robot.
- We have investigated this control principle by using it as a basis for a proposed control strategy for obstacle-aided locomotion.
- The proposed control strategy is a *hybrid controller* aimed at resolving situations where the snake robot is jammed between obstacles.
- In particular, body waves are produced in open-loop in a *leader-follower* scheme as long as the robot is able to move without being jammed between the obstacles. If a *jam* is detected, then a *jam resolution* scheme is carried out in order to effectively 'unlock' the jammed joints.
- We have presented simulation results and experimental results based on the snake robot Kulko, where the hybrid controller was shown to maintain the propulsion of the snake robot in different obstacle environments.
- The high number of jams that occurred during the experiments with the snake robot suggests that the *jam* state of the robot should be treated as a *continuous* rather than a discrete state and that leader-following should not be conducted in open loop, but rather combined with continuous use of the measured contact forces.

Chapter 13
Path Following Control of Snake Robots in Cluttered Environments

With respect to control design, this chapter represents the culmination of this book and can be regarded as a fusion of the control efforts reported in the previous chapters. In particular, forward propulsion of snake robots on planar surfaces based on oscillatory body shape changes was studied in Chap. 4. Subsequently, this oscillatory gait pattern was extended in Chaps. 5 and 8 with directional control capabilities that enabled the snake robot to track straight paths and paths defined by waypoints. Furthermore, the oscillatory gait pattern was extended in Chap. 12 with environment adaptation capabilities that enabled the snake robot to maintain forward propulsion in environments with obstacles. Directional control was, however, not considered in Chap. 12.

In this chapter, we will employ the knowledge gained from these control design efforts to propose a general *framework* for motion control of snake robots. The framework allows the motion of the snake robot to be specified in terms of three separate components, namely a *body wave* component, an *environment adaptation* component, and a *heading control* component. The framework is subsequently used to propose a continuous control strategy for straight line path following control of snake robots in cluttered environments. We will also combine the proposed path following controller with the waypoint guidance strategy from Chap. 8 in order to enable a snake robot to move between waypoints in an obstacle environment. The performance of the path following controller and the waypoint guidance strategy is illustrated with simulation results where a snake robot is successfully steered between waypoints in an environment with obstacles. We will also present experimental results where the snake robot Kulko is successfully propelled through three different obstacle environments with the proposed controller.

An important feature of the path following controller is the idea of a continuous jam resolution action that is performed in parallel with the cyclic wave motion of the robot to continuously adapt the body shape to the environment and prevent the motion from being jammed. This continuous jam resolution action is contained in the *environment adaptation* component of the controller and is based on the jam resolution scheme from Chap. 12. However, whereas a complex hybrid formulation is employed in Chap. 12, the jam resolution action will in this chapter be specified

P. Liljebäck et al., *Snake Robots*, Advances in Industrial Control,
DOI 10.1007/978-1-4471-2996-7_13, © Springer-Verlag London 2013

in terms of simple continuous equations. We consider the step from a *hybrid* to a *continuous* formulation to be important since it makes a formal analysis of the controller more feasible.

The chapter is organised as follows. The relation between this chapter and previous literature is discussed in Sect. 13.1. Section 13.2 presents the general framework for motion control of snake robots. A straight line path following controller based on this framework is proposed in Sect. 13.3 and subsequently fitted within a waypoint guidance strategy in Sect. 13.4. Simulation results and experimental results that illustrate the successful performance of the path following controller are presented in Sects. 13.5 and 13.6, respectively. Finally, the chapter is summarised in Sect. 13.7.

13.1 The Relation Between This Chapter and Previous Literature

This chapter considers path following control of snake robots in cluttered environments, which, to our best knowledge, has not been considered in previous literature. In fact, there are only a few works in previous literature which consider control strategies for snake robots where the surface is no longer assumed to be flat. A review of these works was given in Sect. 12.1.

The first novel result of this chapter in relation to previous literature is the general framework for motion control of snake robots, where the motion is specified in terms of a *body wave* component, an *environment adaptation* component, and a *heading control* component. The second novel result is the control law (based on the controller framework) for straight line path following control of snake robots in environments with obstacles and the fitting of this control law within the waypoint guidance strategy from Chap. 8. A significant contribution of the path following controller is the idea of a continuous jam resolution action that is performed in parallel with the cyclic wave motion of the robot to continuously adapt the body shape to the environment and prevent the motion from being jammed. As a third novel result of this chapter, we present experimental results where the snake robot Kulko is successfully propelled through three different obstacle environments with the proposed controller.

13.2 A Controller Framework for Snake Robot Locomotion

In this section, we propose a general framework for motion control of snake robots. To motivate this framework, we begin by stating a set of claims that we justify with reference to research results reported in the previous chapters of this book:

Claim (The controller should produce body wave motion) During snake robot locomotion on flat surfaces, we showed in Chap. 4 that cyclic body wave motion produces forward propulsion under anisotropic ground friction conditions. The argument for also considering cyclic body wave motion in environments with obstacles was given in the beginning of Sect. 12.5.1. In particular, it was noted in Remark 10.2 in Sect. 10.4 that the phenomenon that propels a snake robot forward due to the anisotropic ground friction properties of the links is similar to the phenomenon that propels a snake robot forward due to the interaction of the links with external obstacles. Since cyclic body wave motion produces forward propulsion under anisotropic ground friction conditions, we can thereby expect this to also be the case in environments with obstacles. Note that the claim is also supported by the motion of biological snakes in nature.

Claim (The controller should continuously perform environment adaptation) The experimental investigation of the hybrid controller presented in Chap. 12 showed that the motion of the physical snake robot was jammed quite frequently. These experimental results indicate that conducting cyclic body wave motion in open loop *will* eventually jam the motion of the robot, which strongly suggests that the cyclic body wave motion should *not* be conducted in open loop, but rather adjusted continuously according to the interaction of the robot with its environment. We therefore claim that environment adaptation should be conducted continuously in parallel with the cyclic body wave motion of the snake robot.

Claim (The controller should steer the heading) This requirement is obvious in order to be able to steer the snake robot to a desired location. We demonstrated path following capabilities of a snake robot in Chaps. 5 and 8.

Based on the above claims, we propose the following general controller framework for snake robots:

Hypothesis 13.1 (The controller framework) *Efficient and intelligent snake robot locomotion in unknown and cluttered environments can be achieved by specifying the reference angles* $\boldsymbol{\phi}_{\text{ref}} = [\phi_{1,\text{ref}}, \dots, \phi_{N-1,\text{ref}}]^T \in \mathbb{R}^{N-1}$ *of the robot as the sum of three individual motion components, namely as*

$$\boldsymbol{\phi}_{\text{ref}} = \boldsymbol{\phi}_{\text{wave}} + \boldsymbol{\phi}_{\text{adapt}} + \boldsymbol{\phi}_{\text{heading}}, \tag{13.1}$$

where $\boldsymbol{\phi}_{\text{wave}}$ *is a* body wave *component that induces propulsive forces on the robot from the environment,* $\boldsymbol{\phi}_{\text{adapt}}$ *is an* environment adaptation *component that adjusts the body shape to the environment, and* $\boldsymbol{\phi}_{\text{heading}}$ *is a* heading control *component that steers the robot according to a specified reference direction.*

Remark 13.1 The lateral undulation gait considered in Part I of this book, which is also considered in the majority of the literature on snake robot locomotion, fits nicely within the framework proposed in (13.1). As defined in (4.32), this gait pattern is achieved by controlling joint i of the snake robot according to

$$\phi_{i,\text{ref}} = \underbrace{\alpha \sin\big(\omega t + (i-1)\delta\big)}_{\boldsymbol{\phi}_\text{wave}} + \underbrace{\phi_o}_{\boldsymbol{\phi}_\text{heading}}, \tag{13.2}$$

where the sinus term constitutes the body wave component, $\boldsymbol{\phi}_\text{wave}$, and ϕ_o, which is an angular offset used to control the direction of the motion, constitutes the heading component, $\boldsymbol{\phi}_\text{heading}$. The gait pattern does not involve adaptation of the body shape to the environment, which means that $\boldsymbol{\phi}_\text{adapt} = \mathbf{0}$.

Remark 13.2 The environment adaptation component $\boldsymbol{\phi}_\text{adapt}$ requires that the snake robot can *sense* its environment in some way. We demonstrated this capability in Chap. 11 by describing the development of a snake robot with contact force sensors installed along its body. Snake robots with discrete contact switches are employed in Bayraktaroglu (2008), Hirose (1993). The environment can also be indirectly sensed through the joint angle measurements, as considered in Andruska and Peterson (2008), Date and Takita (2007).

13.3 Straight Line Path Following Control in Cluttered Environments

In this section, we employ the controller framework presented in Sect. 13.2 to propose a straight line path following controller for snake robots in environments with obstacles. The *body wave* component of the control law is based on a *predecessor-follower* scheme, where each joint follows the angle of the preceding joint ahead of itself. This approach is an improvement over the leader-follower scheme considered in Chap. 12 since leader-following relies on the assumption that the robot moves forward with the same speed as the head angle propagates backward. The *environment adaptation* component is based on the jam resolution principle from Chap. 12. However, whereas a complex hybrid formulation is employed in Chap. 12, the jam resolution motion of the present controller is specified in terms of simple continuous equations. The *heading control* component is similar to a guidance law of the straight line path following controllers proposed in Chaps. 5 and 8.

13.3.1 Control Objective

We choose the control objectives to be identical to the objectives of the path following controller proposed in Chap. 5. The objectives are restated here for completeness.

In order to track the desired straight path, we define the global coordinate system so that the global x axis is aligned with the desired path. The position of the snake robot along the global y axis, p_y, is then the shortest distance from the robot to the desired path (i.e. the cross-track error) and the heading $\overline{\theta}$ of the robot, which was defined in (2.2), is the angle that the robot forms with the desired path. The control objective is thereby to regulate p_y and $\overline{\theta}$ so that they oscillate about zero. We will not attempt to regulate p_y and $\overline{\theta}$ to zero since the heading and position of the robot are expected to display oscillating behaviour during locomotion along the desired path.

Since snake robot locomotion is a slow form of robotic mobility, which is generally employed for traversability purposes, we consider it less important to accurately control the forward velocity of the robot. During path following with a snake robot, it therefore makes sense to focus all the control efforts on converging to the path and subsequently progressing along the path at some non-zero forward velocity $\overline{v}_t(t) > 0$, where $\overline{v}_t(t)$ is the forward velocity of the robot defined in (2.5).

From the above discussion, the control problem is to design a feedback control law such that for all $t > t_c \geq 0$, there exists a $\tau \in [t, t+T]$ satisfying

$$p_y(\tau) = 0, \tag{13.3}$$

$$\overline{\theta}(\tau) = 0, \tag{13.4}$$

$$\overline{v}_t(t) > 0, \tag{13.5}$$

where t_c is some (unknown) finite time duration corresponding to the time it takes the snake robot to converge to the desired straight path, and $T > 0$ is some constant that characterises the time period of the cyclic gait pattern of the snake robot. In other words, we require that p_y and $\overline{\theta}$ are zero at least once *within* each cycle of the locomotion since this means that p_y and $\overline{\theta}$ oscillate about zero. Note that we require $\overline{v}_t(t) > 0$ for all $t > t_c$.

The idea behind the controller proposed in the following is to use the body wave component $\boldsymbol{\phi}_{\text{wave}}$ and the adaptation component $\boldsymbol{\phi}_{\text{adapt}}$ to achieve control objective (13.5), and simultaneously use the heading component $\boldsymbol{\phi}_{\text{heading}}$ to achieve control objectives (13.3) and (13.4).

13.3.2 Notation and Basic Assumptions

So far in this book, we have described the link angles $\boldsymbol{\theta}$ according to Definition 2.1, i.e. with respect to the global x axis. During path following in an obstacle environment, however, we will employ a measure of the link angles with respect to the current heading $\overline{\theta}$ defined in (2.2). As illustrated in Fig. 13.1, we define this measure as follows:

Definition 13.1 (The heading-adjusted link angle) The *heading-adjusted* angle of link $i \in \{1, \dots, N\}$ is denoted by $\widetilde{\theta}_i$ and is given as the angle of link i with respect to the current heading $\overline{\theta}$, i.e. as

$$\widetilde{\theta}_i = \theta_i - \overline{\theta}. \tag{13.6}$$

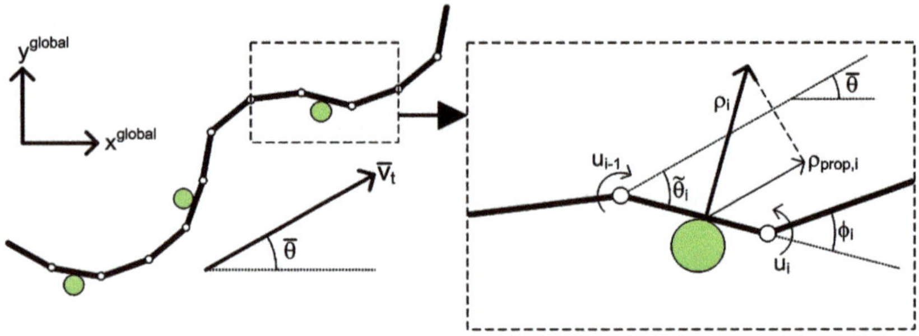

Fig. 13.1 Illustration of the heading-adjusted angle of link i, denoted by $\widetilde{\theta}_i$, and the propulsive component of the contact force on link i, denoted by $\rho_{\text{prop},i}$

Control objective (13.5) simply states that the robot should maintain a positive and non-zero forward velocity. This means that the desired forward direction of motion at any time is along the current heading of the robot. In accordance with Definition 12.2 in Sect. 12.4, the *propulsive component* of the contact force on link $i \in \{1, \dots, N\}$ is therefore given as the component of the constraint force vector $\mathbf{f}_{c,i}$ along the current heading $\overline{\theta}$. This component, which we denote by $\rho_{\text{prop},i} \in \mathbb{R}$, is illustrated in Fig. 13.1 and is easily calculated as

$$\rho_{\text{prop},i} = -\rho_i \sin \widetilde{\theta}_i, \tag{13.7}$$

where ρ_i was defined in Sect. 12.4 as the measured contact force on link $i \in \{1, \dots, N\}$.

Remark 13.3 In Chap. 12, the propulsive component of a contact force was denoted by $\rho_{x,i}$ and given as the force component along the global x axis. As long as the robot is headed along the global x axis, we have that $\rho_{x,i} = \rho_{\text{prop},i}$. In contrast to $\rho_{x,i}$, however, $\rho_{\text{prop},i}$ represents a suitable measure of the propulsive force on link i regardless of the heading of the robot.

We will base the path following controller on the following assumption:

Assumption 13.1 *The control system has access to measurements of the cross-track error p_y, the joint angles $\boldsymbol{\phi}$, the joint angle velocities $\dot{\boldsymbol{\phi}}$, the contact forces $\boldsymbol{\rho}$, and at least one of the absolute link angles θ_i for some $i \in \{1, \dots, N\}$.*

Note that the remaining link angles, and thereby also the heading $\overline{\theta}$, can be calculated from $\boldsymbol{\phi}$ and θ_i.

13.3.3 The Body Wave Component

We begin by presenting the body wave component $\boldsymbol{\phi}_{\text{wave}}$ of the joint reference angles. The hybrid controller proposed in Chap. 12 produces body waves through a leader-follower approach, where the angle of the foremost (head) joint is propagated backwards along the snake body at a constant velocity and used as the reference angle for all subsequent joints. The drawback of this approach is that it relies on the assumption that the snake robot moves forward with the same speed as the head angle propagates backward. If this is not the case, then the head joint angle will generally not be a suitable reference angle for the remaining joints.

In the following, we therefore employ a *predecessor-follower* scheme, where each joint follows the angle of the preceding joint ahead of itself with a specified time delay Δt. The angle of joint i is always a suitable reference angle for joint $i-1$ since the current shape of the snake robot always represents a feasible reference trajectory. The resulting reference angle of joint $i \in \{1, \dots, N-2\}$ in this predecessor-follower scheme can be written

$$\phi_{\text{wave},i}(t) = \phi_{i+1}(t - \Delta t). \tag{13.8}$$

In order to produce body wave motion, we introduce a sinusoidal reference angle for the heading-adjusted angle of the head link, $\widetilde{\theta}_N$, given by

$$\widetilde{\theta}_{N,\text{ref}}(t) = \alpha \sin(\omega t), \tag{13.9}$$

where α and ω are the amplitude and angular frequency, respectively, of the sinusoidal motion. Since the head joint (i.e. joint $N-1$) is at the front of the snake robot, the rotation of the head joint mainly affects the angle of the head link (i.e. link N) and not the angle of the subsequent links. Since it follows from (2.1) that $\phi_{N-1} = \widetilde{\theta}_{N-1} - \widetilde{\theta}_N$, we can track the head link reference angle in (13.9) by controlling the head joint according to the reference

$$\phi_{\text{wave},N-1}(t) = \widetilde{\theta}_{N-1} - \alpha \sin(\omega t). \tag{13.10}$$

From (13.8) and (13.10) we can now write the complete body wave component $\boldsymbol{\phi}_{\text{wave}}$ in matrix form as

$$\boldsymbol{\phi}_{\text{wave}} = \mathbf{S}_{\text{head}}\big(\widetilde{\theta}_{N-1} - \alpha \sin(\omega t)\big) + \mathbf{S}_{\text{joints}}\boldsymbol{\phi}(t - \Delta t), \tag{13.11}$$

where $\boldsymbol{\phi}(t - \Delta t)$ are the measured joint angles at time $t - \Delta t$, and where $\mathbf{S}_{\text{head}}$ and $\mathbf{S}_{\text{joints}}$ are, respectively, a selection vector and a selection matrix defined as

$$\mathbf{S}_{\text{head}} = [0, \dots, 0, 1]^T \in \mathbb{R}^{N-1}, \tag{13.12}$$

$$\mathbf{S}_{\text{joints}} = \begin{bmatrix} 0 & 1 & 0 & & \\ & 0 & 1 & 0 & \\ & & \ddots & \ddots & \\ & & & 0 & 1 \\ & & & & 0 \end{bmatrix} \in \mathbb{R}^{(N-1)\times(N-1)}. \tag{13.13}$$

Remark 13.4 The implementation of (13.11) requires a buffer which keeps track of the angle history of each joint except for the first (tail) joint.

13.3.4 The Environment Adaptation Component

The environment adaptation component $\boldsymbol{\phi}_{\text{adapt}}$ is based on the control principle proposed in Hypothesis 12.1, that was implemented in the jam resolution scheme of the hybrid controller from Chap. 12. However, whereas a complex hybrid formulation is employed in Chap. 12, the jam resolution action of the present control strategy will be specified in terms of simple continuous equations. We consider the step from a *hybrid* to a *continuous* formulation to be important since a continuous formulation makes a formal analysis of the controller more feasible. Unfortunately, a formal analysis of the present control strategy remains an unsolved problem, as explained below in Remark 13.5.

In accordance with the derivation of the jam resolution scheme in Sect. 12.5.3, the idea behind the adaptation strategy is to rotate the links affected by contact forces so that the propulsive component of each contact force increases. Since the propulsive components of the contact forces are what propel the snake robot forward, we conjecture that rotating the contacted links to increase the total propulsive force will adapt the body shape to the environment in a way that maintains or increases the propulsion of the robot. Note that the adaptation strategy only aims at satisfying control objective (13.5), i.e. propelling the snake robot forward in the direction of its current heading.

The change of the propulsive force on link $i \in \{1, \ldots, N\}$ due to a change of the link angle is found by differentiating (13.7) with respect to $\widetilde{\theta}_i$, which gives

$$\frac{\partial \rho_{\text{prop},i}}{\partial \widetilde{\theta}_i} = -\rho_i \cos \widetilde{\theta}_i. \tag{13.14}$$

During adaptation, we choose to rotate links with a high propulsive force gradient with respect to the link angle, which suggests that link i is rotated according to

$$\Delta \widetilde{\theta}_{i,\text{ref}} = k_\rho \frac{\partial \rho_{\text{prop},i}}{\partial \widetilde{\theta}_i} = -k_\rho \rho_i \cos \widetilde{\theta}_i, \tag{13.15}$$

where $k_\rho > 0$ is a controller gain.

Let us now derive how the joint angles ϕ_{i-1} and ϕ_i at each side of link i should be changed to comply with (13.15). We choose that the contact force on link i only should affect the angle of link i, so that $\Delta\widetilde{\theta}_{i-1,\text{ref}} = \Delta\widetilde{\theta}_{i+1,\text{ref}} = 0$. Since we have from (2.1) that $\phi_i = \widetilde{\theta}_i - \widetilde{\theta}_{i+1}$, we can immediately write the desired change of the joint angles ϕ_{i-1} and ϕ_i due to the contact force on link i as

$$\Delta\phi_{i-1,\text{ref}} = \Delta\widetilde{\theta}_{i-1,\text{ref}} - \Delta\widetilde{\theta}_{i,\text{ref}} = k_\rho \rho_i \cos \widetilde{\theta}_i, \tag{13.16}$$

$$\Delta\phi_{i,\text{ref}} = \Delta\widetilde{\theta}_{i,\text{ref}} - \Delta\widetilde{\theta}_{i+1,\text{ref}} = -k_\rho \rho_i \cos \widetilde{\theta}_i. \tag{13.17}$$

By combining the desired change of joint angle ϕ_i due to the measured contact forces on the link at each side of the joint, i.e. the contribution from both ρ_i and ρ_{i+1}, we get that the angle of joint $i \in \{1, \ldots, N-1\}$ in the environment adaptation component $\boldsymbol{\phi}_{\text{adapt}}$ is given by

$$\phi_{\text{adapt},i} = -k_\rho(\rho_i \cos\widetilde{\theta}_i - \rho_{i+1}\cos\widetilde{\theta}_{i+1}). \tag{13.18}$$

The complete environment adaptation component $\boldsymbol{\phi}_{\text{adapt}}$ can thereby be written in matrix form as

$$\boldsymbol{\phi}_{\text{adapt}} = -k_\rho \mathbf{D}\,\text{diag}(\boldsymbol{\rho})\cos\widetilde{\boldsymbol{\theta}}, \tag{13.19}$$

where diag(·) produces a diagonal matrix with the elements of its argument along its diagonal, and where

$$\boldsymbol{\rho} = [\rho_1, \ldots, \rho_N]^T \in \mathbb{R}^N, \tag{13.20}$$

$$\cos\widetilde{\boldsymbol{\theta}} = [\cos\widetilde{\theta}_1, \ldots, \cos\widetilde{\theta}_N]^T \in \mathbb{R}^N, \tag{13.21}$$

$$\mathbf{D} = \begin{bmatrix} 1 & -1 & & \\ & \ddots & \ddots & \\ & & 1 & -1 \end{bmatrix} \in \mathbb{R}^{(N-1)\times N}. \tag{13.22}$$

13.3.5 The Heading Control Component

The heading control component $\boldsymbol{\phi}_{\text{heading}}$ of the joint reference angles is similar to the guidance law of the straight line path following controllers proposed in Chaps. 5 and 8. In particular, we steer the snake robot towards the desired straight path by employing the Line-of-Sight (LOS) guidance law

$$\overline{\theta}_{\text{ref}} = -\arctan\left(\frac{p_y}{\Delta}\right), \tag{13.23}$$

where p_y is the cross-track error, and $\Delta > 0$ is a design parameter referred to as the *look-ahead distance* that influences the rate of convergence to the desired path. As illustrated in Fig. 13.2, the LOS angle $\overline{\theta}_{\text{ref}}$ corresponds to the orientation of the snake robot when it is headed towards the point located a distance Δ ahead of itself along the desired path. To steer the heading $\overline{\theta}$ according to the LOS angle in (13.23), we employ the same approach that was used for controlling the direction of the snake robot with the lateral undulation gait in (4.32), i.e. we offset the reference angle of the head joint according to

$$\phi_{\text{heading},N-1} = k_\theta(\overline{\theta} - \overline{\theta}_{\text{ref}}), \tag{13.24}$$

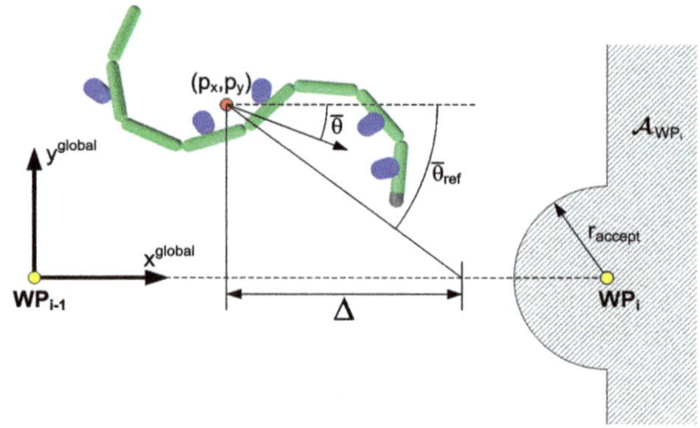

Fig. 13.2 Straight line path following control of the snake robot combined with waypoint guidance in an obstacle environment

where $k_\theta > 0$ is a controller gain. Using (13.12), the heading component can be written in matrix form as

$$\boldsymbol{\phi}_{\text{heading}} = \mathbf{S}_{\text{head}} k_\theta (\overline{\theta} - \overline{\theta}_{\text{ref}}). \tag{13.25}$$

13.3.6 The Joint Angle Controller

In order to make the joint angles $\boldsymbol{\phi}$ track the reference angles given by $\boldsymbol{\phi}_{\text{ref}}$, we set the joint actuator torques $\mathbf{u}$ according to the PD-controller

$$\mathbf{u} = k_p(\boldsymbol{\phi}_{\text{ref}} - \boldsymbol{\phi}) + k_d(\dot{\boldsymbol{\phi}}_{\text{ref}} - \dot{\boldsymbol{\phi}}), \tag{13.26}$$

where $k_p > 0$ and $k_d > 0$ are controller gains.

13.3.7 Summary of the Path Following Controller

The complete straight line path following controller is now summarised. In accordance with the general controller framework defined in (13.1), we conjecture that control objectives (13.3), (13.4), and (13.5) are achieved by employing the PD-controller in (13.26) to control the joint angles of the snake robot according to

$$\boldsymbol{\phi}_{\text{ref}} = \boldsymbol{\phi}_{\text{wave}} + \boldsymbol{\phi}_{\text{adapt}} + \boldsymbol{\phi}_{\text{heading}}, \tag{13.27}$$

where

$$\boldsymbol{\phi}_{\text{wave}} = \mathbf{S}_{\text{head}}\big(\widetilde{\theta}_{N-1} - \alpha\sin(\omega t)\big) + \mathbf{S}_{\text{joints}}\boldsymbol{\phi}(t - \Delta t), \tag{13.28}$$

$$\boldsymbol{\phi}_{\text{adapt}} = -k_\rho \mathbf{D}\,\text{diag}(\boldsymbol{\rho})\cos\widetilde{\boldsymbol{\theta}}, \tag{13.29}$$

$$\boldsymbol{\phi}_{\text{heading}} = \mathbf{S}_{\text{head}} k_\theta (\overline{\theta} - \overline{\theta}_{\text{ref}}). \tag{13.30}$$

Remark 13.5 Due to the complexity of the hybrid model of the snake robot in (10.61), we are currently unable to provide a formal proof of the achievement of objectives (13.3), (13.4), and (13.5) with the proposed controller. It is probably not possible to develop such a proof solely based on the model and control strategy of the robot combined with knowledge of the obstacle locations since it is difficult, if not impossible, to analytically predict the interaction between the robot and the obstacles in advance. However, it may be possible to develop logical arguments regarding the achievement of the control objectives by making assumptions regarding the obstacle interactions. One approach could be to parameterise the contact situation in some way and show that, in certain contact situations, the control action defined by (13.26) and (13.27) will move the snake robot a step closer to the desired path.

13.4 Waypoint Guidance Control in Cluttered Environments

The straight line path following controller defined by (13.26) and (13.27) can easily be combined with the waypoint guidance strategy proposed in Sect. 8.5. We thereby arrive at a complete control strategy for steering a snake robot between waypoints in a cluttered environment, i.e. we obtain a control strategy for obstacle-aided locomotion along arbitrary paths given by waypoints interconnected by straight lines. In the following, we describe the waypoint guidance strategy from Sect. 8.5 within the context of this chapter.

The reason for specifying the path of the robot in terms of waypoints is that future applications of snake robots will generally involve bringing sensors and/or tools to a single or several specified target location(s). In these situations, the exact path taken by the robot as it moves towards the target(s) is generally of less interest as long as the robot reaches the target(s) within a reasonable amount of time. Specifying the motion of a snake robot in terms of waypoints supports this target-oriented control approach.

There are k waypoints, and the ith waypoint is denoted by WP_i, where $i \in \{1, \dots, k\}$. As illustrated in Fig. 13.2, we interconnect the waypoints by straight lines and control the snake robot according to (13.27) in order to steer the robot towards the straight line leading to the next waypoint. The next waypoint is activated as soon as the position of the robot enters inside an *acceptance region* consisting of an *acceptance circle* (with radius r_{accept}) centred in the current waypoint and also the right half-plane of a coordinate system with origo in the current waypoint and x axis pointing away from the previous waypoint (see illustration in Fig. 13.2). The acceptance region of WP_i is denoted by $\mathcal{A}_{WP_i}$. The intended purpose of employing the straight line path following controller is to steer the robot into the acceptance circle of the current waypoint. However, in situations where the obstacle environment prevents the robot from entering inside the acceptance circle, the robot will still proceed towards the next waypoint as soon as the position enters inside the right half-plane contained in the acceptance region. With only acceptance circles enclosing each waypoint, there would be the risk that the robot misses a waypoint, e.g. due to the placement of the obstacles, which would make the robot proceed indefinitely along the path away from the waypoint that was missed.

The above definitions were formalised in Sect. 8.5. We can now state the waypoint guidance strategy as follows:

Algorithm 13.1 (The waypoint guidance strategy)

1. Define the initial position of the snake robot as WP_0.
2. Repeat for all $i \in \{0, \dots, k-1\}$:
   a. Move the origin of the global frame to WP_i and orient the global x axis towards WP_{i+1}.
   b. Conduct path following according to (13.27) until $(p_x, p_y) \in \mathcal{A}_{WP_{i+1}}$.

13.5 Simulation Study: The Performance of the Path Following Controller

This section presents simulation results that illustrate how the path following controller defined by (13.26) and (13.27) performs in combination with the waypoint guidance strategy in Algorithm 13.1.

13.5.1 Simulation Parameters

The hybrid model of the snake robot (10.61) and the guidance strategy in Algorithm 13.1 were implemented in *Matlab R2008b* on a laptop running *Windows XP*. The continuous dynamics of the hybrid model were calculated with the *ode45* solver in Matlab with a relative and absolute error tolerance of 10^{-3}.

The parameters characterising the simulated snake robot were $N = 10$, $l = 0.07$ m, $m = 1$ kg, and $J = 0.0016$ kgm^2. Circular obstacles measuring 10 cm in diameter were placed in a random fashion in the environment of the snake robot. The ground and obstacle friction coefficients were $\mu = 0.3$ and $\mu_o = 0.25$, respectively. The initial link angles and position of the snake robot were $\boldsymbol{\theta} = [-30°, -10°, 30°, 60°, 40°, 0°, -40°, -60°, -30°, 0°]^T$ and $\mathbf{p} = \mathbf{0}_{2\times1}$, respectively.

We defined $k = 5$ waypoints with global frame coordinates $(2.5, 0)$, $(2.5, 1)$, $(0, 1)$, $(1, 2)$, and $(3, 2)$, respectively. The radius of the acceptance circle enclosing each waypoint was $r_{\text{accept}} = 0.5$ m. The remaining controller parameters were $\Delta = 0.7$ m, $k_\theta = 1.3$, $k_\rho = 0.02$, $\Delta t = 0.7$ s, $\alpha = 60°$, $\omega = 40°/\text{s}$, $k_p = 20$, and $k_d = 5$. In order to prevent the measured contact forces in $\boldsymbol{\phi}_{\text{adapt}}$ from producing steps in the joint reference angles $\boldsymbol{\phi}_{\text{ref}}$ in (13.27), the reference angles were filtered using a 2nd-order low-pass filtering reference model (see Appendix C.1). The parameters of the reference model were $\omega = 3\pi/2$ and $\zeta = 1$. This filter also provided

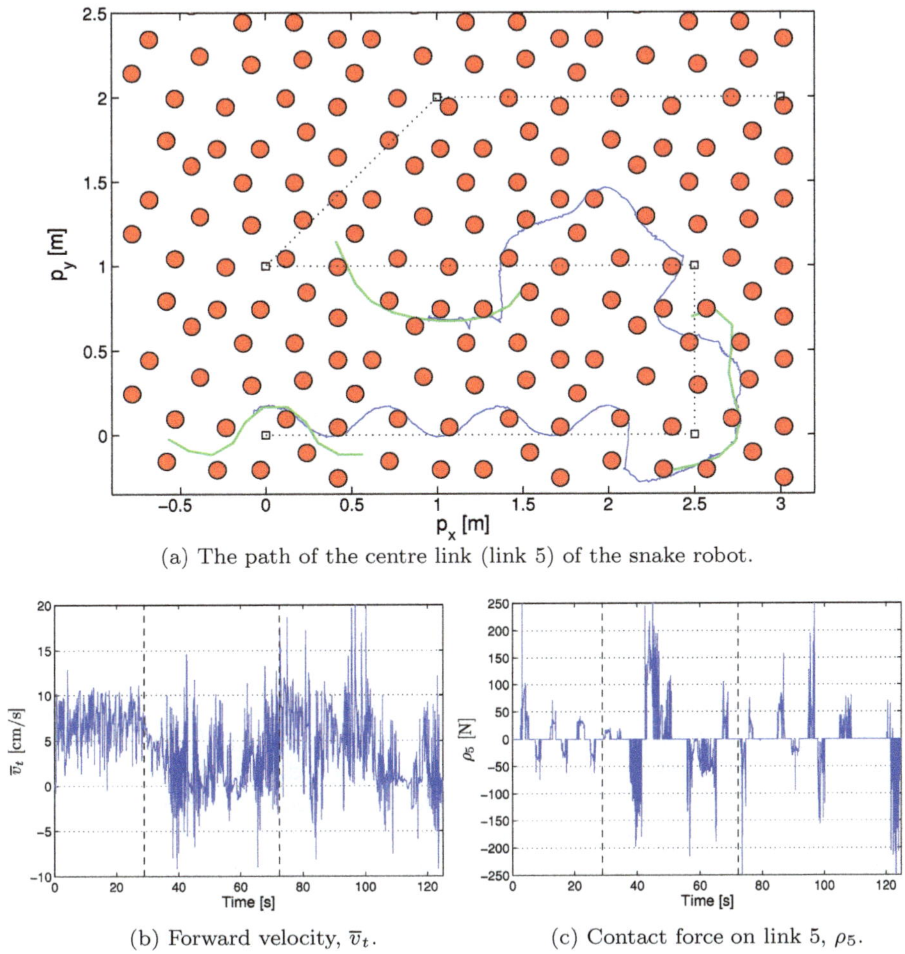

Fig. 13.3 Simulation of the waypoint guidance strategy *without* environment adaptation

the derivative of $\boldsymbol{\phi}_{\text{ref}}$ with respect to time, which is needed by the PD-controller in (13.26).

13.5.2 Simulation Results

To illustrate the importance of environment adaptation, the path following controller of the waypoint guidance strategy was first simulated *without* adaptation ($\boldsymbol{\phi}_{\text{adapt}} = \mathbf{0}$). The path of the centre link of the snake robot (link 5) is shown in blue in Fig. 13.3(a), where black squares indicate the waypoints, the dotted black lines indicate the straight paths between the waypoints, and where the shape and

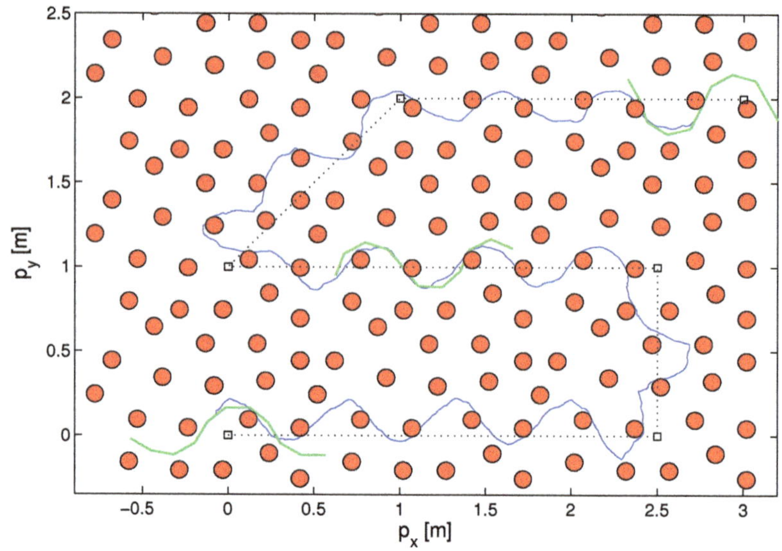

Fig. 13.4 The path of the centre link (link 5) of the snake robot during waypoint guidance *with* environment adaptation

position of the robot are shown in green at $t = 0$ s, $t = 65$ s, and $t = 125$ s, respectively. Furthermore, Fig. 13.3(b)–(c) show the forward velocity, $\overline{v}_t$, and the obstacle constraint force on link 5, ρ_5, respectively. The vertical dashed lines in the plots indicate time instants where the guidance strategy switched to the next waypoint. We see from Fig. 13.3(a) that the robot managed to reach the acceptance region of the two first waypoints. However, the motion was jammed about halfway to the third waypoint, as can be seen from Fig. 13.3(b), which shows that the forward velocity varied around zero after about 110 s. Note that there is a slight overlap between the path of link 5 and some of the obstacles. This issue was commented in Remark 10.4 of Sect. 10.4 and is a consequence of modelling obstacle contact solely by a unilateral force on the contacted link, which means that there is nothing preventing the foremost link (the head) of the snake robot from penetrating an obstacle head-on along its tangential direction.

The same plots for the case where environment adaptation was present (i.e. where $\boldsymbol{\phi}_{\text{adapt}}$ was set according to (13.29)) are shown in Figs. 13.4 and 13.5. In addition to the forward velocity and the obstacle constraint forces, Fig. 13.5 also shows the cross-track error p_y and the heading angle $\overline{\theta}$. With environment adaptation, the propulsion of the robot was maintained through all the waypoints. As seen from Fig. 13.5(c), the forward velocity varied between 5–10 cm/s, which suggests that control objective (13.5) was achieved. Figures 13.5(a)–(b) show that the cross-track error and the heading angle had an oscillatory behaviour around zero after each waypoint switch, which suggests that control objectives (13.3) and (13.4) were also achieved.

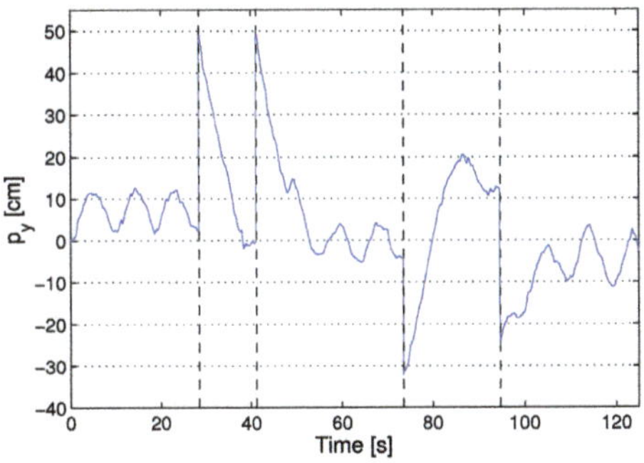

(a) Cross-track error, p_y.

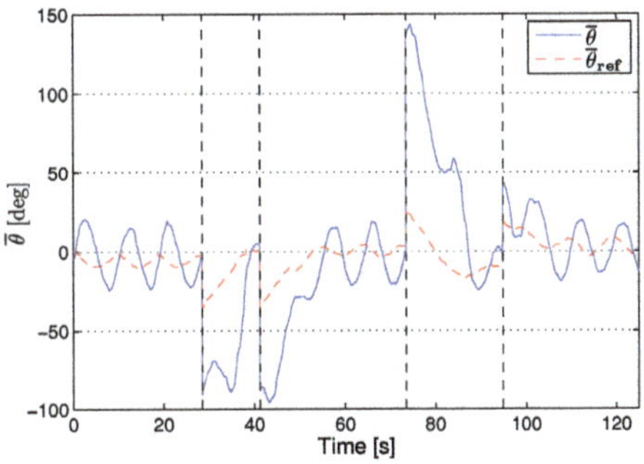

(b) Heading angle, $\overline{\theta}$.

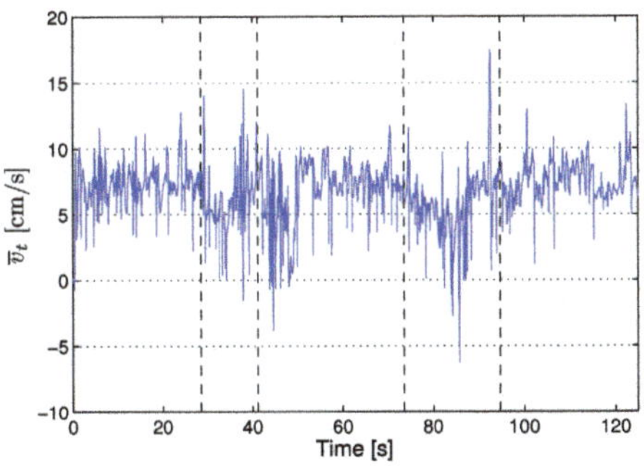

(c) Forward velocity, $\overline{v}_t$.

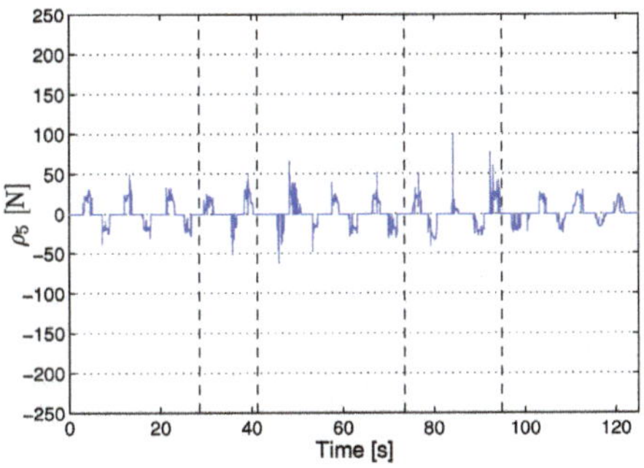

(d) Contact force on link 5, ρ_5.

Fig. 13.5 Simulation of the waypoint guidance strategy *with* environment adaptation

It is interesting to note that, in addition to improving the propulsion of the robot, the environment adaptation strategy reduces the constraint forces on the robot significantly, which is seen by comparing Fig. 13.5(d) with Fig. 13.3(c). This is also expected since the environment contact forces opposing the motion will naturally be larger when the motion is performed without considering the environment.

In summary, the path following controller defined by (13.26) and (13.27) combined with the waypoint guidance strategy in Algorithm 13.1 maintained the propulsion and steered the snake robot to the acceptance region of each waypoint in the obstacle environment.

13.6 Experimental Study: The Performance of the Environment Adaptation Strategy

In this section, we present experimental results in order to demonstrate that a snake robot is propelled forward in an obstacle environment when the joints are controlled according to (13.27). In particular, the results demonstrate the propulsion produced by the body wave component in (13.28) and the environment adaptation component in (13.29). Due to the limited range of the obstacle course, we did not consider heading control during the experiments, which means that the heading control component in (13.30) was set to zero.

13.6.1 Experimental Setup

We employed the same experimental setup that was used to investigate the jam resolution controller in Sect. 12.8. In particular, the experimental setup consisted of the snake robot Kulko and the camera-based position measurement system presented in Chap. 11. The joint reference angles defined by (13.27) were calculated on an external computer with the parameters $N = 10$, $l = 0.07$ m, $k_\rho = 0.01$, $\Delta t = 0.9$ s, $\alpha = 60°$, and $\omega = 30°/\text{s}$, and with $\boldsymbol{\phi}_{\text{heading}} = 0$. The joint torque controller given by (13.26) was not implemented since accurate torque control is not supported by the servo motors installed in the snake robot. The joint angles were instead controlled according to a proportional controller implemented in the microcontroller of each joint module.

We considered three obstacle environments with similar obstacle configuration as in the experiments described in Sect. 12.8. The first obstacle environment contained five obstacles with x coordinates $(-123.9, -89.6, -48.4, -8.2, -0.6)$ cm, y coordinates $(20.2, -15.7, 13.2, -23.5, 24.8)$ cm, and diameters $(30, 20, 30, 30, 20)$ cm, respectively. The second obstacle environment contained four obstacles with x coordinates $(-90.9, -35.5, 5.1, 31.7)$ cm, y coordinates $(-20.3, 4.2, -28.9, 15.9)$ cm, and diameters $(30, 30, 30, 30)$ cm, respectively. The third and final obstacle environment contained five obstacles with x coordinates $(-93.1, -79.4, -45.4, -17.4, 14.6)$ cm, y coordinates $(-61.7, -6.3, 29.4, -18.9, 24.3)$ cm, and diameters $(30, 30, 20, 30, 30)$ cm, respectively.

The initial link angles in the first, second, and third environments were $\boldsymbol{\theta} = [49°, 43°, 6°, 14°, -19°, -20°, -3°, 16°, 11°, 1°]^T$, $\boldsymbol{\theta} = [58°, 47°, 25°, -14°, -35°, -27°, -12°, 3°, 28°, 25°]^T$, and $\boldsymbol{\theta} = [-4°, -4°, -2°, 15°, 30°, 65°, 40°, 29°, -3°, -22°]^T$, respectively, and the initial position of the head link was $(x_N = 0, y_N = 0)$.

13.6.2 Experimental Results

The experimental results from the three obstacle environments are shown in Figs. 13.6 and 13.7, Figs. 13.8 and 13.9, and Figs. 13.10 and 13.11, respectively. As seen by the plots of the head position along the forward direction in Figs. 13.6(a), 13.8(a), and 13.10(a), the overall forward propulsion of the robot was maintained throughout all three trials. This was also the main goal of the experiments. In other words, using the same controller with the same set of controller parameters, the snake robot was able to move through three different obstacle environments, which

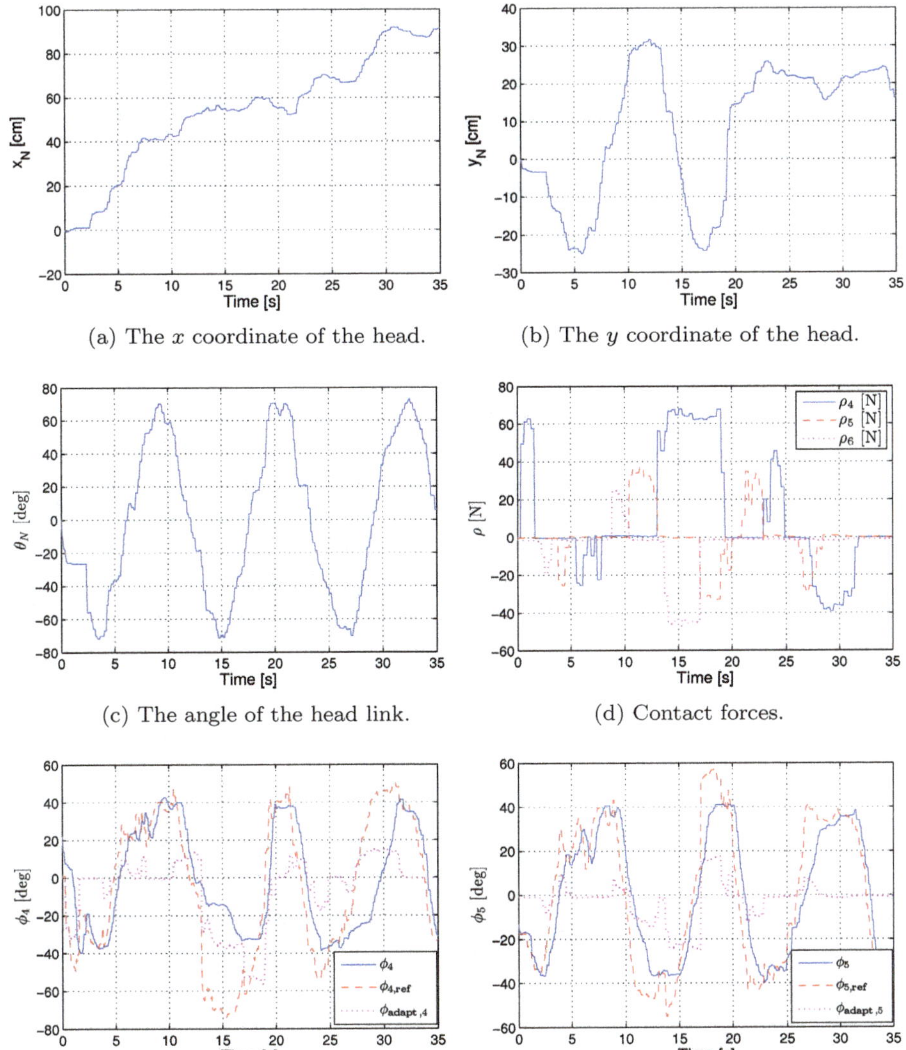

(a) The x coordinate of the head.

(b) The y coordinate of the head.

(c) The angle of the head link.

(d) Contact forces.

(e) The angle of joint 4.

(f) The angle of joint 5.

Fig. 13.6 Experimental results of obstacle-aided locomotion in the first obstacle environment

we consider to be evidence that the proposed control strategy provides a snake robot with environment adaptation skills.

As explained in conjunction with the experimental results of Sect. 12.8, the forward speed of the robot was relatively slow in all trials mainly due to the limited torque of the joints of the snake robot compared to the rather large ground and obstacle friction forces opposing the motion. In particular, the snake robot is rather heavy (about 10 kg) compared to its maximum actuator torque (about 4 Nm). We

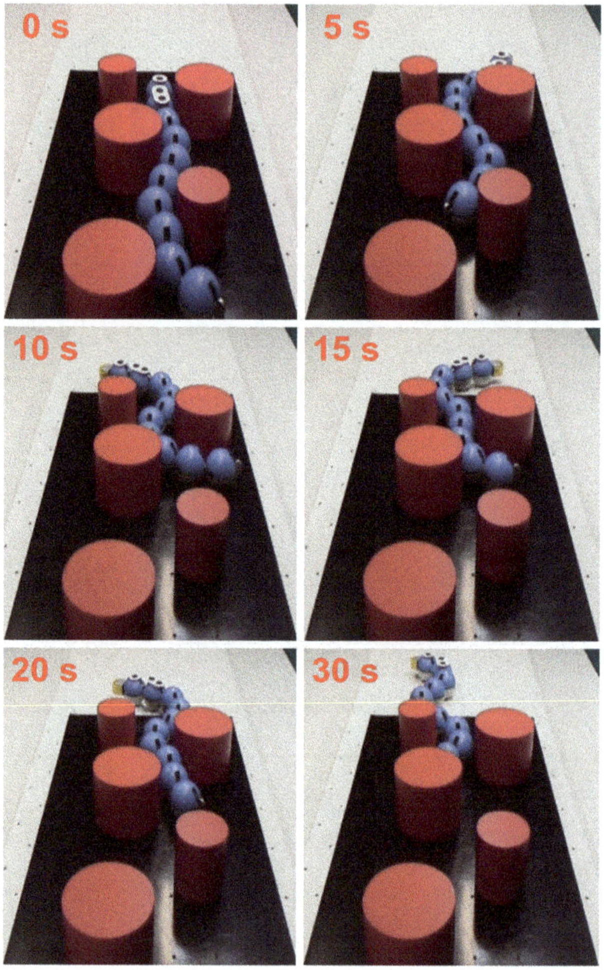

Fig. 13.7 The motion of the snake robot in the first obstacle environment

therefore claim that the limited speed during the experiments was caused by limitations of the physical snake robot and is not a general property of the proposed control strategy. Had the experiments been carried out using a snake robot with a larger actuator strength to weight ratio, then the controller parameters could have been adjusted to increase the forward speed significantly.

To give an idea of the forces needed to propel the robot forward in the three environments, the measured contact forces on joint modules 4, 5, and 6 (module 1 is the tail) are shown in Figs. 13.6(d), 13.8(d), and 13.10(d), respectively. We see that contact forces in the range 30–60 N occurred during the motion.

The motion of two joints of the snake robot during the three trials is shown in Figs. 13.6(e)–(f), 13.8(e)–(f), and 13.10(e)–(f), respectively. These figures show the measured (solid line) and the reference angles (dashed line) of the two joints interconnecting modules 4, 5, and 6, namely joints 4 and 5. In accordance with the

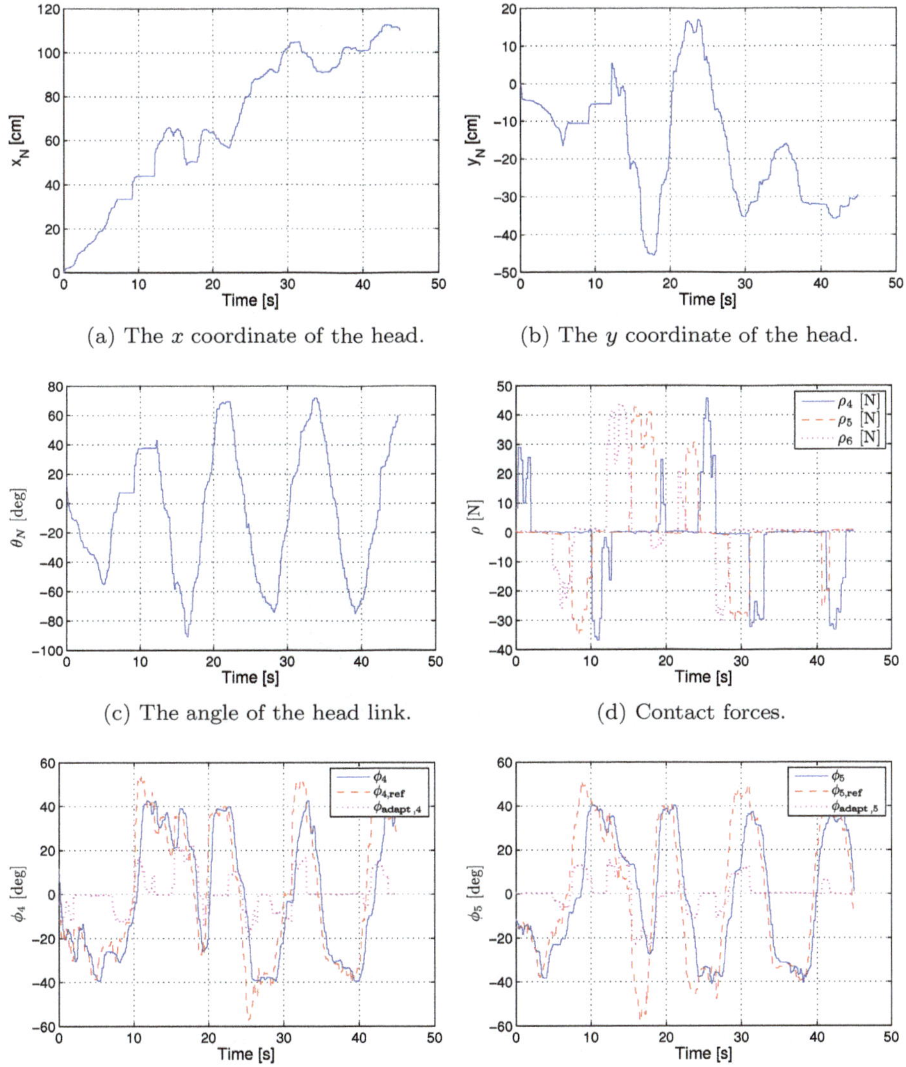

(a) The x coordinate of the head.

(b) The y coordinate of the head.

(c) The angle of the head link.

(d) Contact forces.

(e) The angle of joint 4.

(f) The angle of joint 5.

Fig. 13.8 Experimental results of obstacle-aided locomotion in the second obstacle environment

environment adaptation strategy in (13.29), the measured contact forces on modules 4, 5, and 6 affect the reference angles of joints 4 and 5. To illustrate this effect, we have plotted the environment adaptation components of the reference angles, i.e. $\phi_{\mathrm{adapt},4}$ and $\phi_{\mathrm{adapt},5}$, with a dotted line together with the reference angles. It was clearly observed during the experiments that the adaptation component of the joint reference angles serves as a 'curvature generator' at contacted points along the body of the snake robot. In other words, whenever the snake robot makes contact with an

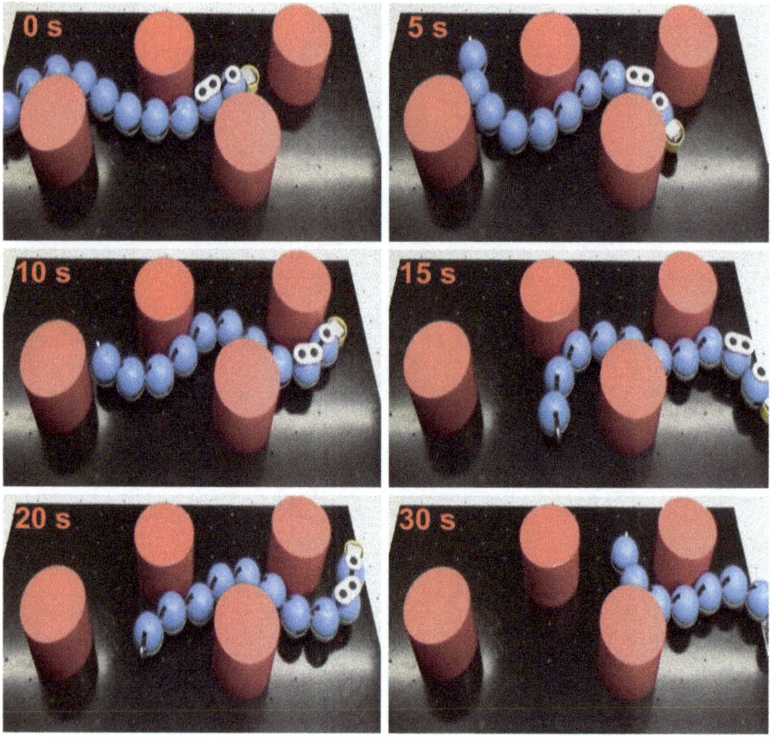

Fig. 13.9 The motion of the snake robot in the second obstacle environment

obstacle, the environment adaptation component in (13.29) produces more body curvature at this location. This curvature is subsequently propagated backwards by the body wave component in (13.28), which generates a push against the contacted obstacle.

It is interesting to compare the experimental results of this section with the experimental results from the jam resolution controller in Sect. 12.8. In particular, the forward velocity of the snake robot in the experimental results from this section is higher than in the experimental results from Sect. 12.8, which can be verified by comparing the plots of the forward position of the robot during the experiments. This difference in forward velocity is not surprising since the jam resolution controller from Chap. 12 spends much time resolving jams that occur during the motion. With the control strategy proposed in this chapter, jams are resolved continuously since there is a continuous adaptation of the body shape to the environment, which is clearly a more efficient strategy than employing an explicit jam resolution scheme. Moreover, the control strategy proposed in this chapter is a lot easier to implement than the jam resolution controller from Chap. 12.

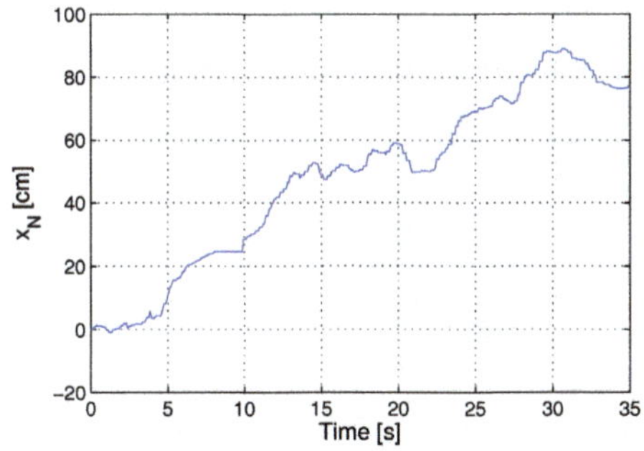

(a) The x coordinate of the head.

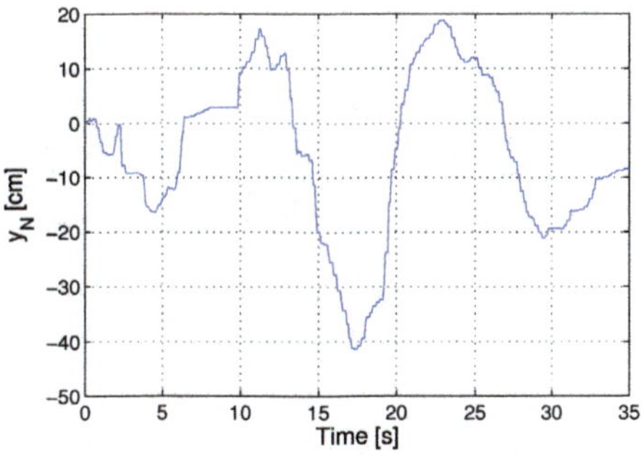

(b) The y coordinate of the head.

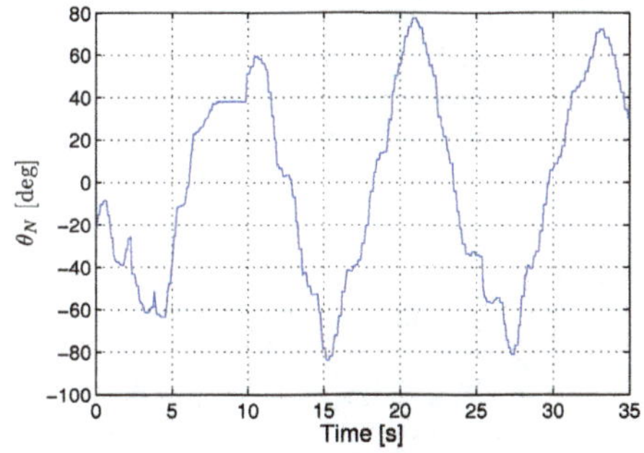

(c) The angle of the head link.

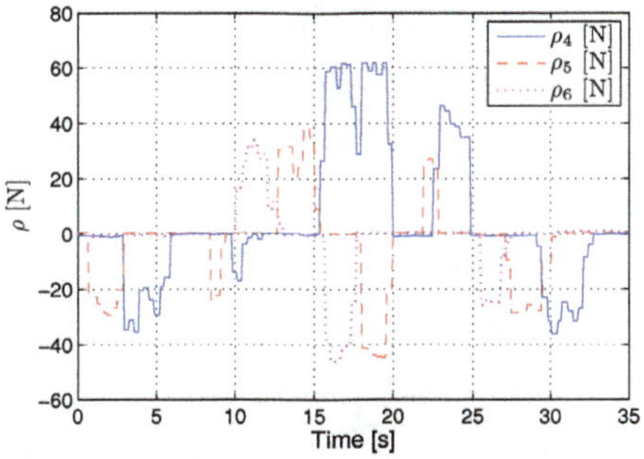

(d) Contact forces.

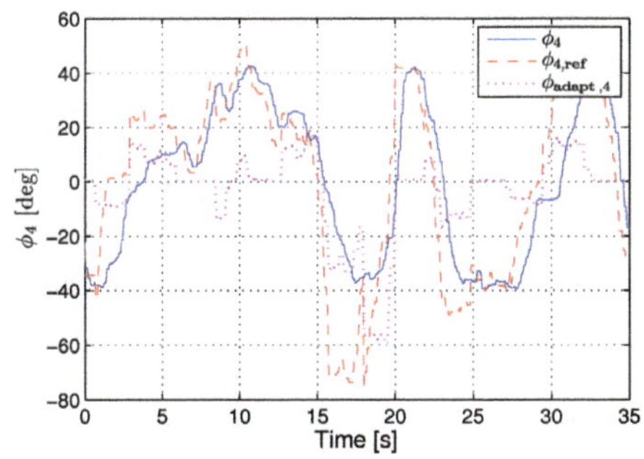

(e) The angle of joint 4.

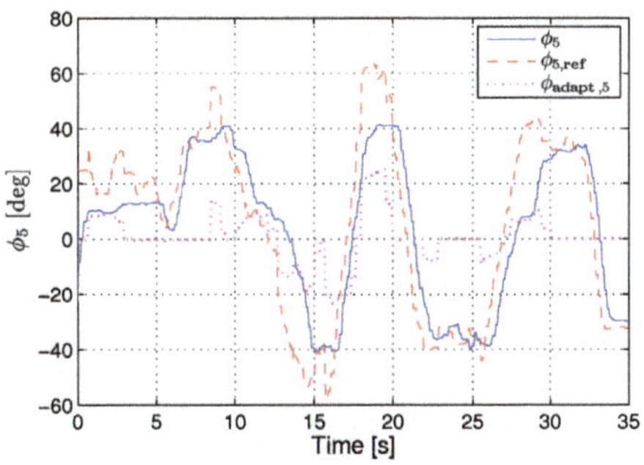

(f) The angle of joint 5.

Fig. 13.10 Experimental results of obstacle-aided locomotion in the third obstacle environment

13.7 Chapter Summary

This chapter is summarised as follows:

- We have proposed a general framework in (13.1) for motion control of snake robots, where the motion is specified in terms of a *body wave* component, an *environment adaptation* component, and a *heading control* component.

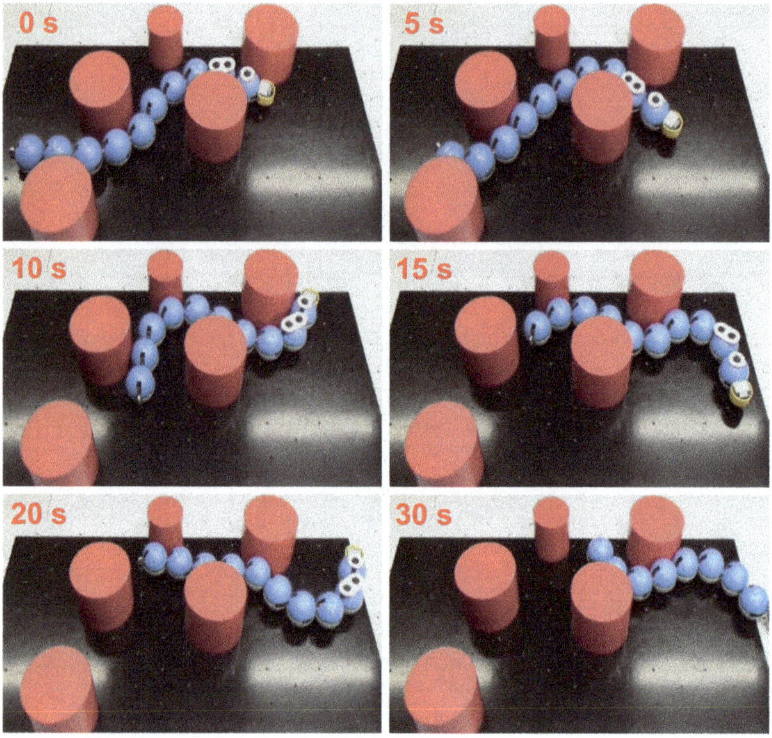

Fig. 13.11 The motion of the snake robot in the third obstacle environment

- We have employed the controller framework to propose a continuous control strategy, defined by (13.26) and (13.27), for straight line path following control of snake robots in cluttered environments.
- We have fitted the path following controller within the waypoint guidance strategy from Chap. 8 (see Algorithm 13.1).
- We have presented simulation results where the path following controller, in combination with the waypoint guidance strategy, was seen to successfully steer the snake robot between waypoints in an obstacle environment.
- We have presented experimental results where the snake robot Kulko was successfully propelled through three different obstacle environments when the joints were controlled according to (13.27).

Chapter 14
Future Research Challenges of Snake Robot Locomotion

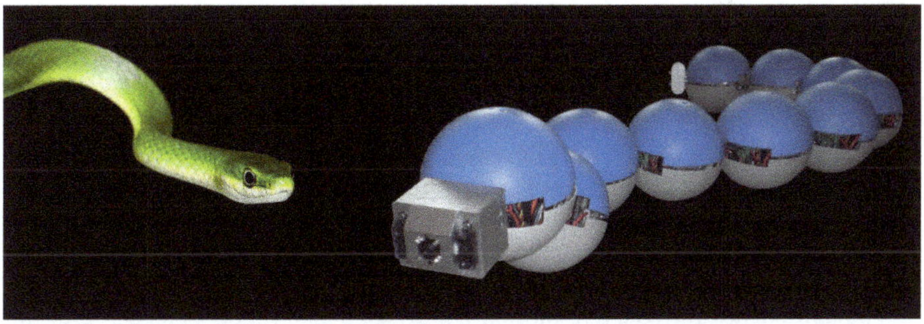

We end this book with a discussion which points out important research challenges that must be addressed before we will ever see useful snake robots outside the laboratory. This book has presented experimental results that demonstrate *planar* snake robot locomotion in cluttered environments based on environment sensing and body shape adaptation. However, to our best knowledge, the fact still remains that *non-planar* (3D) locomotion in cluttered environments based on environment sensing and body shape adaptation has *not* yet been demonstrated. Our primary claim is therefore that future applications of snake robots require significantly more research on adaptive behaviour during motion in unknown and cluttered environments.

14.1 Control Design Challenges

Analysable Mathematical Models Future control design efforts for adaptive motion of snake robots should go beyond pure heuristics and instead base the controllers on analysable mathematical models and well-established control design techniques. This will allow for a more generic and general understanding of the properties and control challenges of snake robots. Model-based control design for

P. Liljebäck et al., *Snake Robots*, Advances in Industrial Control,
DOI 10.1007/978-1-4471-2996-7_14, © Springer-Verlag London 2013

snake robots is, however, a major challenge. As shown in this book, a mathematical model of the dynamics of a snake robot on a flat surface is very complex due to the many degrees of freedom of the robot. When contact forces from a cluttered environment are included, the model becomes even more complex because the discrete nature of the contact forces turns the model of the robot into a hybrid system. However, model-based control design can be achieved by pursuing simplified mathematical descriptions of the interaction between a snake robot and its environment that can be analysed from a control perspective. In particular, a simple relationship between body shape changes of a snake robot during environment contact and the resulting translational and rotational motion of the robot could enable an analytical derivation of the joint torques that will produce the desired motion. The controller development in this book based on the simplified model of snake robot locomotion is an example of how a simplified modelling approach can be employed to derive model-based control strategies for these systems.

Feedback Control Laws Based on Environment Sensing Environment sensing is a requirement for efficient snake robot locomotion in unknown and cluttered environments. The challenge of utilising this sensor information intelligently to maintain the propulsion of the robot is closely related to the challenge of developing analysable models of the robot. With a suitable description of how the environment interaction affects the motion, it is possible to analytically derive the control action that, in a given environment, will propel the robot in a desired direction. Control design for snake robots is also challenging because these mechanisms are generally underactuated, i.e. they have more degrees of freedom than actuators.

The approaches for adaptive motion control of snake robots considered in this book are *local* approaches since the body shape is adjusted locally at each contact point to increase the propulsive force without considering how the adjustment affects the remaining contact forces on the robot. More efficient control strategies are, however, likely to be obtained by considering *global* approaches to environment adaption, i.e. approaches where the body shape adjustments are made by considering the overall interaction of the robot with its environment. In particular, situations where body shape adjustments act 'against' each other are likely to occur with a local approach but may be prevented by a global approach. Such global approaches represent a significant control design challenge.

SLAM Enabling a mobile robot to generate a map of its own environment and simultaneously determine its own position in this map is called *simultaneous localisation and mapping* (SLAM). SLAM represents an extensive and important research area today since such capabilities are generally essential for autonomous operations of mobile robots. SLAM is also very relevant for snake robots since these mechanisms will typically be employed in situations where a map of the environment is not available in advance. To our best knowledge, previous literature has not considered SLAM explicitly in the context of snake robot locomotion.

Although SLAM is, in many ways, independent of the specific propulsion mechanism of the robot, there are features of snake robots which make SLAM for these mechanisms particularly interesting. In particular, most applications of SLAM involve map creation based on different types of vision sensors. While such sensors are also relevant to snake robots, these robots will generally also be able to extract information about their environment from the contact sensing capabilities along their long and slender body. We therefore claim that future research on SLAM for snake robots should focus on map creation by combining data from vision sensors with data from the contact force sensors of the robot. Moreover, since snake robot locomotion is highly dependent on the interaction between the robot and its environment, SLAM for snake robots should not only focus on mapping the specific geometries of the environment, but also on mapping other properties of the environment which influence the motion. In particular, the contact force sensing capabilities of a snake robot can for instance be used to extract information about the friction coefficients of surfaces and objects which come into contact with the snake robot. The elasticity and plasticity of external objects are also important parameters which should be mapped since these parameters determine if an object is completely rigid, or if there is a spring effect in the interaction with the object, or if the object is displaced when the robot makes contact with it.

Motion Planning Strategies With an efficient system for SLAM in place, a snake robot may be able to make intelligent decisions about where to go in order to solve a specific task. Motion planning for snake robots is in many ways similar to motion planning for mobile robots in general. However, the unique features of snake robot locomotion suggest that motion planning for these mechanisms should be attacked with a somewhat different approach than motion planning for more conventional mobile robots. In particular, while *obstacle avoidance* is an important topic for wheeled, tracked, and legged robots, a goal of snake robot locomotion is rather *obstacle utilisation* since objects in the environment of a snake robot represent push points that the robot can use for propulsion. While conventional wheeled, tracked, and legged robots will usually try to find the *shortest* path to a given location, a snake robot should rather seek out the most *efficient* path, which for instance may involve taking small detours in order to reach locations with push points that can be used for propulsion.

14.2 Hardware Design Challenges

Environment Sensing Measuring external contact forces on the snake robot is a natural approach for sensing the environment. The force sensing system of a snake robot is, however, particularly challenging since the robot is articulated, which introduces the challenge of preventing the joint motion from interfering with the measurements of the external forces. Measuring forces directly along the body of a snake robot is, in other words, a significant design challenge. The design of the snake robot *Kulko*, which was presented in this book, represents a proposed solution to this challenge. However, the robot cannot measure external forces very well at the connection point between two modules.

An alternative approach, which was proposed in this book, is to estimate the external forces acting on the robot solely through force measurements at each articulation point along the robot. The instrumentation system of this solution is significantly simpler than the instrumentation required to measure external forces on the robot directly. Further investigations of this approach represent an interesting and important topic of future work.

Research on environment sensing for snake robots is also highly relevant to many other application areas within robotics, which suggests that researchers working with snake robots should identify and pursue synergies with other robotic research areas where environment adaptation is important.

Vision The above discussion regarding SLAM and efficient motion planning strategies suggests that future snake robots need to be equipped with a suitable vision system. Vision for mobile robots is a large and active area of research where progress is continuously being made. Although there has been very limited focus in previous literature on vision specifically for snake robots, the available hardware that can be employed to implement such a system is continuously being improved. Note that a snake robot has limited payload capabilities, which means that the hardware of the vision system should be both small and of limited weight.

Solutions for Untethered Operations In many future applications of snake robots, a tethered connection between the robot and the human operator will not be possible. Consequently, the robot must carry its own power supply and also communicate with some human operator through a wireless connection. The onboard power supply of a snake robot represents a significant design challenge since snake robot locomotion is an energy-demanding form of propulsion, and at the same time, a snake robot will generally have limited payload capabilities. Moreover, since many future applications of snake robots involve motion in environments which are inaccessible by humans, it will usually be impossible to retrieve the robot if it runs out of power before it can make its way back to the human operator.

Operations in inaccessible environments also introduce challenges related to the wireless communication between the snake robot and the human operator. In particular, maintaining a reliable wireless link in such environments may often be difficult. Moreover, in situations where the communication link with the human operator is lost, the snake robot must be able to operate autonomously until the communication link is re-established.

Ground Friction Force Limitation If the propulsion of the snake robot is based on forward gliding motion similar to the motion of biological snakes, then a sufficiently smooth exterior surface is very important since any irregularities along the body may potentially induce large obstructive friction forces on the robot. Obtaining a smooth surface combined with contact force sensing at articulated parts of the robot represents a significant design challenge. The snake robot Kulko presented in this book has a smooth surface due to spherical shells that cover each joint module and can measure external forces due to force sensors mounted underneath the shells.

The friction forces opposing the motion of a snake robot can also be limited by introducing active propulsion along the body. This approach is employed by the snake robot with active tracks presented in Granosik et al. (2006) and by the skin drive mechanism described in McKenna et al. (2008). The drawback of active propulsion along the body of a snake robot is that the mechanical complexity of the robot is significantly increased. In our opinion, the ideal solution is a snake robot with a passive and smooth tactile skin that can glide forward like a biological snake. Mechanism simplicity is important to the future use of snake robots since this increases the reliability and reduces the development cost of the robots.

Robust and Strong Actuation Mechanisms In order to move in challenging environments, the snake robot must generally be able to lift parts of its body. This means that there is some lower limit to the ratio between the strength of the actuators and the weight of the robot. Developing joint mechanisms for snake robots where this ratio is maximised is an important design challenge that must be addressed. Furthermore, locomotion in cluttered environments generally requires that the actuators can work against environment contact forces over time without overheating. The modified servo motor for snake robots described in Wright et al. (2007) targets this design challenge. A compliant joint mechanism is advantageous during locomotion in cluttered environments, which is the motivation for the use of pneumatic actuators in the snake robot with active tracks presented in Granosik et al. (2006). However, compliance can also be enforced by the controller of the robot if the contact forces along the body are measured.

Dustproofing and Waterproofing In order to make use of snake robots outside the generally clean lab environments, the robots must be able to operate despite mud and dirt in their environment. Water resistance is also generally a great advantage. Both dustproofing and waterproofing a snake robot is challenging, in particular when we also require force sensing capabilities and a smooth exterior surface.

Appendix A
Proof of Lemma 8.2

The proof of Lemma 8.2 has previously been presented in Pavlov et al. (2007) and is included here for completeness. The lemma is proved by showing that a quadratic Lyapunov function candidate of the system (8.40) is negative definite, thereby implying that (8.40) is UGAS.

The system (8.40) can be written as

$$\begin{bmatrix} \dot{\overline{p}}_y \\ \dot{\overline{v}}_n \end{bmatrix} = \begin{bmatrix} -\frac{v_t}{\sqrt{\overline{p}_y^2+\Delta^2}} & \frac{\Delta}{\sqrt{\overline{p}_y^2+\Delta^2}} \\ \frac{X\Delta v_t}{(\sqrt{\overline{p}_y^2+\Delta^2})^3} & (Y - \frac{X\Delta^2}{(\sqrt{\overline{p}_y^2+\Delta^2})^3}) \end{bmatrix} \begin{bmatrix} \overline{p}_y \\ \overline{v}_n \end{bmatrix}. \tag{A.1}$$

Consider the quadratic Lyapunov function candidate $V = 1/2\overline{p}_y^2 + \kappa/2\overline{v}_n^2$ with $\kappa > 0$. The derivative of V along the solutions of (A.1) is given by

$$\begin{aligned} \dot{V} &= \overline{p}_y \dot{\overline{p}}_y + \kappa \overline{v}_n \dot{\overline{v}}_n \\ &= -\frac{v_t \overline{p}_y^2}{\sqrt{\overline{p}_y^2 + \Delta^2}} + \frac{\Delta \overline{p}_y \overline{v}_n}{\sqrt{\overline{p}_y^2 + \Delta^2}} + \kappa \frac{X\Delta v_t \overline{p}_y \overline{v}_n}{(\sqrt{\overline{p}_y^2 + \Delta^2})^3} \\ &\quad + \kappa \left(Y - \frac{X\Delta^2}{(\sqrt{\overline{p}_y^2 + \Delta^2})^3} \right) \overline{v}_n^2. \end{aligned} \tag{A.2}$$

Since $v_t \in [V_{\min}, V_{\max}]$ by Assumption 8.1, and since $X \leq |X|$, we can estimate $\dot{V}$ as

$$\begin{aligned} \dot{V} &\leq -\frac{V_{\min} \overline{p}_y^2}{\sqrt{\overline{p}_y^2 + \Delta^2}} + \frac{\Delta \overline{p}_y \overline{v}_n}{\sqrt{\overline{p}_y^2 + \Delta^2}} + \kappa \frac{|X|\Delta V_{\max} \overline{p}_y \overline{v}_n}{(\sqrt{\overline{p}_y^2 + \Delta^2})^3} \\ &\quad + \kappa \left(Y + \frac{|X|\Delta^2}{(\sqrt{\overline{p}_y^2 + \Delta^2})^3} \right) \overline{v}_n^2. \end{aligned} \tag{A.3}$$

P. Liljebäck et al., *Snake Robots*, Advances in Industrial Control,
DOI 10.1007/978-1-4471-2996-7, © Springer-Verlag London 2013

By introducing the variable $z = |\overline{p}_y|/\sqrt{\overline{p}_y^2 + \Delta^2}$, this estimate can be written as

$$\begin{aligned}\dot{V} &\le -V_{\min} z^2 \sqrt{\overline{p}_y^2 + \Delta^2} \\ &\quad + \left(\Delta + \kappa \frac{|X| \Delta V_{\max}}{\overline{p}_y^2 + \Delta^2}\right) z |\overline{v}_n| + \kappa \left(Y + \frac{|X| \Delta^2}{(\sqrt{\overline{p}_y^2 + \Delta^2})^3}\right) \overline{v}_n^2. \end{aligned} \tag{A.4}$$

Finally, using the inequalities $-\sqrt{\overline{p}_y^2 + \Delta^2} \le -\Delta$ and $1/(\overline{p}_y^2 + \Delta^2) \le 1/\Delta^2$, we obtain

$$\dot{V} \le -V_{\min} \Delta z^2 + \left(\Delta + \kappa \frac{|X| V_{\max}}{\Delta}\right) z |\overline{v}_n| + \kappa \left(Y + \frac{|X|}{\Delta}\right) \overline{v}_n^2. \tag{A.5}$$

We now choose $\kappa = \Delta^2 (2\beta - 1)/(|X| V_{\max})$, where

$$\beta = \frac{V_{\min}(-\Delta Y - |X|)}{V_{\max} |X|}. \tag{A.6}$$

It is straightforward to show that condition (8.28) of Theorem 8.2 is equivalent to $\beta > 1$. The chosen value of κ is therefore strictly positive. Substituting this κ into (A.5) gives

$$\begin{aligned}\dot{V} &\le -V_{\min} \Delta z^2 + 2\beta \Delta z |\overline{v}_n| - \frac{\Delta(2\beta - 1)\beta}{V_{\min}} \overline{v}_n^2 \\ &= -\Delta \left(\sqrt{V_{\min}} z - \frac{\beta |\overline{v}_n|}{\sqrt{V_{\min}}}\right)^2 - \frac{\Delta(\beta - 1)\beta}{V_{\min}} \overline{v}_n^2. \end{aligned} \tag{A.7}$$

Finally, substituting the expression for z into this estimate gives

$$\dot{V} \le -\Delta \left(\frac{\sqrt{V_{\min}} |\overline{p}_y|}{\sqrt{\overline{p}_y^2 + \Delta^2}} - \frac{\beta |\overline{v}_n|}{\sqrt{V_{\min}}}\right)^2 - \frac{\Delta(\beta - 1)\beta}{V_{\min}} \overline{v}_n^2. \tag{A.8}$$

Since condition (8.28) guarantees that $\beta > 1$, we can conclude that $\dot{V} < 0$, which implies that the origin of the system (8.40) is UGAS (see Khalil 2002). This completes the proof of Lemma 8.2.

Appendix B
Proof of Lemma 8.3

The lemma is proved by showing that system (8.40) is ULES (uniformly locally exponentially stable), which, together with the UGAS property established by Lemma 8.2, implies that (8.40) is globally $\mathcal{K}$-exponentially stable according to Corollary 8.1.

The linearisation of system (8.40) about the origin is easily calculated as

$$\begin{bmatrix} \dot{\overline{p}}_y \\ \dot{\overline{v}}_n \end{bmatrix} = \begin{bmatrix} -\frac{v_t}{\Delta} & 1 \\ \frac{X v_t}{\Delta^2} & Y - \frac{X}{\Delta} \end{bmatrix} \begin{bmatrix} \overline{p}_y \\ \overline{v}_n \end{bmatrix}. \tag{B.1}$$

Denoting the system matrix of (B.1) by $\mathbf{W}$, we can calculate the eigenvalues of $\mathbf{W}$ from its characteristic equation

$$\lambda^2 - \mathrm{tr}(\mathbf{W})\lambda + \det(\mathbf{W}) = 0, \tag{B.2}$$

where $\mathrm{tr}(\mathbf{W})$ and $\det(\mathbf{W})$ are the trace and the determinant of $\mathbf{W}$, respectively. $\mathbf{W}$ is Hurwitz (see e.g. Khalil 2002) if the coefficients of this characteristic equation are strictly positive, i.e. if $\mathrm{tr}(\mathbf{W}) < 0$ and $\det(\mathbf{W}) > 0$. Since $v_t > 0$, $Y < 0$, and $\Delta > 2|X|/|Y|$ (this follows from (8.28)), the trace of $\mathbf{W}$ satisfies

$$\begin{aligned} \mathrm{tr}(\mathbf{W}) &= -\frac{v_t}{\Delta} + Y - \frac{X}{\Delta} \leq -\frac{v_t}{\Delta} - |Y| + \frac{|X|}{\Delta} \\ &\leq -\frac{v_t}{\Delta} - |Y| + \frac{1}{2}|Y| = -\frac{v_t}{\Delta} - \frac{1}{2}|Y| < 0, \end{aligned} \tag{B.3}$$

and the determinant of $\mathbf{W}$ satisfies

$$\det(\mathbf{W}) = -\frac{v_t}{\Delta}\left(Y - \frac{X}{\Delta}\right) - \frac{X v_t}{\Delta^2} = -\frac{v_t}{\Delta} Y > 0. \tag{B.4}$$

The system matrix $\mathbf{W}$ of the linearised system (B.1) is therefore Hurwitz, which implies that the origin of system (8.40) is ULES (see Khalil 2002, Corollary 4.3). Since, by Lemma 8.2, the origin of (8.40) is also UGAS, Corollary 8.1 implies that the origin of (8.40) is globally $\mathcal{K}$-exponentially stable. This completes the proof of Lemma 8.3.

P. Liljebäck et al., *Snake Robots*, Advances in Industrial Control,
DOI 10.1007/978-1-4471-2996-7, © Springer-Verlag London 2013

Appendix C
Low-Pass Filtering Reference Models

In order to ensure that the state reference of a control system complies with the dynamical capabilities of the system, the commanded state of the system can be passed through a low-pass filter. The filter, which is called a *reference model*, keeps the output from the control system within the physical capabilities of the actuators of the system. In addition to the filtered state reference, the output from the filter typically also includes the derivatives of the state reference with respect to time, which are often needed in the control law of the system. A low-pass filtering reference model can for example be used in combination with PD-control of the angle of a mechanical joint. Since the derivative part of the PD-controller includes the derivative of the reference angle with respect to time, any steps in the reference angle would make the output from the PD-controller infinitely large. This infinite control output is avoided by using a reference model to low-pass filter the reference angle.

In the following, we present a *2nd-order* and a *3rd-order low-pass filtering reference model* for a control system with a single scalar state value. The commanded state reference is denoted by $r \in \mathbb{R}$, and the filtered state reference is denoted by $x_{\text{ref}} \in \mathbb{R}$. In other words, we want the system to reach the state r, but we apply the filtered state reference x_{ref} to the controller in order to ensure that the system is brought to r through smooth control actions. For control systems with multiple states, the reference model can be applied individually to each state. The following material is based on Chap. 5 in Fossen (2002).

C.1 A 2nd-Order Low-Pass Filtering Reference Model

A reference model of 2nd-order ensures smooth reference signals for x_{ref} and $\dot{x}_{\text{ref}}$, but not for $\ddot{x}_{\text{ref}}$. We define the reference model as a *mass-damper-spring* system with the transfer function

$$\frac{x_{\text{ref}}}{r} = \frac{\omega^2}{s^2 + 2\zeta\omega s + \omega^2}, \tag{C.1}$$

P. Liljebäck et al., *Snake Robots*, Advances in Industrial Control,
DOI 10.1007/978-1-4471-2996-7, © Springer-Verlag London 2013

where $\omega > 0$ and $\zeta > 0$ are, respectively, the *natural frequency* and the *relative damping ratio* of the mass-damper-spring system. The reference model, which satisfies

$$\lim_{t\to\infty} x_{\text{ref}} = r, \tag{C.2}$$

can be written

$$\ddot{x}_{\text{ref}} + 2\zeta\omega\dot{x}_{\text{ref}} + \omega^2 x_{\text{ref}} = \omega^2 r \tag{C.3}$$

and can be implemented in a control system by defining the state vector

$$\mathbf{w} = \begin{bmatrix} x_{\text{ref}} \\ \dot{x}_{\text{ref}} \end{bmatrix} \tag{C.4}$$

and calculating the dynamics of this state according to

$$\dot{\mathbf{w}} = \begin{bmatrix} 0 & 1 \\ -\omega^2 & -2\zeta\omega \end{bmatrix} \mathbf{w} + \begin{bmatrix} 0 \\ \omega^2 \end{bmatrix} r. \tag{C.5}$$

C.2 A 3rd-Order Low-Pass Filtering Reference Model

In order to ensure that x_{ref}, $\dot{x}_{\text{ref}}$, and $\ddot{x}_{\text{ref}}$ are all sufficiently smooth, the filter should be of 3rd order. This is achieved by cascading the mass-damper-spring system (C.1) with a 1st-order low-pass filter, which gives the transfer function

$$\frac{x_{\text{ref}}}{r} = \frac{\omega^2}{(1+Ts)(s^2+2\zeta\omega s+\omega^2)}, \tag{C.6}$$

where $T = 1/\omega$ is the time constant of the 1st-order low-pass filter. This transfer function can be written

$$\frac{x_{\text{ref}}}{r} = \frac{\omega^3}{s^3 + (2\zeta+1)\omega s^2 + (2\zeta+1)\omega^2 s + \omega^3}. \tag{C.7}$$

The reference model, which satisfies

$$\lim_{t\to\infty} x_{\text{ref}} = r, \tag{C.8}$$

can be written

$$x_{\text{ref}}^{(3)} + (2\zeta+1)\omega\ddot{x}_{\text{ref}} + (2\zeta+1)\omega^2\dot{x}_{\text{ref}} + \omega^3 x_{\text{ref}} = \omega^3 r \tag{C.9}$$

and can be implemented in a control system by defining the state vector

$$\mathbf{w} = \begin{bmatrix} x_{\text{ref}} \\ \dot{x}_{\text{ref}} \\ \ddot{x}_{\text{ref}} \end{bmatrix} \tag{C.10}$$

and calculating the dynamics of this state according to

$$\dot{\mathbf{w}} = \begin{bmatrix} 0 & 1 & 0 \\ 0 & 0 & 1 \\ -\omega^3 & -(2\zeta+1)\omega^2 & -(2\zeta+1)\omega \end{bmatrix} \mathbf{w} + \begin{bmatrix} 0 \\ 0 \\ \omega^3 \end{bmatrix} r. \tag{C.11}$$

Glossary[1]

Anisotropic ground friction properties Properties of the links of the snake robot such that the friction coefficient describing the ground friction force parallel to and perpendicular to a link, respectively, are different. Such properties are advantageous for efficient snake robot locomotion.

Environment adaptation The use of environment sensing information in the control system of the snake robot in order to adapt the motion of the robot to its environment in some appropriate manner.

Environment sensing The use of sensors to extract information about the interaction between the snake robot and its environment.

Flat surface An environment which is completely flat without any protrusions. The majority of previous research considers motion of snake robots on flat surfaces.

Flat surface locomotion Snake robot locomotion across a flat surface. Note that this is not the same as planar locomotion.

Head link The foremost link of the snake robot. This is link number N of a snake robot consisting of N links.

Heading-adjusted link angle The orientation of a link with respect to the current heading of the snake robot. The heading is always referred to some global coordinate system.

Isotropic ground friction properties The opposite of anisotropic ground friction properties, i.e. properties of the links of the snake robot such that the ground friction force on each link is independent of the link orientation with respect to its direction of motion.

Joint actuator The mechanism which induces a controlled relative motion between two interconnected links (usually an electrical motor). A planar snake robot with N links will generally have $N - 1$ joint actuators.

Joint angle The relative angle between two interconnected links.

[1]The following glossary lists frequently used terms and phrases along with their corresponding meaning in this book.

P. Liljebäck et al., *Snake Robots*, Advances in Industrial Control,
DOI 10.1007/978-1-4471-2996-7,

Lateral undulation The fastest and most common form of snake locomotion. The gait pattern consists of continuous body waves that are propagated backwards from the head to the tail in order to push the robot forward.

Link angle The orientation of a link with respect to some global coordinate system.

Obstacle An object or an irregular surface in the environment of the snake robot.

Obstacle constraint force The component of an obstacle contact force in the direction perpendicular to the link.

Obstacle contact force The force acting on a link due to the interaction between the link and an obstacle.

Obstacle detachment The event when the contact between a link of the snake robot and an obstacle is broken.

Obstacle friction force The component of an obstacle contact force in the direction parallel to the link.

Obstacle impact The event when a link of the snake robot comes into contact with an obstacle.

Obstacle-aided locomotion Locomotion where a snake robot is propelled forward due to interaction with obstacles in its environment.

Planar locomotion Snake robot locomotion which only consists of link motion components in some common plane (typically purely horizontal motion where none of the links are lifted from the ground).

Planar snake robot A snake robot which only displays planar locomotion.

Snake robot A robotic mechanism designed to move like a biological snake.

Snake robot joint The mechanical connection between two links.

Snake robot link A rigid body segment of the snake robot with a joint in one or both ends.

Snake robot locomotion A general term denoting the motion of a snake robot during forward progression.

Tail link The backmost link of the snake robot. This is link number 1.

Uneven and cluttered environment Any environment with geometries beyond merely a flat surface (i.e. an environment with obstacles).

Unilateral velocity constraint A mathematical motion constraint on a link which prevents sideways motion in one direction but allows sideways motion in the opposite direction. The interaction between a link and an obstacle can be modelled mathematically with a unilateral velocity constraint.

Wheel-less snake robot locomotion Snake robot locomotion which is *not* based on passive wheels mounted along the body of the robot.

Wheeled snake robot locomotion Locomotion where a snake robot is propelled forward due to passive (not motorised) wheels mounted along its body. Passive wheels represent a simple way of implementing anisotropic ground friction properties on a physical snake robot and will usually introduce sideslip constraints on the links, i.e. constraints which prevent the links from moving sideways.

References

Andruska, A.M., Peterson, K.S.: Control of a snake-like robot in an elastically deformable channel. IEEE/ASME Trans. Mechatron. **13**(2), 219–227 (2008)

Bauchot, R.: Snakes: a Natural History. Sterling Publishing Company, New York (1994)

Bayraktaroglu, Z.Y.: Snake-like locomotion: experimentations with a biologically inspired wheel-less snake robot. Mech. Mach. Theory **44**(3), 591–602 (2008)

Bayraktaroglu, Z.Y., Blazevic, P.: Understanding snakelike locomotion through a novel push-point approach. J. Dyn. Syst. Meas. Control **127**(1), 146–152 (2005)

Bernstein, D.S.: Matrix Mathematics: Theory, Facts, and Formulas, 2nd edn. Princeton University Press, Princeton (2009)

Bianchini, R.M., Stefani, G.: Graded approximations and controllability along a trajectory. SIAM J. Control Optim. **28**(4), 903–924 (1990)

Bloch, A.M., Baillieul, J., Crouch, P., Marsden, J.: Nonholonomic Mechanics and Control. Springer, Berlin (2003)

Børhaug, E.: Nonlinear control and synchronization of mechanical systems. Ph.D. thesis, Norwegian University of Science and Technology (2008)

Boyer, F., Porez, M., Khalil, W.: Macro-continuous computed torque algorithm for a three-dimensional eel-like robot. IEEE Trans. Robot. **22**(4), 763–775 (2006)

Brockett, R.: Asymptotic stability and feedback stabilization. In: Differential Geometric Control Theory, pp. 181–191 (1983)

Brogliato, B.: Nonsmooth Mechanics, 2nd edn. Springer, London (1999)

Brunete, A., Gambao, E., Torres, J.E., Hernando, M.: A 2 DoF servomotor-based module for pipe inspection modular micro-robots. In: IEEE/RSJ Int. Conf. Intelligent Robots and Systems, pp. 1329–1334 (2006)

Burdick, J.W., Radford, J., Chirikjian, G.S.: A sidewinding locomotion gait for hyper-redundant robots. Adv. Robot. **9**(3), 195–216 (1995)

Chen, L., Wang, Y., Ma, S., Li, B.: Studies on lateral rolling locomotion of a snake robot. In: Proc. IEEE Int. Conf. Robotics and Automation, pp. 5070–5074 (2004)

Chen, L., Wang, Y., Li, B., Ma, S., Duan, D.: Study on locomotion of a crawling robot for adaptation to the environment. In: Bioinspiration and Robotics: Walking and Climbing Robots, pp. 301–316. I-Tech Education and Publishing, Vienna (2007)

Chen, T.L.T., Liu, S., Yen, J.: A bio-mimetic snake-like robot: sensor based gait control. In: IEEE Workshop on Advanced Robotics and Its Social Impacts, 2008. ARSO 2008, pp. 1–6 (2008)

Chernousko, F.: Snake-like locomotions of multilink mechanisms. J. Vib. Control **9**(1–2), 235–256 (2003)

Chernousko, F.: Modelling of snake-like locomotion. Appl. Math. Comput. **164**(2), 415–434 (2005)

P. Liljebäck et al., *Snake Robots*, Advances in Industrial Control,
DOI 10.1007/978-1-4471-2996-7,

Chirikjian, G.S.: Theory and applications of hyper-redundant robotic manipulators. Ph.D. thesis, California Institute of Technology, Pasadena, CA (1992)

Chirikjian, G.S.: Design and analysis of some nonanthropomorphic, biologically inspired robots: an overview. J. Robot. Syst. **18**(12), 701–713 (2001)

Chirikjian, G.S., Burdick, J.W.: The kinematics of hyper-redundant robot locomotion. IEEE Trans. Robot. Autom. **11**(6), 781–793 (1995)

Coron, J.-M., Rosier, L.: A relation between continuous time-varying and discontinuous feedback stabilization. J. Math. Syst. Estim. Control **4**(1), 67–84 (1994)

Cottle, R.W., Pang, J.S., Stone, R.E.: The Linear Complementarity Problem. Academic Press, San Diego (1992)

Crespi, A., Ijspeert, A.J.: Online optimization of swimming and crawling in an amphibious snake robot. IEEE Trans. Robot. **24**(1), 75–87 (2008)

Date, H., Takita, Y.: Control of 3D snake-like locomotive mechanism based on continuum modeling. In: Proc. ASME 2005 Int. Design Engineering Technical Conf., pp. 1351–1359 (2005)

Date, H., Takita, Y.: Adaptive locomotion of a snake like robot based on curvature derivatives. In: Proc. IEEE/RSJ Int. Conf. Intelligent Robots and Systems, San Diego, CA, USA, pp. 3554–3559 (2007)

Date, H., Hoshi, Y., Sampei, M.: Locomotion control of a snake-like robot based on dynamic manipulability. In: Proc. IEEE/RSJ Int. Conf. Intelligent Robots and Systems (2000)

Date, H., Hoshi, Y., Sampei, M., Shigeki, N.: Locomotion control of a snake robot with constraint force attenuation. In: Proc. American Control Conf., pp. 113–118 (2001a)

Date, H., Sampei, M., Nakaura, S.: Control of a snake robot in consideration of constraint force. In: Proc. IEEE Int. Conf. Control Applications, pp. 966–971 (2001b)

Do, K.D., Pan, J.: Global tracking control of underactuated ships with off-diagonal terms. In: IEEE Conf. Decision and Control, vol. 2, pp. 1250–1255 (2003)

Dowling, K.: Limbless locomotion: learning to crawl. In: Proc. IEEE Int. Conf. Robotics and Automation, vol. 4 (1999)

Dowling, K.J.: Limbless locomotion: learning to crawl with a snake robot. Ph.D. thesis, The Robotics Institute, Carnegie Mellon University (1997)

Egeland, O., Gravdahl, J.T.: Modeling and Simulation for Automatic Control. Marine Cybernetics, Trondheim (2002)

Endo, G., Togawa, K., Hirose, S.: Study on self-contained and terrain adaptive active cord mechanism. In: Proc. IEEE/RSJ Int. Conf. Intelligent Robots and Systems, vol. 3, pp. 1399–1405 (1999)

EUROP/CARE: Robotic visions to 2020 and beyond—the strategic research agenda for robotics in Europe. Technical report (2009)

Fjerdingen, S.A., Mathiassen, J.R., Schumann-Olsen, H., Kyrkjebø, E.: Adaptive snake robot locomotion: a benchmarking facility for experiments. In: European Robotics Symp. 2008, vol. 44, pp. 13–22 (2008)

Fjerdingen, S.A., Liljebäck, P., Transeth, A.A.: A snake-like robot for internal inspection of complex pipe structures (PIKo). In: Proc. IEEE/RSJ Int. Conf. Intelligent Robots and Systems, St. Louis, MO, USA, pp. 5665–5671 (2009)

Fossen, T.I.: Marine Control Systems: Guidance, Navigation and Control of Ships, Rigs and Underwater Vehicles. Marine Cybernetics, Trondheim (2002)

Fredriksen, E., Pettersen, K.Y.: Global κ-exponential way-point maneuvering of ships: theory and experiments. Automatica **42**, 677–687 (2006)

Gao, J., Gao, X., Zhu, W., Zhu, J., Wei, B.: Design and research of a new structure rescue snake robot with all body drive system. In: IEEE Int. Conf. Mechatronics and Automation, pp. 119–124 (2008)

Goebel, R., Sanfelice, R., Teel, A.: Hybrid dynamical systems. IEEE Control Syst. Mag. **29**(2), 28–93 (2009)

Goldstein, H., Poole, C., Safko, J.: Classical Mechanics, 3rd edn. Addison-Wesley, Reading (2002)

Gonzalez-Gomez, J., Zhang, H., Boemo, E.: Locomotion principles of 1D topology pitch and pitch-yaw-connecting modular robots. In: Bioinspiration and Robotics: Walking and Climbing Robots, pp. 403–428. I-Tech Education and Publishing, Vienna (2007)

Gonzalez-Gomez, J., Gonzalez-Quijano, J., Zhang, H., Abderrahim, M.: Toward the sense of touch in snake modular robots for search and rescue operations. In: Proc. ICRA 2010 Workshop "Modular Robots: State of the Art", pp. 63–68 (2010)

Grabec, I.: Control of a creeping snake-like robot. In: Proc. 7th Int. Workshop on Advanced Motion Control, pp. 526–531 (2002)

Granosik, G., Borenstein, J., Hansen, M.G.: Serpentine robots for industrial inspection and surveillance. In: Industrial Robotics: Programming, Simulation and Applications, pp. 633–662. Pro Literatur Verlag, Mammendorf (2006)

Gray, J.: The mechanism of locomotion in snakes. J. Exp. Biol. **23**(2), 101–120 (1946)

Greenfield, A., Rizzi, A.A., Choset, H.: Dynamic ambiguities in frictional rigid-body systems with application to climbing via bracing. In: Proc. IEEE Int. Conf. Robotics and Automation, pp. 1947–1952 (2005)

Gu, Y.-L., Xu, Y.: A normal form augmentation approach to adaptive control of space robot systems. In: Proc. IEEE Int. Conf. Robotics and Automation, vol. 2, pp. 731–737 (1993)

Hara, M., Satomura, S., Fukushima, H., Kamegawa, T., Igarashi, H., Matsuno, F.: Control of a snake-like robot using the screw drive mechanism. In: IEEE Int. Conf. Robotics and Automation, pp. 3883–3888 (2007)

Harville, D.A.: Matrix Algebra from a Statistician's Perspective. Springer, Berlin (2000)

Hatton, R.L., Choset, H.: Approximating displacement with the body velocity integral. In: Proc. Robotics: Science and Systems (2009a)

Hatton, R.L., Choset, H.: Generating gaits for snake robots by annealed chain fitting and keyframe wave extraction. In: Proc. IEEE/RSJ Int. Conf. Intelligent Robots and Systems, pp. 840–845 (2009b)

Hatton, R.L., Choset, H.: Sidewinding on slopes. In: IEEE Int. Conf. Robotics and Automation, pp. 691–696 (2010)

Hicks, G., Ito, K.: A method for determination of optimal gaits with application to a snake-like serial-link structure. IEEE Trans. Autom. Control **50**(9), 1291–1306 (2005)

Hicks, G.P.: Modeling and control of a snake-like serial-link structure. Ph.D. thesis, North Carolina State University (2003)

Hirose, S.: Biologically Inspired Robots: Snake-Like Locomotors and Manipulators. Oxford University Press, Oxford (1993)

Hu, D., Nirody, J., Scott, T., Shelley, M.: The mechanics of slithering locomotion. In: Proc. National Academy of Sciences, USA, vol. 106, pp. 10081–10085 (2009)

Ijspeert, A.J., Crespi, A., Ryczko, D., Cabelguen, J.-M.: From swimming to walking with a salamander robot driven by a spinal cord model. Science **315**(5817), 1416–1420 (2007)

Ishikawa, M.: Iterative feedback control of snake-like robot based on principal fiber bundle modeling. Int. J. Adv. Mechatron. Syst. **1**(3), 175–182 (2009)

Ishikawa, M., Owaki, K., Shinagawa, M., Sugie, T.: Control of snake-like robot based on nonlinear controllability analysis. In: IEEE Int. Conf. Control Applications, pp. 1134–1139 (2010)

Jones, B.A., Walker, I.D.: Kinematics for multisection continuum robots. IEEE Trans. Robot. **22**(1), 43–55 (2006)

Kamegawa, T., Yarnasaki, T., Igarashi, H., Matsuno, F.: Development of the snake-like rescue robot 'Kohga'. In: Proc. IEEE Int. Conf. Robotics and Automation, vol. 5, pp. 5081–5086 (2004)

Kamegawa, T., Harada, T., Gofuku, A.: Realization of cylinder climbing locomotion with helical form by a snake robot with passive wheels. In: Proc. IEEE Int. Conf. Robotics and Automation, pp. 3067–3072 (2009)

Kane, T.R., Lecison, D.A.: Locomotion of snakes: a mechanical 'explanation'. Int. J. Solids Struct. **37**(41), 5829–5837 (2000)

Kanso, E., Marsden, J.E., Rowley, C.W., Melli-Huber, J.B.: Locomotion of articulated bodies in a perfect fluid. J. Nonlinear Sci. **15**, 255–289 (2005)

Kelly, S.D., Murray, R.M.: Geometric phases and robotic locomotion. J. Robot. Syst. **12**(6), 417–431 (1995)

Khalil, H.K.: Nonlinear Systems, 3rd edn. Prentice Hall, New York (2002)

Kimura, H., Hirose, S.: Development of Genbu: active wheel passive joint articulated mobile robot. In: IEEE/RSJ Int. Conf. Intelligent Robots and Systems, vol. 1, pp. 823–828 (2002)

Krishnaprasad, P.S., Tsakiris, D.P.: G-snakes: nonholonomic kinematic chains on lie groups. In: Proc. 33rd IEEE Conf. Decision and Control, Lake Buena Vista, FL, USA, vol. 3, pp. 2955–2960 (1994)

Kulali, G., Gevher, M., Erkmen, A., Erkmen, I.: Intelligent gait synthesizer for serpentine robots. In: Proc. IEEE Int. Conf. Robotics and Automation, vol. 2, pp. 1513–1518 (2002)

Kuwada, A., Wakimoto, S., Suzumori, K., Adomi, Y.: Automatic pipe negotiation control for snake-like robot. In: Proc. IEEE/ASME Int. Conf. Advanced Intelligent Mechatronics, pp. 558–563 (2008)

Lefeber, E.: Tracking control of nonlinear mechanical systems. Ph.D. thesis, Department of Applied Mathematics, University of Twente, The Netherlands (2000)

Li, J., Shan, J.: Passivity control of underactuated snake-like robots. In: Proc. 7th World Congress on Intelligent Control and Automation, pp. 485–490 (2008)

Liljebäck, P., Pettersen, K.Y.: Waypoint guidance control of snake robots. In: Proc. IEEE Int. Conf. Robotics and Automation, Shanghai, China, pp. 937–944 (2011)

Liljebäck, P., Stavdahl, Ø., Pettersen, K.Y.: Modular pneumatic snake robot: 3D modelling, implementation and control. In: Proc. 16th IFAC World Congress, Prague, Czech Republic (2005)

Liljebäck, P., Stavdahl, Ø., Beitnes, A.: SnakeFighter—development of a water hydraulic fire fighting snake robot. In: Proc. IEEE Int. Conf. Control, Automation, Robotics, and Vision (ICARCV), Singapore (2006)

Liljebäck, P., Stavdahl, Ø., Pettersen, K.Y.: Modular pneumatic snake robot: 3D modelling, implementation and control. Model. Identif. Control **29**(1), 21–28 (2008)

Liljebäck, P., Fjerdingen, S., Pettersen, K.Y., Stavdahl, Ø.: A snake robot joint mechanism with a contact force measurement system. In: Proc. IEEE Int. Conf. Robotics and Automation, Kobe, Japan, pp. 3815–3820 (2009a)

Liljebäck, P., Pettersen, K.Y., Stavdahl, Ø.: Modelling and control of obstacle-aided snake robot locomotion based on jam resolution. In: Proc. IEEE Int. Conf. Robotics and Automation, Kobe, Japan, pp. 3807–3814 (2009b)

Liljebäck, P., Pettersen, K.Y., Stavdahl, Ø., Gravdahl, J.T.: Controllability analysis of planar snake robots influenced by viscous ground friction. In: Proc. IEEE/RSJ Int. Conf. Intelligent Robots and Systems, St. Louis, MO, USA, pp. 3615–3622 (2009c)

Liljebäck, P., Pettersen, K.Y., Stavdahl, Ø., Gravdahl, J.T.: Stability analysis of snake robot locomotion based on Poincaré maps. In: Proc. IEEE/RSJ Int. Conf. Intelligent Robots and Systems, St. Louis, MO, USA, pp. 3623–3630 (2009d)

Liljebäck, P., Haugstuen, I.U., Pettersen, K.Y.: Experimental investigation of a path following controller for planar snake robots. In: Proc. IEEE Int. Conf. Control, Automation, Robotics, and Vision (ICARCV), Singapore, pp. 2325–2332 (2010a)

Liljebäck, P., Haugstuen, I.U., Pettersen, K.Y.: Path following control of planar snake robots using a cascaded approach. In: Proc. IEEE Conf. Decision and Control, Atlanta, GA, USA, pp. 1969–1976 (2010b)

Liljebäck, P., Pettersen, K.Y., Stavdahl, Ø.: A snake robot with a contact force measurement system for obstacle-aided locomotion. In: Proc. IEEE Int. Conf. Robotics and Automation, Anchorage, AK, USA, pp. 683–690 (2010c)

Liljebäck, P., Pettersen, K.Y., Stavdahl, Ø., Gravdahl, J.T.: Stability analysis of snake robot locomotion based on averaging theory. In: Proc. IEEE Conf. Decision and Control, Atlanta, GA, USA, pp. 1977–1984 (2010d)

Liljebäck, P., Pettersen, K.Y., Stavdahl, Ø., Gravdahl, J.T.: Experimental investigation of fundamental properties of snake robot locomotion. In: Proc. IEEE Int. Conf. Control, Automation, Robotics, and Vision (ICARCV), Singapore, pp. 187–194 (2010e). Finalist for the Best Paper Award

Liljebäck, P., Pettersen, K.Y., Stavdahl, Ø., Gravdahl, J.T.: Fundamental properties of snake robot locomotion. In: Proc. IEEE/RSJ Int. Conf. Intelligent Robots and Systems, Taipei, Taiwan, pp. 2876–2883 (2010f)

Liljebäck, P., Pettersen, K.Y., Stavdahl, Ø., Gravdahl, J.T.: A hybrid model of obstacle-aided snake robot locomotion. In: Proc. IEEE Int. Conf. Robotics and Automation, Anchorage, AK, USA, pp. 675–682 (2010g)

Liljebäck, P., Pettersen, K.Y., Stavdahl, Ø., Gravdahl, J.T.: A simplified model of planar snake robot locomotion. In: Proc. IEEE/RSJ Int. Conf. Intelligent Robots and Systems, Taipei, Taiwan, pp. 2868–2875 (2010h)

Liljebäck, P., Pettersen, K.Y., Stavdahl, Ø., Gravdahl, J.T.: Hybrid modelling and control of obstacle-aided snake robot locomotion. IEEE Trans. Robot. **26**(5), 781–799 (2010i)

Liljebäck, P., Stavdahl, Ø., Pettersen, K.Y., Gravdahl, J.T.: Two new design concepts for snake robot locomotion in unstructured environments. Paladyn J. Behav. Robot. **1**(3), 154–159 (2010j)

Liljebäck, P., Haugstuen, I.U., Pettersen, K.Y.: Path following control of planar snake robots using a cascaded approach. IEEE Trans. Control Syst. Technol. **20**, 111–126 (2011a). doi:10.1109/TCST.2011.2107516

Liljebäck, P., Pettersen, K.Y., Stavdahl, Ø., Gravdahl, J.T.: Path following control of snake robots in unstructured environments. In: Proc. IEEE Int. Conf. Robotics and Automation, Shanghai, China, pp. 503–510 (2011b)

Liljebäck, P., Pettersen, K.Y., Stavdahl, Ø., Gravdahl, J.T.: Snake robot locomotion in environments with obstacles. IEEE Trans. Mechatron. (2011c). doi:10.1109/TMECH.2011.2159863

Liljebäck, P., Pettersen, K.Y., Stavdahl, Ø., Gravdahl, J.T.: A review on modelling, implementation, and control of snake robots. Robot. Auton. Syst. **60**, 29–40 (2011d). doi:10.1016/j.robot.2011.08.010

Liljebäck, P., Pettersen, K.Y., Stavdahl, Ø., Gravdahl, J.T.: Controllability and stability analysis of planar snake robot locomotion. IEEE Trans. Autom. Control **56**(6), 1365–1380 (2011f)

Liljebäck, P., Pettersen, K.Y., Stavdahl, Ø., Gravdahl, J.T.: Experimental investigation of obstacle-aided locomotion with a snake robot. IEEE Trans. Robot. **27**(4), 792–800 (2011g)

Liljebäck, P., Pettersen, K.Y., Stavdahl, Ø., Gravdahl, J.T.: Lateral undulation of snake robots: a simplified model and fundamental properties. Robotica (2012). Submitted

Linnemann, R., Paap, K.L., Klaassen, B., Vollmer, J.: Motion control of a snakelike robot. In: Proc. Third European Workshop on Advanced Mobile Robots, pp. 1–8 (1999)

Lipkin, K., Brown, I., Choset, H., Rembisz, J., Gianfortoni, P., Naaktgeboren, A.: Differentiable and piecewise differentiable gaits for snake robots. In: Proc. IEEE/RSJ Int. Conf. Intelligent Robots and Systems, San Diego, CA, USA, pp. 1864–1869 (2007)

Lochmatter, T., Roduit, P., Cianci, C., Correll, N., Jacot, J., Martinoli, A.: Swistrack—a flexible open source tracking software for multi-agent systems. In: IEEE/RSJ Int. Conf. Intelligent Robots and Systems, pp. 4004–4010 (2008)

Lötstedt, P.: Coulomb friction in two-dimensional rigid body systems. Z. Angew. Math. Mech. **61**, 605–615 (1981)

Lötstedt, P.: Mechanical systems of rigid bodies subject to unilateral constraints. SIAM J. Appl. Math. **42**(2), 281–296 (1982)

Ma, S.: Analysis of snake movement forms for realization of snake-like robots. In: Proc. IEEE Int. Conf. Robotics and Automation, Detroit, MI, USA, vol. 4, pp. 3007–3013 (1999)

Ma, S.: Analysis of creeping locomotion of a snake-like robot. Adv. Robot. **15**(2), 205–224 (2001)

Ma, S., Tadokoro, N.: Analysis of creeping locomotion of a snake-like robot on a slope. Auton. Robots **20**, 15–23 (2006)

Ma, S., Araya, H., Li, L.: Development of a creeping snake-robot. In: Proc. IEEE Int. Symp. Computational Intelligence in Robotics and Automation, pp. 77–82 (2001)

Ma, S., Ohmameuda, Y., Inoue, K., Li, B.: Control of a 3-dimensional snake-like robot. In: Proc. IEEE Int. Conf. Robotics and Automation, Taipei, Taiwan, vol. 2, pp. 2067–2072 (2003)

Ma, S., Ohmameuda, Y., Inoue, K.: Dynamic analysis of 3-dimensional snake robots. In: Proc. IEEE/RSJ Int. Conf. Intelligent Robots and Systems, pp. 767–772 (2004)

Masayuki, A., Takayama, T., Hirose, S.: Development of "Souryu-III": connected crawler vehicle for inspection inside narrow and winding spaces. In: Proc. IEEE Int. Conf. Intelligent Robots and Systems, vol. 1, pp. 52–57 (2004)

Mason, M.T., Wang, Y.: On the inconsistency of rigid-body frictional planar mechanics. In: Proc. IEEE Int. Conf. Robotics and Automation, vol. 1, pp. 524–528 (1988)

Matsuno, F., Mogi, K.: Redundancy controllable system and control of snake robots based on kinematic model. In: Proc. IEEE Int. Conf. Decision and Control, vol. 5, pp. 4791–4796 (2000)

Matsuno, F., Sato, H.: Trajectory tracking control of snake robots based on dynamic model. In: Proc. IEEE Int. Conf. Robotics and Automation, pp. 3029–3034 (2005)

Matsuno, F., Suenaga, K.: Control of redundant 3D snake robot based on kinematic model. In: Proc. IEEE Int. Conf. Robotics and Automation, pp. 2061–2066 (2003)

Mattison, C.: The Encyclopedia of Snakes. Cassell, London (2002)

McIsaac, K.A., Ostrowski, J.P.: Motion planning for anguilliform locomotion. IEEE Trans. Robot. Autom. **19**(4), 637–652 (2003a)

McIsaac, K.A., Ostrowski, J.P.: A framework for steering dynamic robotic locomotion systems. Int. J. Robot. Res. **22**(2), 83–97 (2003b)

McKenna, J.C., Anhalt, D.J., Bronson, F.M., Brown, H.B., Schwerin, M., Shammas, E., Choset, H.: Toroidal skin drive for snake robot locomotion. In: Proc. IEEE Int. Conf. Robotics and Automation, pp. 1150–1155 (2008)

Mehta, V., Brennan, S., Gandhi, F.: Experimentally verified optimal serpentine gait and hyperredundancy of a rigid-link snake robot. IEEE Trans. Robot. **24**(2), 348–360 (2008)

Melli, J.B., Rowley, C.W., Rufat, D.S.: Motion planning for an articulated body in a perfect planar fluid. SIAM J. Appl. Dyn. Syst. **5**(4), 650–669 (2006)

Miller, G.: Snake robots for search and rescue. In: Neurotechnology for Biomimetic Robots, pp. 271–284. MIT Press, Cambridge (2002)

Moon, B.R., Gans, C.: Kinematics, muscular activity and propulsion in gopher snakes. J. Exp. Biol. **201**, 2669–2684 (1998)

Morgansen, K.A., Duidam, V., Mason, R.J., Burdick, J.W., Murray, R.M.: Nonlinear control methods for planar carangiform robot fish locomotion. In: Proc. IEEE Int. Conf. Robotics and Automation, vol. 1, pp. 427–434 (2001)

Morgansen, K.A., Vela, P.A., Burdick, J.W.: Trajectory stabilization for a planar carangiform robot fish. In: Proc. IEEE Int. Conf. Robotics and Automation, vol. 1, pp. 756–762 (2002)

Morgansen, K.A., Triplett, B.I., Klein, D.J.: Geometric methods for modeling and control of free-swimming fin-actuated underwater vehicles. IEEE Trans. Robot. **23**(6), 1184–1199 (2007)

Mori, M., Hirose, S.: Three-dimensional serpentine motion and lateral rolling by active cord mechanism ACM-R3. In: Proc. IEEE/RSJ Int. Conf. Intelligent Robots and Systems, pp. 829–834 (2002)

Murugendran, B., Transeth, A.A., Fjerdingen, S.A.: Modeling and path-following for a snake robot with active wheels. In: Proc. IEEE/RSJ Int. Conf. Intelligent Robots and Systems, pp. 3643–3650 (2009)

Nijmeijer, H., van der Schaft, A.: Nonlinear Dynamical Control Systems. Springer, New York (1990)

Nilsson, M.: Ripple and roll: slip-free snake robot locomotion. In: Proc. Mechatronical Computer Systems for Perception and Action, Piza, Italy (1997)

Nilsson, M.: Snake robot—free climbing. IEEE Control Syst. Mag. **18**(1), 21–26 (1998)

Nilsson, M.: Serpentine locomotion on surfaces with uniform friction. In: Proc. IEEE/RSJ Int. Conf. Intelligent Robots and Systems, pp. 1751–1755 (2004)

Ohashi, H., Yamada, T., Hirose, S.: Loop forming snake-like robot ACM-R7 and its Serpenoid Oval control. In: Proc. IEEE/RSJ Int. Conf. Intelligent Robots and Systems, pp. 413–418 (2010)

Ohno, H., Hirose, S.: Design of slim slime robot and its gait of locomotion. In: Proc. IEEE/RSJ Int. Conf. Intelligent Robots and Systems, vol. 2, pp. 707–715 (2001)

Ostrowski, J., Burdick, J.: The geometric mechanics of undulatory robotic locomotion. Int. J. Robot. Res. **17**(7), 683–701 (1998)

Ostrowski, J.P.: The mechanics and control of undulatory robotic locomotion. Ph.D. thesis, California Institute of Technology (1996)

Paap, K.L., Kirchner, F., Klaassen, B.: Motion control scheme for a snake-like robot. In: Proc. IEEE Int. Symp. Computational Intelligence in Robotics and Automation, pp. 59–63 (1999)

Panteley, E., Loria, A.: On global uniform asymptotic stability of nonlinear time-varying systems in cascade. Syst. Control Lett. **33**(2), 131–138 (1998)

Panteley, E., Lefeber, E., Loria, A., Nijmeijer, H.: Exponential tracking control of a mobile car using a cascaded approach. In: Proc. IFAC Workshop on Motion Control, pp. 221–226 (1998)

Parker, T.S., Chua, L.O.: Practical Numerical Algorithms for Chaotic Systems. Springer, Berlin (1989)

Pavlov, A., Børhaug, E., Panteley, E., Pettersen, K.Y.: Straight line path following for formations of underactuated surface vessels. In: Proc. IFAC NOLCOS (2007)

Pettersen, K.Y., Egeland, O.: Exponential stabilization of an underactuated surface vessel. In: Proc. 35th IEEE Conf. Decision and Control, vol. 1, pp. 967–972 (1996)

Pettersen, K.Y., Lefeber, E.: Way-point tracking control of ships. In: Proc. IEEE Conf. Decision and Control, pp. 940–945 (2001)

Poi, G., Scarabeo, C., Allotta, B.: Traveling wave locomotion hyper-redundant mobile robot. In: Proc. IEEE Int. Conf. Robotics and Automation, vol. 1, pp. 418–423 (1998)

Prautsch, P., Mita, T.: Control and analysis of the gait of snake robots. In: Proc. IEEE Int. Conf. Control Applications, Kohala Coast, HI, USA, pp. 502–507 (1999)

Prautsch, P., Mita, T., Iwasaki, T.: Analysis and control of a gait of snake robot. Trans. Inst. Electr. Eng. Jpn., Sect. D **120-D**(3), 372–381 (2000)

Reyhanoglu, M., van der Schaft, A., McClamroch, N.H., Kolmanovsky, I.: Dynamics and control of a class of underactuated mechanical systems. IEEE Trans. Autom. Control **44**(9), 1663–1671 (1999)

Rincon, D.M., Sotelo, J.: Ver-Vite: dynamic and experimental analysis for inchwormlike biomimetic robots. IEEE Robot. Autom. Mag. **10**(4), 53–57 (2003)

Saito, M., Fukaya, M., Iwasaki, T.: Serpentine locomotion with robotic snakes. IEEE Control Syst. Mag. **22**(1), 64–81 (2002)

Sanders, J.A., Verhulst, F., Murdock, J.: Averaging Methods in Nonlinear Dynamical Systems, 2nd edn. Applied Mathematical Sciences, vol. 59. Springer, Berlin (2007)

Sato, T., Watanabe, W., Ishiguro, A.: An adaptive decentralized control of a serpentine robot based on the discrepancy between body, brain and environment. In: Proc. IEEE Int. Conf. Robotics and Automation, pp. 709–714 (2010)

Sfakiotakis, M., Tsakiris, D.: Biomimetic centering for undulatory robots. Int. J. Robot. Res. **26**, 1267–1282 (2007)

Shan, Y., Koren, Y.: Design and motion planning of a mechanical snake. IEEE Trans. Syst. Man Cybern. **23**(4), 1091–1100 (1993)

Shan, Y., Koren, Y.: Obstacle accommodation motion planning. IEEE Trans. Robot. Autom. **11**(1), 36–49 (1995)

Shapiro, A., Greenfield, A., Choset, H.: Frictional compliance model development and experiments for snake robot climbing. In: Proc. IEEE Int. Conf. Robotics and Automation, pp. 574–579 (2007)

Song, P., Kraus, P., Kumar, V., Dupont, P.: Analysis of rigid body dynamic models for simulation of systems with frictional contacts. J. Appl. Mech. **68**, 118–128 (2000)

Sørdalen, O.J., Egeland, O.: Exponential stabilization of nonholonomic chained systems. IEEE Trans. Autom. Control **40**(1), 35–49 (1995)

Spong, M.W.: Partial feedback linearization of underactuated mechanical systems. In: Proc. IEEE/RSJ/GI Int. Conf. Intelligent Robots and Systems, vol. 1, pp. 314–321 (1994)

Sussmann, H.J.: A general theorem on local controllability. SIAM J. Control Optim. **25**(1), 158–194 (1987)

Taal, S.R., Yamada, H., Hirose, S.: 3 axial force sensor for a semi-autonomous snake robot. In: Proc. IEEE Int. Conf. Robotics and Automation, pp. 4057–4062 (2009)

Tanaka, M., Matsuno, F.: Control of 3-dimensional snake robots by using redundancy. In: IEEE Int. Conf. Robotics and Automation, pp. 1156–1161 (2008a)

Tanaka, M., Matsuno, F.: Modeling and control of a snake robot with switching constraints. In: SICE Annual Conf. (2008b)

Tanaka, M., Matsuno, F.: A study on sinus-lifting motion of a snake robot with switching constraints. In: Proc. IEEE Int. Conf. Robotics and Automation, pp. 2270–2275 (2009)

Tanev, I., Ray, T., Buller, A.: Automated evolutionary design, robustness, and adaptation of sidewinding locomotion of a simulated snake-like robot. IEEE Trans. Robot. **21**(4), 632–645 (2005)

Togawa, K., Mori, M., Hirose, S.: Study on three-dimensional active cord mechanism: development of ACM-R2. In: Proc. IEEE/RSJ Int. Conf. Intelligent Robots and Systems, vol. 3, pp. 2242–2247 (2000)

Transeth, A.A., van de Wouw, N., Pavlov, A., Hespanha, J.P., Pettersen, K.Y.: Tracking control for snake robot joints. In: Proc. IEEE/RSJ Int. Conf. Intelligent Robots and Systems, San Diego, CA, USA, pp. 3539–3546 (2007b)

Transeth, A.A., Leine, R.I., Glocker, C., Pettersen, K.Y.: 3D snake robot motion: non-smooth modeling, simulations, and experiments. IEEE Trans. Robot. **24**(2), 361–376 (2008a)

Transeth, A.A., Leine, R.I., Glocker, C., Pettersen, K.Y., Liljebäck, P.: Snake robot obstacle aided locomotion: modeling, simulations, and experiments. IEEE Trans. Robot. **24**(1), 88–104 (2008b)

Transeth, A.A., Pettersen, K.Y., Liljebäck, P.: A survey on snake robot modeling and locomotion. Robotica **27**, 999–1015 (2008c)

Transeth, A.A., Liljebäck, P., Fjerdingen, S., Kyrkjebø, E., Mugaas, T.: New possibilities—next generation robotic systems for inspection and maintenance operations. In: Proc. EuroMaintenance, Fiera di Verona, Italy, pp. 229–232 (2010)

Trinkle, J., Pang, J.S., Sudarsky, S., Lo, G.: On dynamic multi-rigid-body contact problems with coulomb friction. Z. Angew. Math. Mech. **77**, 267–280 (1997)

Tsatsomeros, M.J., Li, L.: A recursive test for P-matrices. BIT Numer. Math. **40**, 410–414 (2000)

Ute, J., Ono, K.: Fast and efficient locomotion of a snake robot based on self-excitation principle. In: Proc. 7th Int. Workshop on Advanced Motion Control, pp. 532–539 (2002)

van der Schaft, A.J., Schumacher, J.M.: An Introduction to Hybrid Dynamical System. Springer, Berlin (2000)

Vela, P.A., Morgansen, K.A., Burdick, J.W.: Underwater locomotion from oscillatory shape deformations. In: Proc. IEEE Conf. Decision and Control, vol. 2, pp. 2074–2080 (2002a)

Vela, P.A., Morgansen, K.A., Burdick, J.W.: Second order averaging methods for oscillatory control of underactuated mechanical systems. In: Proc. American Control Conf., vol. 6, pp. 4672–4677 (2002b)

Wang, Z., Ma, S., Li, B., Wang, Y.: Stability and adaptability of passive creeping of a snake-like robot. In: Proc. IEEE/RSJ Int. Conf. Intelligent Robots and Systems, pp. 395–400 (2010)

Watanabe, K., Iwase, M., Hatakeyama, S., Maruyama, T.: Control strategy for a snake-like robot based on constraint force and verification by experiment. In: IEEE/RSJ Int. Conf. Intelligent Robots and Systems, pp. 1618–1623 (2008)

Westervelt, E.R., Grizzle, J.W., Chevallereau, C., Choi, J.H., Morris, B.: Feedback Control of Dynamic Bipedal Robot Locomotion. CRC Press, Boca Raton (2007)

Wiriyacharoensunthorn, P., Laowattana, S.: Analysis and design of a multi-link mobile robot (serpentine). In: Proc. IEEE Int. Conf. Robotics, Intelligent Systems and Signal Processing, vol. 2, pp. 694–699 (2002)

Worst, R., Linnemann, R.: Construction and operation of a snake-like robot. In: Proc. IEEE Int. Joint Symp. Intelligence and Systems, Rockville, MD, USA, pp. 164–169 (1996)

Wright, C., Johnson, A., Peck, A., McCord, Z., Naaktgeboren, A., Gianfortoni, P., Gonzalez-Rivero, M., Hatton, R., Choset, H.: Design of a modular snake robot. In: Proc. IEEE/RSJ Int. Conf. Intelligent Robots and Systems, pp. 2609–2614 (2007)

Yamada, H., Hirose, S.: Study on the 3D shape of active cord mechanism. In: Proc. IEEE Int. Conf. Robotics and Automation, pp. 2890–2895 (2006a)

Yamada, H., Hirose, S.: Development of practical 3-dimensional active cord mechanism ACM-R4. J. Robot. Mechatron. **18**(3), 1–7 (2006b)

Yamada, H., Hirose, S.: Study of active cord mechanism—generalized basic equations of the locomotive dynamics of acm and analysis of sinus lifting. J. Robot. Soc. Jpn. **26**(7), 801–811 (2008). In Japanese

Yamada, H., Hirose, S.: Study of a 2-DOF joint for the small active cord mechanism. In: Proc. IEEE Int. Conf. Robotics and Automation, pp. 3827–3832 (2009)

Yamada, H., Hirose, S.: Steering of pedal wave of a snake-like robot by superposition of curvatures. In: Proc. IEEE/RSJ Int. Conf. Intelligent Robots and Systems, pp. 419–424 (2010)

Yamada, H., Chigisaki, S., Mori, M., Takita, K., Ogami, K., Hirose, S.: Development of amphibious snake-like robot ACM-R5. In: Proc. 36th Int. Symp. Robotics (2005)

Yamakita, M., Hashimoto, M., Yamada, T.: Control of locomotion and head configuration of 3D snake robot (SMA). In: Proc. IEEE Int. Conf. Robotics and Automation, vol. 2, pp. 2055–2060 (2003)

Ye, C., Ma, S., Li, B., Wang, Y.: Turning and side motion of snake-like robot. In: Proc. IEEE Int. Conf. Robotics and Automation, vol. 5, pp. 5075–5080 (2004a)

Ye, C., Ma, S., Li, B., Wang, Y.: Locomotion control of a novel snake-like robot. In: Proc. IEEE/RSJ Int. Conf. Intelligent Robots and Systems, vol. 1, pp. 925–930 (2004b)

Ye, C., Ma, S., Li, B., Liu, H., Wang, H.: Development of a 3D snake-like robot: Perambulator-II. In: Int. Conf. Mechatronics and Automation, pp. 117–122 (2007)

Yim, M.: New locomotion gaits. In: Proc. IEEE Int. Conf. on Robotics and Automation, vol. 3, pp. 2508–2514 (1994)

Yim, M., Duff, D.G., Roufas, K.D.: Walk on the wild side. IEEE Robot. Autom. Mag. **9**(4), 49–53 (2002)

Yu, S., Ma, S., Li, B., Wang, Y.: Analysis of helical gait of a snake-like robot. In: IEEE/ASME Int. Conf. Advanced Intelligent Mechatronics, pp. 1183–1188 (2008)

Yu, S., Ma, S., Li, B., Wang, Y.: An amphibious snake-like robot: design and motion experiments on ground and in water. In: Int. Conf. Information and Automation, pp. 500–505 (2009)

Zarrouk, D., Sharf, I., Shoham, M.: Analysis of earthworm-like robotic locomotion on compliant surfaces. In: Proc. IEEE Int. Conf. Robotics and Automation, pp. 1574–1579 (2010)

Zuo, Z., Wang, Z., Li, B., Ma, S.: Serpentine locomotion of a snake-like robot in water environment. In: IEEE Int. Conf. Robotics and Biomimetics, pp. 25–30 (2008)

Index

P. Liljebäck et al., *Snake Robots*, Advances in Industrial Control,
DOI 10.1007/978-1-4471-2996-7, © Springer-Verlag London 2013

www.ingramcontent.com/pod-product-compliance
Ingram Content Group Australia Pty Ltd
76 Discovery Rd, Dandenong South VIC 3175, AU
AUHW011923051225
420567AU00001B/1